Solar Energy Applications to Dwellings

Solar Energy R&D in the European Community

Series A:

Solar Energy Applications to Dwellings

Volume 2

Publication arrangements: D. NICOLAY

Solar Energy R&D
in the European Community

Series A Volume 2

Solar Energy Applications to Dwellings

Proceedings of the
EC Contractors' Meeting held in
Meersburg (F.R.G.), 14-16 June 1982

edited by

W. PALZ
Commission of the European Communities

and

C. DEN OUDEN
Institute of Applied Physics TNO-TH

D. REIDEL PUBLISHING COMPANY

Dordrecht, Holland / Boston, U.S.A. / London, England

for the Commission of the European Communities

Library of Congress Cataloging in Publication Data

Main entry under title:

Solar energy applications to dwellings.

(Solar Energy R&D in the European Community. Series A: v. 2)
1. Solar houses–Europe, Western. 2. Solar energy research–Europe, Western. I. Palz, Wolfgang. II. Ouden, C. den III. Commission of the European Communities. IV. Series.
TH7414.S653 1982 697.78 82-20545
ISBN 90-277-1491-6

Organization of the Contractors' meeting by
Commission of the European Communities
Directorate-General Science, Research and Development, Brussels

Publication arrangements by
Commission of the European Communities
Directorate-General Information Market and Innovation, Luxembourg

EUR 8046

Published by D. Reidel Publishing Company
P.O. Box 17, 3300 AA Dordrecht, Holland

Sold and distributed in the U.S.A. and Canada
by Kluwer Boston Inc.,
190 Old Derby Street, Hingham, MA 02043, U.S.A.

In all other countries, sold and distributed
by Kluwer Academic Publishers Group,
P.O. Box 322, 3300 AH Dordrecht, Holland

D. Reidel Publishing Company is a member of the Kluwer Group

Printed in The Netherlands

C O N T E N T S

SECTION 2 - PASSIVE SOLAR COMPONENTS

A. Subsection : walls, façades, etc.

B. Subsection : Shutters, window systems, etc.

C. Subsection : Control devices, etc.

SECTION 3 - HIGH PERFORMANCE COLLECTORS

SECTION 4 - SOLAR COOLING MACHINES

SECTION 5 - HEAT STORAGE

INTRODUCTION

This volume 2 gives again a broad overview of the Commission's activities in the development of solar heating and cooling which are being published in Series A. We are glad to be in a position to provide here a report of high quality which will no doubt be of value for all those interested in solar energy.

The Commission of the European Communities has at this time concluded more than a hundred contracts on solar heating and cooling on a cost-sharing basis. Contractors' reports which are presented in this book give a full view of their activities. It is important to note that the various aspects of this programme are being carried out by working groups in which, by and large, all member countries of the European Communities are involved and work together towards a common strategy. The spirit of cooperation which finds here its expression has so far led to many concrete results which I will try to summarise in the following.

The key elements of an active solar heating system are the solar collector and the storage unit which have been the subject of an intense R+D effort within the Commission's programme. Collectors with selective absorbers are now very successful on the market place while high performance collectors of a more sophisticated type e.g. those employing vacuum did not yet get beyond the prototype stage. The work towards the establishment of generally recognised methods for the performance rating of solar collectors - it is of paramount importance for the European industry working in this field - has become close to successful completion. Future work has to concentrate on quality, durability and lifetime criteria for collectors. In the field of heat storage much emphasis is given to seasonal storage by means of huge water reservoirs. Water is either stored at high temperature for direct use or at low temperature involving heat pumps. Large-scale experiments are now under way and first conclusions can be expected within 1 or 2 years from now. New work has been started on performance rating of solar heat storage and first criteria are being set up.

The programme in this sector includes also promising work on chemical heat storage and low-cost devices are now being developed.

Integration of collectors and storage devices into an operational solar heating system needs careful optimisation. The Commission has reviewed within this programme many active systems in Europe and had to conclude that as a rule, performances were lower than anticipated. It must be recognised that there remains much room for improvement of efficiency and cost-effectiveness. The Commission has taken up this subject as a major research area and has given it a lot of attention. European modellers and designers from all member countries are working actively together and have by now agreed upon a general model which is able to cope with all design variation of active heating systems. A simplified version of the model is also being developed. It can be expected that this European model will meet the designer's needs of a reliable method of general value as compared to the confusingly large number of models, all more or less perfect and suitable as they have been available until now. The European model for active space heating is presently being finalised and fully validated for all climates with the help of the European test facilities.

Turning now to solar cooling, it is obvious that the European market here is much smaller than for heating. Some of the work which is reported in this book gives a good idea of the technical complexity cooling systems may imply. There is new work however in particular a chemical adsorption system developed by a French contractor demonstrating that simple and cheap design for ice production by solar energy is possible.

For passive solar heating and cooling the Commission's R+D activities are more recent than for the active systems discussed here above. Hence only preliminary results are available at this time. This holds in particular true for passive solar components. Quite a few interesting concepts have been presented nevertheless by the Commission's contractors be it a transparent Trombe wall, evacuated glass tubes, or others.

One contractor reported that he is marketing prefabricated passive solar houses designed for the two most typical climates in France. Overcost is such that it pays off in not more than 10 years. The Commission's programme on passive solar systems adresses itself to a lot more than just component development. Concerted activities involving all member countries are now in progress in particular on design tools and simulation models. A second European architectural competition for passive solar energy use has been started successfully.

As a general conclusion we can take note of the fact that these days the technological basis for solar heating including software has become much broader. With 250.000 solar water heaters already installed throughout the EEC, with hundreds of companies involved and thousands of new employments, solar heating is now well rooted in Europe. The Commission's activities contributed a reasonable part to this success.

For the appraisal of future needs, the Commission has initiated an assessment study which will lay down guidelines for the most promising options. It will be published in full within a few months in this Series A.

W. PALZ
Head of the Solar Energy R+D Programme

SECTION I - OVERVIEWS

OVERVIEW OF PAPERS PRESENTING CONTRACT RESEARCH WITHIN CONCERTED ACTIONS

Author: C. den Ouden, Projectleader Project A

1. Introduction

In the session 'Concerted Actions' 10 papers were presented by the co-ordinators of these collaborative actions, in which, in general, from each member-country within the EC a participating contractor is represented.

The structure of Project A, and especially of the concerted actions, reflects the fact that there are National Research Programmes on Solar Energy in the various member-countries. Therefore, the topics studied within these collaborative actions are in general directed towards:

- the development of European testprocedures for solar components (collectors, heat storage systems, etc.),
- the development of uniform approaches for total system design, monitoring and evaluation of total system performance and validation of computer simulation models for different solar systems, both for active and passive solar designs, etc.

The relationship between these concerted actions is shown schematically in figure 1.
In this figure all aspects of the Research and Development activities on Solar Thermal are indicated. As can be seen from this figure all activities are directed towards design, development and improvements of solar systems (passive and active) in order to stimulate the implementation of these systems within Europe. Development of solar components, which gets enough attention within the various national programmes only gets little support from the EC programme and this support is mainly directed towards standard approach for testing and testing as component of total systems.

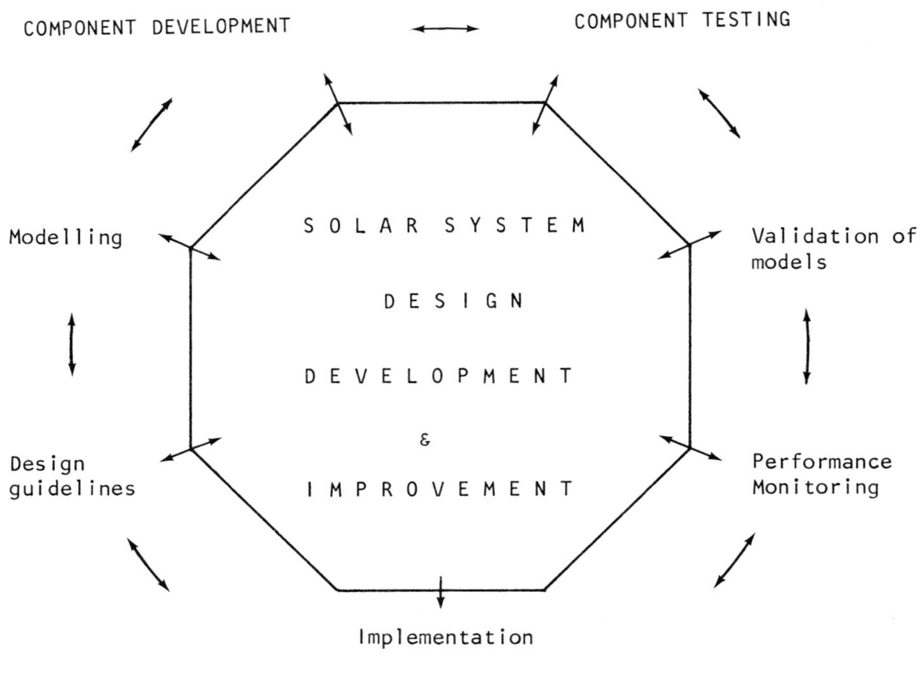

2. Review of the presentations

The first presentation in the session of the concerted actions was presented by D. Turrent and concerned the results obtained in an Assessment Study covering the Solar Thermal Applications. This study was carried out with 10 sub-contractors of all member countries in order to find out in which direction the main R & D effort for the next coming programme should be directed.
Within this study also the state of the technology as far as the use of collectors, storages, solar water heating, active space heating and passive space heating etc. were discussed.
A survey of the European solar industry was given and finally the assessment itself was presented by N. Backer. It was shown that the potential for solar energy in Europe is considerably larger than previously thought. In the domestic sector only, solar energy could provide in between 35 mtoe and 50 mtoe, depending on the successful use of interseasonal heat storage. A detailed economic evaluation of various solar systems in the different climatic regions in Europe was presented as well.

The second presentation of R. Lebens gave an overview of the progress made with the preparation of the 'Second Passive Solar Competition' organized to stimulate architects and schools of archetecture to make use of the sun in the best possible way to reduce energy consumption of houses etc.

The third paper, also presented by R. Lebens, the work carried out in the 'European Passive Solar Working Group' was discussed.
This group has several tasks and subjects to study, such as:
- the production of a European Passive Solar Handbook,
- the preparation of a European Passive Solar Products Catalogue, - to compile information to be used for Planning and Urban Design, - to study the potential for passive solar retrofitting, - the preparation of workshops on passive solar to serve as an educational support in the various member countries, - to study various special technical subjects including air movement and heat distribution as well as comfort aspects in passive solar houses, etc.

The fourth paper, presented by J.M. Hauglustaine, dealt with the work of the 'Passive Solar Modelling Sub-group' a task which is devoted to the selection and improvement of a reliable computer programme for simulating the performance of passive solar buildings.
Furthermore, this group will select a number of simplified models suitable for mini and microcomputers and also programmable desk calculators. The validity of these simplified methods will be calibrated against the large computer model and possibly to emperical data before they can be documented for the Passive Solar Handbook and successfully used by architects and designers of passive solar houses.

In the fifth presentation, by Prof. W. Dutré, the progress of the work of the 'European Modelling Group for Space Heating Systems and Domestic Hot Water Systems' was discussed. Within this task a detailed simulation programme (EMGP-2) for active space heating and domestic hot water systems is under development. This programme is modularly structured and has been extended with a large number of subroutines such that it can be used to simulate a large variety of different solar systems. An essential part of this concerted action is the validation of this model by means of experimental data from the Solar Pilot Test Facilities (see 7th presentation). Another aspect of this collaborative action is the prediction of the annual performance of various solar systems in different climatic zones of the European Communities and to perform an economic

evaluation of the considered class of solar systems in Europe.
Another task of this EMG-group is the development and validation of simplified computation methods, which was presented by J. Adnot and B. Bourges. Within this subtask of the European Modelling Group half a dozen simplified methods, such as: accumulated frequency curve methods, - f-chart and ø-f-chart methods etc. will be compared systematically not only with the validated large programme (EMGP-2), but also with measured results obtained from the Performance Monitoring Group (see presentation R. Ferraro).

The seventh presentation of G. Olive gave the progress of the experimental work of the European Solar Pilot Testfacilities. With these testfacilities, consisting of real solar systems with collectors, storage, controls and associated piping, but with the dwelling's thermal distribution system replaced by a microprocessor-controlled physical load simulator, capable of producing a typical thermal load for a house, interactive with the actual weather and taking into account the effect of occupancy etc., the European Modelling Group can be supplied with reliable data to validate its models. In the last heating season 1981/1982 a large number of system variations have been carefully monitored and the measured results have been stored on magnetic tapes to serve the Modelling Group.
Several other important studies are carried out by the eight participants in the SPTF-group, such as: - extrapolation of measured results to other locations (climates), - investigation of the period of time to measure within a certain system configuration to be able to conclude certain results,- reliability studies etc.
Above all these experimental groups try to measure all relevant system parameters as accurately as possible in order to get sufficient confidence in the measured results, which is not always possible in monitored solar houses with occupants, etc.

The next presentation, the eighth, by R. Ferraro, gave an overview of the work within the Performance Monitoring Group. The overall objective of this group is to collect data from monitored solar heating systems in order to draw conclusions about their performance in European climates. This group has been very successful in the past in producing formats to enable uniform presentation of system performance data for domestic hot water systems, active and passive space heating systems. Currently, two books are being published with the reported and analysed results of the functioning of 28 solar water heaters, and 33 solar heating systems (both active and passive).
Moreover, a book will soon be published with guidelines for monitoring solar heating systems, to promote the use of uniform monitoring procedures and to improve the current state of knowledge about instrumentation and the use of data acquisition systems.
An important part of the presentation of the work of the PMG-participants for the next year was devoted to the recently initiated work of design teams, which have to perform a study with input from PMG-experience, system modellers, industry, architects and costing experience, resulting in exemplary designs of second generation systems. These systems will show a higher useful energy output per invested amount of money in comparison with the systems analysed before.

In the ninth presentation of J.E. Moon the progress of the work within the 'Collector Testing Group' was reported of. In this collaborative action more than 20 laboratories within Europe work together on testprocedures for solar collectors and related research topics.
Currently the work concentrates on testing air collectors and developing

standard test procedures for these collector types. Also results of the latest Round Robin test of a high performance collector were presented and particularly encouraging results obtained with solar simulator measurements were reported from a workshop held in ISPRA on this topic, attended by many participants of the Collector Testing Group. Other work within this group is: - collector reliability and durability aspects, Round Robin tests of optical properties of absorber surface samples and the reproducibility tests.

The last presentation within the session of concerted actions, by E. van Galen, was devoted to the work of the 'Solar Storage Testing Group'. In this group, initiated early 1982, 5 laboratories and 2 observing members are participating. The overall objective of this group is to draw up recommendations for testprocedures for solar storage systems, which must provide useful tools for modellers, system designers and industry. These testprocedures must be as simple as possible, applicable over a wide range of storage systems and the validity must be proven. The objectives of the work for the period till June 1983 are to draw up draft recommendations and guidelines for the design of testfacilities. An attempt will be made to present the results of the testprocedure in the form of a simple model expressing the 'storage efficiency' over a certain timestep as a function of the relevant parameters, such as flow rate, storage temperature etc. and furtheron in simple figures and tables.

SOLAR ENERGY ASSESSMENT STUDY (PROJECT A)

Authors	:	D. TURRENT, N. BAKER	
Contract Number	:	ESA-G-111-81-UK (H)	
Duration	:	9 months	1st October 1981 - 30th June 1982
Total budget	:	£ 54 950	CEC contribution: 100%
Head of project	:	D. Turrent	
Contractor	:	Energy Conscious Design	
Address	:	44 Earlham Street London WC2	
Subcontractors	:	F. Jäger, Germany G. Olive, France J. de Grijs, Netherlands C. Boffa, Italy J.O. Lewis, Ireland	P. Kristensen, Denmark D. Vokaer, Belgium G. Leach, UK T. Tsingas, Greece

Summary

This paper describes progress made on the Solar Energy Assessment Study during the last nine months. It begins by stating the aims and objectives of the Study and then goes on to describe the method of working and the basic approach to the assessment. The structure of the final report is outlined, covering four main areas: the context for solar R & D, the current state of the technology, applications and quantitative assessment and a profile of the solar industry. Within each of these areas a summary is then given of the main findings and these are followed by a description of the main problems encountered and some of the provisional conclusions reached. In summary, the main purpose of the Assessment Study is to draw together the practical conclusions drawn so far in the area of R & D, to assess the potential of the technology in terms of displacing conventional fuels in the longer term and to establish priorities for future R & D work.

1.1 Aims of the Study

The main aims of the Assessment Study are; to review current progress in solar R & D, assess the potential market for low temperature solar applications in Europe, and define future R & D requirements to ensure that it is technically feasible to reach this potential.

Rather than concentrating solely on the Commissions' own Solar Energy R & D Programme, the brief asked for a wide ranging study, to include a review of current and planned National Solar R & D programmes as well as R & D activities outside Europe. It is therefore intended that the final report will provide a base document to help the Commission plan their next 4 years programme starting in July 1983. At the same time it should provide individual governments with a means of measuring their own progress in this field. The anticipated readership for the report includes policy makers as well as members of the research community, the solar industry and hopefully, members of the building industry, including architects and engineers.

As shown in Fig (i), the Assessment study for solar energy applications in dwellings comes at an appropriate moment, after nearly a decade of gradually intensifying R & D activity and at a time when many countries in Europe are about to embark on a new phase of work. With only one more year to run in the Commissions' current 4 year programme this is a good time to take stock, to make an objective assessment of the current state of the technology and to assess its future prospects.

1.2 Methods of Working

The study is being carried out by Energy Conscious Design, London under the direction of T.C. Steemers, DG XII and C den Ouden, Leader of Project A. Additionally, an Advisory Group has been appointed to assist in the study, consisting of representatives nominated by members of expert working Group A.

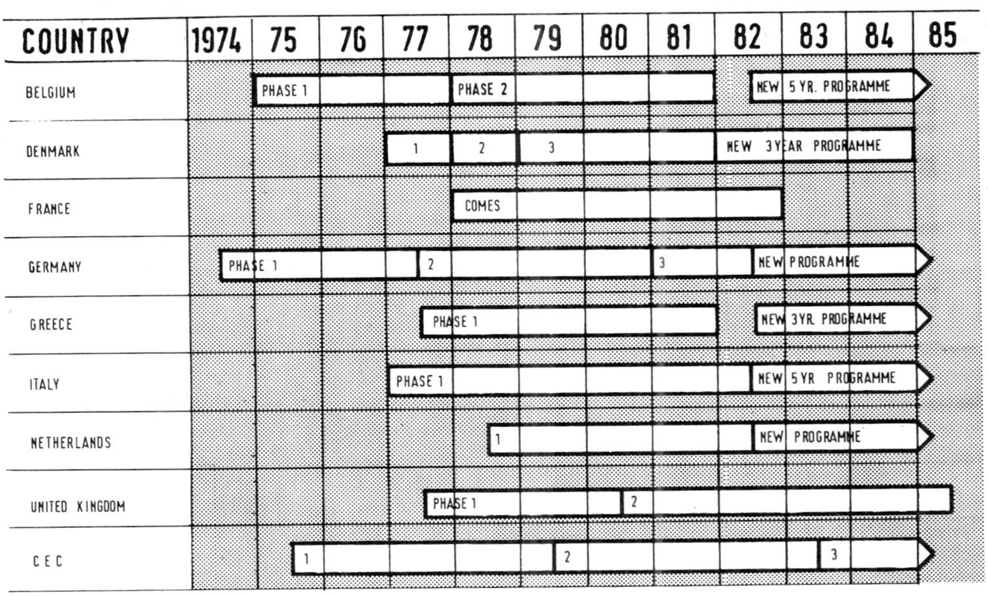

Fig (i). Timescales of National R & D Programmes.

From the outset, it has been agreed that the study must take account of, and embody the views of, leading experts in the solar R & D field. One central objective of the study is to achieve a degree of concensus within the R & D community about some of the key issues affecting progress in the development and implementation of solar technology. Through a series of structured interviews and meetings with the advisory group it has been possible to identify some of the major constraints and uncertainties currently affecting the technology, as well as define some of the more promising areas for future development. To date, the views and opinions of some sixty leading experts in the field have been taken into account. It is intended that the general conclusions of the study should reflect this concensus, although of course opinions will vary on matters of detail. The main body of evidence leading up to these conclusions is structured into four main sections:

- The Context for Solar R & D
- The State of the Technology
- An Assessment of Market Potential
- A Profile of the Solar Industry.

In the following chapters we will describe our approach to each of these sections before finally outlining some provisional conclusions.

1.3 The Context for Solar R & D

Ten years ago, solar energy research was confined to a very specialised area of scientific experimentation. Today there are several thousand researchers active in the field and a small industry is beginning to develop. This rapid development of a new technology has taken place against a background of rising energy prices and a growing awareness that conventional forms of energy supply are finite, and must be conserved. In particular, Europe has become aware of its heavy dependence on imported oil, currently about 55% of gross primary energy consumption, most of which comes from the Middle East. Reduction of imported oil is therefore a central objective of European energy policy. Clearly, energy conservation will play a major role in this policy as will the development of renewable energy sources, such as solar, wind, wave, biomass etc. Many experts believe that the contribution of all the renewables to Europe's gross energy demand could be in excess of 10% by the year 2020. Solar energy in particular could have an important role to play in the future as a source of low temperature energy supply. It is estimated that 40-50% of Europe's primary energy is consumed in buildings, mostly to provide low temperature space and water heating.

Historically, most of the R & D carried out to date has been on active solar systems for water and space heating. Recently there has been a gradual shift towards passive solar space heating, especially in France and the United Kingdom. Interest in passive solar technology is also growing in Italy and Greece, but in contrast in Germany, 95% of the national solar R & D budget is likely to be allocated to active solar R & D.

The purpose of the Assessment Study is to provide a context for the whole range of R & D activities being carried out in Europe. It allows the individual researcher working in a particular aspect such as say, temperature stratification in water stores, to see how this work fits in to the whole development of solar space heating systems. It also enables him to see his effort in relation to other people's work and to get a feel for the longer terms quantitative potential that solar technology has to offer. Most importantly however, it provides a goal to work towards and a better understanding of the actions needed by both the research community and the industry to achieve the goal.

1.4 The State of the Technology

An assessment of the current state of solar technology forms a large part of the study. The section begins by discussing recent developments in component design, solar collectors, short term heat storage and long term heat storage, before going on to consider various approaches to system design. Current costs and performance of active and passive systems are outlined and the prospects for further system optimisation and cost reduction are discussed. Attention is also paid to recent developments in hybrid systems, incorporating elements of both active and passive approaches, and to solar assisted heat pump systems and larger scale grouped systems. Finally, progress in the field of simulation modelling is reviewed, together with developments towards more simplified design methods.

In the case of solar collectors, whilst the operating characteristics and performance of flat plate designs are now well known, there is a need for much more feedback on long term durability and reliability. In terms of collector performance, there is still scope for improvement, especially in improved cover design and selective absorber surfaces. High performance evacuated tube collectors are now being commercially manufactured by several companies. The first results of round-robin indoor tests show that the U-value is reduced to less than 2 Watts/^{o}C. Most of the technical problems affecting the production of evacuated tube collectors now appear to have been resolved, but unit costs are still high. With an expanding market however, some manufacturers believe that costs could become competitive with flat plate collectors.

Research on heat storage has resulted in a good understanding of the behaviour of water tanks to provide short term heat storage in active systems, although there is still some disagreement on the benefits to be gained from thermal stratification. There is also disagreement about the use of phase change materials for latent heat storage and the possibility for volume reduction. Standard test procedures are clearly needed and have now been initiated as part of a new concerted action in this area. One conclusion drawn, is that the heat store cannot be considered in isolation, but must be designed and evaluated as part of a whole system, including the heat distribution and auxiliary system.

Long term heat storage has received less attention but is now recognised as a key area for future research, especially for active systems serving groups of houses. A major effort is needed in this area to develop low cost solutions. Various approaches look promising at this stage, including heat storage in the ground, acquifiers and solar ponds.

Turning to systems, solar water heating is now a proven technology, although there is still some uncertainty about system lifetime. Average annual system efficiencies of 25% have been measured, although this could be increased to 40% with some technical improvements. However, solar water heating systems are still not ecomonic at current fuel prices. System costs need to be reduced by a factor of 2 to make solar water heating competitive with electricity. Nevertheless there is evidence of some cheaper systems coming on the market and simpler thermosyphon systems also deserve more attention.

Results to date for active solar space heating in individual houses have not been very encouraging. The first generation of liquid based systems were over-sized, over complicated, over-expensive and over optimistic in performance terms. Measured performance figures for monitored solar houses show an average output of around 160 kWh/m^2 p.a. with system efficiencies of less than 20%. Despite this rather disappointing performance, a great deal of valuable practical experience has been gained. There is now a strong concensus that the next generation of active systems should be simpler and smaller, designed to maximise the useful solar output per m^2, for a minimum investment cost. Preliminary optimisation studies have pointed to 'over-sized' water heating systems with collector areas in the region of 10-15m^2 as showing some real promise for new houses and also possibly for the retrofit market. These 'second generation' systems will have a higher output, possibly in excess of 250kWh/m^2 p.a.

Passive solar space heating has been gaining increasing interest in the last few years, particularly since the simpler forms of direct gain passive solar design appear to be cost effective now. Nevertheless there is still a considerable degree of uncertainty concerning the cost and performance of other forms of passive solar design and more R & D is needed to reduce these. In particular there is a need for a validated mathematical model and for more experimental data from monitored houses. There is also a need for more education of design professionals and the public to increase awareness in this area.

The state of the art in solar cooling is considerably less advanced than it is for solar space heating. Most of the R & D carried out to date has been on active systems based on the absorption cycle, but most designs are still only at laboratory-prototype stage. Passive cooling systems have received very little attention so far, although there are a number of options which look interesting for southern Europe.

1.5 The Assessment Methodology

One of the major objectives of the Study is to assess the size of the solar heating resource for Europe. This is important for two reasons. First it provides a means of comparing the size of the resource with the potential from other sources and second, it establishes a goal for the R & D community to work towards.

Making predictions of the likely contribution from solar energy in the future is very difficult. Conventional scenarios have been based on market penetration factors which are based in turn on assumptions about future energy prices, social attitudes, institutional factors and so on.

To avoid the obvious pitfalls of this approach or the alternative approach of making extrapolations based on present trends, we have chosen to quantify the potential contribution from solar energy by the application of reference systems to the European housing stock. These originate from plausible and familiar present day systems for which we have a firm knowledge of performance and rather less firm but useful knowledge of cost. The Reference Systems do not have to be highly technically specific, provided that their performance lies within technically feasible limits. Furthermore, whilst reference systems can be initially defined in terms of current technical performance, projections can be made for increased system efficiency for future date-line predictions. One interesting case would be the incorporation of inter-seasonal storage which would lead to a sudden increase in annual system efficiency. A reference system reflecting this improvement could be used when and where appropriate. The structure of the quantitative assessment is shown in Fig (ii).

The logical starting point is the Total Resource which we define as the annual total solar radiation falling on Europe. This is a large number and is of academic interest only since the gulf between the Total Resource and its utilisation is enormous. The next step is to define the Ultimate Potential which is the total demand for low temperature heat.

In defining the Ultimate Potential for any future time horizons we have to take into account the influence of conservation measures on present day building stock and its incorporation into new build (this is indicated on the diagram). The establishment of a conservation scenario is difficult in that it contains implicit assumptions about economic forces which we are trying to postpone until after the application of our reference systems. However, for all but present day potentials, we shall have to make conservation scenario predictions.

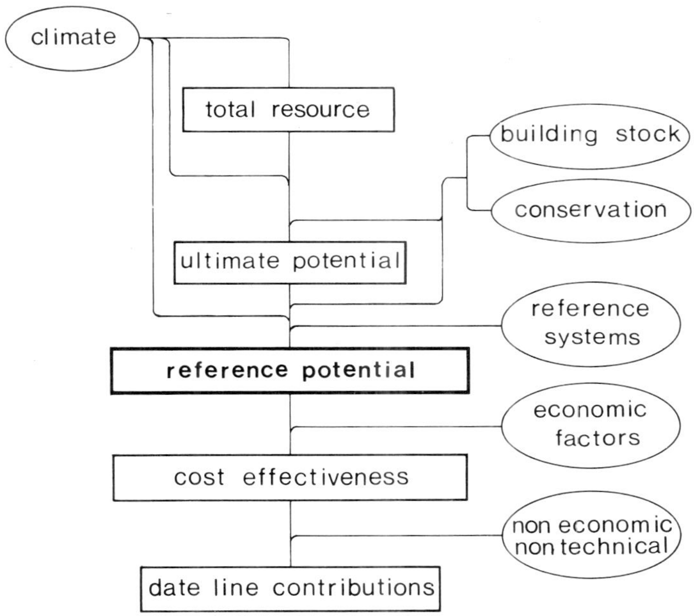

Fig (ii) Schematic illustration of the Assessment Methodology.

The performance of the reference system will vary for different sites in Europe due to the different availability of solar radiation and the different ambient temperatures. Clearly, it is not possible to ascribe a single absolute performance or even system efficiency to cover all locations.

In order to assess the performance of the reference systems under various combinations of climate and load, a performance modelling system will be used. In the case of active systems, this will be the F-chart method and for passive systems the Solar/Load Ratio method will be applied. This method, similar to the F-chart approach predicts a useful solar fraction as a function of the solar gain to load ratio, on a monthly basis. The reference systems wil be applied to 23 sites in Europe for which we have the necessary monthly weather data. National averages for performance will then be derived and applied to the appropriate proportion of the national building stock in order to evaluate a National Reference Potential.

The influence of conservation will be accounted by defining two different reference building loss co-efficients. The first will be derived from estimates of the current national average, and the second, a lower value, assumes the takeup of a package of conservation measures. The mix of these two will alter for different time horizons, as the high conservation proportion increases. The proportion of the building stock for which a particular solar application is appropriate will be established by considerations of overshadowing, orientation etc. Also, the rate of new build will be accounted for in the assumption that all new build can incorporate the high level of conservation, and only new build can incorporate 'total' as distinct from 'retrofit' passive solar design.

The application of Reference Systems will then lead to Reference Potentials. These will include Present-day Potentials reflecting current system performance for systems which can be retrofitted to existing buildings, and Future Potentials, which will allow for technological improvement and incorporation of 'total' passive systems. One such hypothetical improvement which will be considered, is the solution of the inter-seasonal storage problem.

The Reference Potentials take no account of the economic viability of the systems, only technical limitations. The next stage in the analysis discusses the economic factors which may cause the realisable contribution of solar applications to fall short of the Potential. The economic viability will be dependent upon the following factors; system cost, conventional fuel cost, and the cost of borrowing money. Present day system costs are relatively easy to establish, at least for active systems. Future viability will of course be influenced by future costs, and these are less certain.

Conventional energy costs must then be considered and the sensitivity of economic viability to real price increases explored. Clearly, the returns from a solar system are measured in terms of the value of the conventional energy displaced; thus even if energy prices are known, viability will differ according to the particular type of energy being displaced.

To be considered cost effective, the solar investment must be tested against the returns that could be realised from other kinds of investment. For example, if a householder could obtain a higher rate of interest (net of tax) by investing his 2500 ECU in a Building Society than the value of the energy saved by a 4 sq.m. SWH system, then the latter cannot be regarded as a good investment. For this evaluation the appropriate Discount Rate must be applied.

Even when solar measures are cost effective, their adoption is not immediate. The number of applications with time will show an 'S-curve' rather than a 'step function' characteristic. The rate at which the take up occurs is known as the market diffusion rate and is sensitive to a number of factors, most of which are not strictly economic factors, but social, environmental and institutional.

It is clearly outside the scope of this paper to go into more detail on the non-technical issues, so we shall confine ourselves to a description of the reference systems and the preliminary estimates of reference potential for Europe.

1.6 Reference Systems and Reference Potentials

In this chapter we describe the reference systems used for solar water heating, active space heating and passive space heating. We then go on to explain the application of these systems to the housing stock, and finally we present some provisional results.

For solar water heating we have used two systems: a $4m^2$ system for northern Europe and a $2m^2$ for southern Europe. Both have single glazed selective surface collectors and a load of 100 litres per day is assumed for calculation purposes. A fixed annual system efficiency of 25% is also used. In the case of active solar space heating, two systems are again employed. The first is a $15m^2$ system with short term heat storage providing both water heating and some space heating. The second is a $25m^2$ system with interseasonal heat storage covering approximately 80% of the annual load in northern Europe.

The final two reference systems are for passive solar space heating. One has a solar aperture area of $10m^2$, which in practice would be a lean-to sunspace for retrofit applications. The other has a $15m^2$ aperture area and is used as a direct gain system in new build houses.

In applying the reference systems to the building stock there will in some cases be a choice between different systems for the same building. For example for the current (1980) date line some houses could adopt the passive retrofit systems, some the solar water heaters and others, the active space heating system. To a large extent these populations will overlap, i.e. all buildings that can adopt active space heating will include water heating, and most buildings that could adopt passive retrofit could also adopt solar water heating. In principle some buildings could adopt both active space heating and passive retrofit. Also it may be that some buildings, due to orientation, overshadowing, or even 'non-technical' reasons, could adopt active space heating but not passive retrofit, and vice versa. Since the performance of space heating systems is dependent upon the building load, we also have to take account of the take up of conservation measures; i.e. the proportion of systems applied to houses with or without energy conservation measures.

We now bring together the simplified conservation scenario, the percentage suitability from the building stock characteristics, and reference systems to propose scenarios for the three datelines 1980, 2000 and 2020, and calculate the resulting energy saved at each dateline. The percentage of existing houses suitable for each reference system, or combination of systems, are based on limited survey data. This data suggests that 60% of houses can be fitted with solar water heating systems, 30% with active space heating and 45% with passive space heating. However, depending on the varying proportions of houses and flats in each country, these percentages are then weighted to give a percentage suitability.

Preliminary results show that the amount of conventional energy displaced by solar energy could be in excess of 50 m.t.o.e. per annum by the year 2020. Moreover, it is estimated that a further 10 m.t.o.e. could be saved in the non domestic sector. This latter figure is a very crude estimate however, since very little work has been done on solar energy applications in the non-domestic sector and there is very little data available on the non-domestic building stock. At this stage there appear to be some promising applications for solar water heating in hospitals and for passive solar space heating in schools and office buildings.

1.7 The European Solar Industry

The European solar industry now has a business value of about 100 million ECU per annum, with the potential for becoming a 5000 million ECU p.a. business by the year 2000, assuming a growth rate of 25% per annum. In fact, sales have been growing at a rate of just over 30% p.a. between 1976-1980, although there was a fall-off in 1981, as shown in Fig (iii). In general, the industry has developed on the basis of domestic solar water heating systems, although solar heated swimming pools are an important sector, representing perhaps 50% of the market. According to our survey, some quarter of a million solar water heating systems have been installed in Europe since 1974. Most of these are in single family houses. A high proportion of the collectors installed were manufactured in the country where the installation was carried out, although a small quantity is imported, either from other European countries or from the USA, Japan, Australia or Israel.

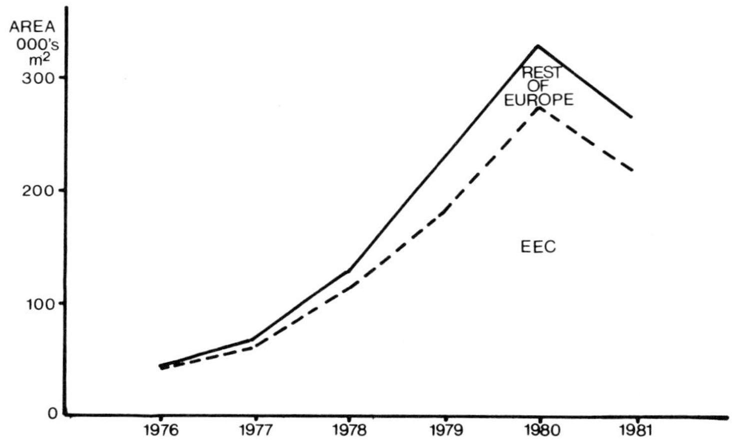

Fig (iii) Area of solar collectors installed 1976-1981

The typical solar DHW system installed to date has a collector area between 5-6m^2, a storage volume of 270 litres and has a total installed price of 490 ECU/m^2. According to current knowlege however, this is probably oversized for northern Europe. Prices certainly vary widely across Europe, from 260 ECU/m^2 in Greece to 580 ECU/m^2 in Ireland.

Concerning the industry itself, this is made up of between 400-500 mainly small to medium size business organisations, including manufacturers, importers, distributors and installers. Of the suppliers of collectors, it is estimated that probably 50 or 60 firms supply as much as 75% of the total sales volume. Although the majority of firms in the industry are small, having moved into the solar field from metal fabrication or heating and ventilating, in some countries, the main suppliers are large household name firms.

In most countries, Trade Associations have now been formed to represent the industry in negotiations with governments on, for example, the setting of standards or introduction of solar legislation. Other activities include promotion and marketing through publications and use of the media. Most of the Trade Associations, together with national standards organisations have been working on the preparation of codes of practice, specifications and test methods.

Looking to the future, the long term viability of the industry is dependent on many factors, not least of which is the future price of conventional fuels. We have already discounted the 'scenario' approach to predictions of the future because it is over-simplistic. Nevertheless it is interesting to examine the figures resulting from this approach to assessing future sales of equipment. Assuming a 'no growth' scenario, the number of installed systems would increase from 250,000 in 1981 to about 1.8 million by the year 2000. This represents a market penetration of only 1%. However, if the number of systems installed increase by 25% each year, the number of installations increases to 20 million by 2000, representing a 15% market penetration. These calculations assume that saturation is reached in any particular country with a 70% market penetration. If this were reached in practice it would result in over 90 million installations in Europe.

Clearly these scenarios constitute a considerable business potential with good prospects for increasing employment as well. There is a general concensus within the industry however, that this potential will not be reached without financial support, either from the financial institutions or from government direct.

The solar industry is a new venture, and like all new ventures it is likely to make a loss for a number of years until the potential customer has become fully aware of the product. There is a need therefore, for increased financial support to the industry and for increased effort on marketing the product and educating potential consumers. On the subject of research and development, opinions from private industry varied, although on the whole there was a feeling that current programmes are just about adequate. The major gaps are seen to be in the areas of long term storage and performance results from existing systems. Awareness of real commercial and industrial needs were felt to be somewhat inadequate and it was also considered that links between the industry and the research community could be improved.

1.8 Main Problems Encountered

A large proportion of the technical assessment has now been carried out, although it has proved time consuming because of the large scope of work to be covered. Likewise, the collection and analysis of data from the solar industry has proved to be a laborious process, mainly because of the need to cross check and compare data from several different sources in each country. However, the main problems have been concerned with the actual assessment of solar potential.

Although we have opted for a fairly simple methodology in conceptual terms, the implementation of this methodology has been more difficult than anticipated. Briefly, the acquisition of suitable data about the building stock, both now and in the future, has proved to be a stumbling block. So too, has the question of establishing national average house heat loss coefficients, both current and in the future. It is very difficult to predict the rate of take-up of energy conservation measures and their overall effect on heat loss. To what extent, for example, will conservation measures merely result in improved comfort conditions.

There is also the problem of establishing percentages of new and existing buildings suitable for solar measures either active or passive. Again there is virtually no knowledge on this subject available at the present time. Whilst carrying out the assessment, the importance of these issues has become increasingly clear. In particular, the long term potential has proved to be particularly sensitive to the rate of new build construction assumed.

Finally, in order to arrive at date-line solar contributions it is necessary to introduce market penetration factors, and particularly economics. There are also many uncertainties in this area, not least of which is the question of future fuel prices, system costs, system lifetimes and most importantly, which discount rates to apply in the economic calculations.

1.9 Provisional Conclusions

It is premature at this stage in the study to draw firm conclusions, but a number of points can be mentioned, having emerged as being of some particular significance.

The technical viability of domestic solar heating technology is now well established. R & D activity during the past decade has created a sound technical base from which to develop economically viable systems. A major effort is needed, however, to find effective ways to disseminate the results of research and development work to industry and to architects and engineers 'in the field'. This process of technology transfer is a major task in the immediate future. At the same time potential users need to be informed about the benefits of solar energy. More effort is also needed on the marketing of solar technology. This point has been stressed as the single, most important requirement for the future develpment of the technology by both researchers and industrialists. There is also a need for more field trials and demonstration projects to inform both the public and the design professions about the benefits of solar technology. However, these demonstration projects need to be properly monitored in order that they can be of value to future designers. The level of monitoring and the size of the statistical sample are important factors to be considered.

Solar water heating is now a proven technology, but the market must be stimulated to achieve further developments. At present most of the effort is going into reducing total system costs and developing optimised designs. Recently there have been some interesting developments in 'packaged' systems which considerably reduce installation times. Thermosyphon systems also appear to offer some significant advantages. In the case of active space heating, some promising '2nd generation' systems are beginning to emerge, consisting mainly of smaller, better optimised systems capable of higher outputs per unit collector area. It is generally acknowledged however, that the full potential of active space heating will not be reached until long term heat storage systems are developed.

Passive designs look particularly promising for both retrofit and new build aplications. The quantitative assessment shows that passive systems can make a contribution of the same order of magnitude as active systems, with the possiblity of very low costs. However, passive solar design is at an earlier stage of development and further R & D is necessary to reduce some current uncertainties about the potential benefits.

Finally, it has been noted that most of the R & D carried out to date has been concerned with solar energy applications in the domestic sector. The work now needs to be extended to include non-domestic buildings such as schools, factories and offices.

SECOND EUROPEAN PASSIVE SOLAR COMPETITION 1982

Authors: R Lebens, D Clarke

Contract No: ESA-A-059-UK (G)

Duration: April 1981 - 31 March 1983

Total Budget: £103,645 CEC contribution 100%

Head of Project: R Lebens

Contractor: Ralph Lebens Associates

Address: 4 Tottenham Mews, London W1P 9PJ UK

Summary

Following the success of the 1st European Passive Solar Competition it was decided to hold a second competition. The Aim of the competition is to disseminate information and knowledge of passive solar design by providing the opportunity for innovatory housing designs based on passive solar principles. The competition is structured to encourage entries from designers with no previous experience of passive solar design. Each entrant is provided with a guidelines booklet, a step by step calculation method and a booklet of climatic data. These documents have been prepared in the seven major languages of the EC. Special arrangements have been made to encourage participation of schools of architecture.

Applications have closed at 2156

The final submission date for completed entries is 27 August with judging in September.

The prize giving and exhibition of entries will now be during the International Conference 'Passive Solar from Research to Design' to be held in Cannes from 13-16 December 1982.

1.0 AIMS

This competition provides the opportunity for innovatory housing designs which combine solutions to the usual architectural constraints and to the energy problems of the selected climate, without the use of complex or elaborate mechanical equipment.

The competition is being structured to encourage entries from designers with no previous experience in passive solar design. A booklet of passive design guidelines, included in the entry forms package, provides detailed information. It is extensively illustrated with cartoons and diagrams. The competition is open to all architects and students of architecture in the EC.

1.1 BACKGROUND

Promotion and coordination of energy research is one of the tasks pursued by the Commission of the European Communities in the framework of its energy policy. In this context it is the aim of the Commission to assist in the stimulation and dissemination of new technological developments which show potential benefits to the economies of the EC countries either by energy saving or by calling upon new non-fossil energy sources. Fossil fuels are becoming increasingly scarce and expensive. They are vital to the health of the european economies in general, but their use for certain forms of low grade energy may not be necessary. Some architects and engineers are beginning to think about these new problems and are developing cost-effective climate-responsive designs which can significantly reduce energy consumptions. "Passive Solar Design" is the universally accepted term used to describe this climate-responsive architecture. Such a departure from existing architectural practice involves the innovation of inexpensive climate-sensitive living systems and the re-evaluation of existing methods of designing and detailing buildings.

The "Second European Passive Solar Competition" is a housing ideas competition, launched by the Commission in response to the growing needs for dissemination of the philosophy and principles of passive solar design within the European Community. It is thought that such a competition will be a most effective method of disseminating the concepts of passive solar design to the architectural and mechanical engineering professions.

1.2 PROGRAMME

The main change in the programme has been the alteration in the exhibition date to coincide with the Passive Solar Conference in December in Cannes. The only other alteration was a printing delay which meant that the calculations booklets were only available from the end, not the beginning, of January 1982.

2.0 PROGRESS REPORT

2.1 CHOICE OF ASSESSORS

A shortlist of assessors was drawn up in conjunction with the Commission the final list is:-

Technical Assessors:

- Nick Baker, Cambridge University, UK
- Orazio Barra, Universita di Calabria, Cosenza, Italy
- Paul Caluwaerts, Centre Scientifique et Technique de la Construction, Brussels, Belgium
- Albert Dupagne, Universite de Liege, Belgium

PROGRAMME FOR SECOND EUROPEAN PASSIVE SOLAR COMPETITION

	1981 A	M	J	J	A	S	O	N	D	1982 J	F	M	A	M	J	J	A	S	O	N	D	1983 J	F	M
Preparation of Brief	>>>>>>>>>>>>																							
Judges Meeting			*																					
Invitation to Schools				*																				
Document Preparation					>>>>>>>>>>>>>>																			
Main Announcement						*																		
Printing								>>>>>>>>>>																
Despatch of Documents										>>>>>>>>>>>>>>>>>>>														
Sorting Entries																	>>>>>>							
Last date for Entries																	*							
Technical Assessment																		>>						
Main Assessment																		>>						
Press Release																			>>>>					
Models																			>>>>>>>>					
Prize Giving																					*			
Exhibition																					>>			
Final Report																						>>>>>>>>>		

Aldo Fanchiotti, Istituto Universitario di Architettura di Venezia
Dean Hawkes, Cambridge University, UK
Chris Kupke, Fraunhofer-Institut fur Bauphysik, Stuttgart, FDR.
J Owen Lewis, University College, Dublin, Eire
Michel Raoust, Consulting Engineer, Paris, France
Michel Schneider, CNRS, Nice, France
Hasso Schreck, Technische Universitat, Berlin, Deutschland
Alexandros Tombazis, Architect, Athens, Greece
Martin de Wit, Technische Hogeschool, Eindhoven, Nederland

Architectural Assessors:
James Barrett, Architect, Dublin, Ireland
Edward Cullinan, Architect, London, UK
Michel Gerber, Architecte, Perpignan, France
Sergio Los, Architetto e Professore, Venezia, Italia
Jaques Michel, Architecte, Paris, France

2.2 PREPARATION OF BRIEF

The assessors met on 26 June 1981 to discuss the proposed design categories and documentation for the competition.

It was felt that the single house category had attracted a very good response in the first european passive solar competition and possibly to the detriment of the other categories. After some discussion it was agreed for the second competition to limit the choices to two categories:-

Category A: Low rise high density housing
Category B: Retrofit and rehabilitation of dwellings

It was decided to encourage greater participation from schools of architecture by giving them special entry arrangements and advance publicity so that the competition could be included in their programmes.

2.3 LANGUAGES

All documentation had to be produced in the seven principal languages of the European Community. It was thought that ample time had been allowed but despite the use of air courier for all technical proofreading it was not possible to achieve the programme deadlines for translated material. Fortunately some of this lost time was made up and the principal information was available at the stated date of 1 January 1982.

2.4 DOCUMENTS

There are two main types of printed documents - publicity material for the competition and documents for the use of the entrants.

Each entrant receives the following papers:
- an illustrated guidelines booklet in editions for each of 7 languages
- a Calculations booklet of 24 pages in editions for each of 7 languages
- a Data booklet of 16 pages in 2 editions of 4 languages in each
- an entry form in 2 editions of 4 languages each
- a Conditions and Instructions booklet in 2 editions of 4 languages each
- a separate official envelope to contain the entry form
- self addressed label for entries

It was impossible to estimate accurately the number of entrants requiring each language so an allowance for reprinting was made and was used for German and English.

2.5 CALCULATION METHOD

The calculation method which entrants have to fill out is based on the Solar Load Ratio development at the University of California, Los Alamos scientific Laboratories. The method had to be converted into metric and put into a form with a step by step guide to take the reader through. A manual method such as this is necessarily lengthy and although it may appear daunting at first it was felt important to communicate the idea that passive solar design must be based on a clear understanding of principles and thorough estimation of performance.

2.6 SCHOOLS OF ARCHITECTURE

The response from schools of architecture had varied widely in different member countries, see 2.7, but it is thought that the early announcement by letter to all schools of architecture has increased the response.

The arrangement for schools is that they receive up to 40 sets of documents for £25 (compared to £5 for one set for other entrants) and are allowed to submit up to 5 designs from these.

2.7 ENTRANTS

The applications for documents have arrived consistently up to the closing date of 18 May 1982. It has been decided, in view of the extreme postal delays from some parts of the EC to send documents for applications postmarked before 18 May, even if they arrive after that date. The provisional list for applications is:-

DOCUMENTS SENT OUT

Country	Individual Applications	Schools of Architecture
BELGIUM	84	200
DENMARK	75	-
FRANCE	366	40
GERMANY	376	80
GREECE	27	-
IRELAND	35	70
ITALY	170	66
LUXEMBOURG	3	-
NETHERLANDS	140	-
UK	119	305
TOTAL	1395	761

Potential entries individuals : 1395
from Schools of Architecture: 110

The likely number of entrants will be approximately 25%

LIST OF DOCUMENTS USED BY LANGUAGE

	Danish	Dutch	English	French	German	Greek	Italian
To individual applicants	75	156	166	412	389	27	170
To schools of architecture	-	-	375	240	106	-	40
To CEC	55	55	55	55	55	55	55
To judges, translators	2	5	6	6	3	2	4
Lost in post	-	3	40	-	1	-	-
TOTAL	132	219	642	713	554	84	269

A number of applications were received from outside the EC due to inaccurate reporting of the press release. These were all returned.

2.8 PUBLICITY

Individual letters were sent out on 9 July 1981 to all schools of architecture in the EC. 2,500 press releases were sent out in August 1981 to specialist publications, interested institutions and individuals throughout the community. The address list from the first competition applicants was also used.

3.0 FUTURE WORK

3.1 PRIZE GIVING AND EXHIBITION

The site for the prize giving and exhibition has been moved from Brussels to Cannes to coincide with the International Conference 'Passive Solar From Research to Design', 13 to 16 December 1982. As in the first competition the winners will be asked to make models of their schemes for the exhibition. It is hoped that the exhibition will be sent to various countries in the community during 1983.

3.2 ASSESSMENT

Space has been reserved at the Royal Institute of British Architects in London for assessment. Technical assessment will be from 6 to 10 September 1982, architectural assessment will be on 15 to 17 September.

3.3 PUBLICITY

To maximise interest in the competition a press release including illustrations of winning schemes will be distributed to come out simultaneously with the announcement of results. Invitations to the prize giving and exhibition will also be widely distributed.

3.4 FINAL REPORT

It is intended that the final report will be published to provide a permanent record of the competitions achievements. The main content will be the drawings and models of the winning and commended designs together with the performance predictions and description of the organisation of the competition.

CO-ORDINATING CONTRACT FOR THE EUROPEAN PASSIVE SOLAR WORKING GROUP AND MODELLING SUB-GROUP

CONTRACT No. : ESA-PW-108-UK

DURATION : from: 1 July 1981 to: 30 June 1983

TOTAL BUDGET : £99,526 CEC contribution: 100%

HEAD OF PROJECT: R Lebens

ORGANIZATION : Ralph Lebens Associates,

ADDRESS : 4 Tottenham Mews, London W1, UK

SUMMARY

The Passive Solar Working Group (PSWG) and Modelling Sub Group (MSG) were established by the Commission to promote and improve passive solar design in EC countries. Although only recently started, Actions its proposed tasks are ambitious for such a short time-span. The brief is to carry out research and to recommend work to be included in the next four year programme within the following subject areas:

- Simulation and simple design models
- Design Guidelines and Educational Support
- Planning and Urban Design
- Retrofit and Rehabilitation
- Passive Solar Components
- Air Movement and Heat Distribution
- Thermal Comfort
- Test Facilities

The task of investigating passive solar simulation tools requires a group of modelling experts and has therefore lead to the formation of the Modelling Sub-Group. The detailed work of this Modelling Group is described by its Chairman, Prof A Dupagne, in the paper following this one.

The major tasks of the PSWG are described in this paper.

1.0 PARTICIPANTS OF THE PASSIVE SOLAR WORKING GROUP

Organization:

DG XII of the CEC - T C Steemers, 200 Rue de la Loi, 1049 Brussels, Belgium. Tel.02/235 6878
Project A Co-ordinator - C den Ouden, Institute of Applied Physics TNO-TH, 2600 AD Delft, The Netherlands.
Co-ordinating Contractor - R Lebens (as on first page)

Representatives from participating EC countries were recommended by the ACPM members from each country:

Belgium - P Caluwaerts, Belg. Bldg. Res. Inst., Lombardstraat 41, B-1000 Brussels, Tel.02/653.88.01

Denmark - L Olsen - Thermal Insul. Lab., Tech. Univ. of Denmark, Bldg. 118, Dk - 2800 Lyngby, Tel.02-88.35.11 x5228

France - M Bellanger - SOREIB, 27 rue de Longchamp, 92200 Neuilly-sur-Seine, Paris CEDEX 16, Tel. 1 722 8349

F R Germany - K. Bertsch - Fraunhofer-Institut fur Bauphysik, Koningstrasse 74, 700 Stuttgart Tel. 49 711 765008

Greece - E Tsingas, P O Box 8, Thessaloniki, Tel.(31) 20.07.02

Ireland - J O Lewis, Energy Research Group, School of Arch., Univ. College, Richview, Clonskeagh, Dublin 14, Tel.01-696416

Italy - S Los, Via Boschetto 9, Bassano del Grappa, Tel.0424/21/046

The Netherlands - R Kiel - Bouwcentrum, Weena 700, Postbus 299, NL-3000 AG Rotterdam, Tel. 010 3110/11.61.81.

United Kingdom - D Hawkes, Martin Centre for Arch. & Urban Studies Dept. of Architecture, Univ. of Cambridge, 6 Chaucer Rd., Cambridge. Tel. 0223/69501

Observers:

Modelling Sub Group Chairman: A Dupagne
Performance Monitoring Group Co-ordinators: Energy Conscious Design (R Godoy)

The PSWG receives 100% funding from the CEC.

2.0 PARTICIPANTS OF THE MODELLING SUB GROUP

Organisation:

DG XII of the CEC - T C Steemers, 200 Rue de la Loi, 1049 Brussels, Belgium. Tel 02/235 6878
Project A Co-ordinator - C den Ouden, Institute of Applied Physics TNO-TH, 2600 AD Delft, The Netherlands

Co-ordinating Contractor - R Lebens, Ralph Lebens Associates, 4 Tottenham Mews, London W1, UK Tel. 01 636 7172

Participants:

Belgium - A Dupagne (Chairman), Laboratoire de Physique du Batiment, 15 Ave des Tilleuls, Bat D1, B-4000 Liege, Tel 041/52 01 80 ext 367

Denmark - H Lund, Tech, Univ. of Denmark, Thermal Insul. Lab., Bldg. 118. DK-2800 Lyngby, Tel. 288 35 11

France - M Raoust, SCPA, CLAUX-PESSO-RAOUST, 70 Bd de Magenta, 75010 Paris, Tel. 205 53 02

F R of Germany - Chr Kupke, Fraunhofer-Institut fur Bauphysik, 7 Stuttgart - 70 Konigstrasse 70, Tel. 0711/76 50 08

Ireland - J Cash, Dublin Inst. of Tech., Bolton Street, Dublin 1 Tel. 01/74 99 13

Italy - G P Alpa, Polytechnic School of Turin, C Duca degli Abruzzi 24, 10129 Torino

The Netherlands - G Pernot, Samen-werkingsverband FAGO/TPD, T H Eindhoven, Postbus 513, 5600 MB Eindhoven, Tel 040/ 47 26 78

United Kingdom - J Littler, Bldg. Unit, Polytechnic of Central London, 35 Marylebone Rd, London NW1 5LS, Tel 01 /486 58 11 x345

Observer:

Performance Monitoring Group Co-ordinators: Energy Conscious Design (R Godoy)

The work undertaken by the participants of the Group is being partially funded by the Commission through research contracts which will pay the following percentages of costs:

Research time	50%
Computer costs	50%
Travel costs	100%

The coordination work of the Chairman is 100% funded by the Commission.

3.0 AIMS

The aims of the two groups were established at the very outset. These were as follows:

3.1 THE CEC PASSIVE SOLAR HANDBOOK

Assembling information from several exisiting European and American guidelines, the PSWG would write an extensive European passive solar handbook, providing useful design methods and data for building professionals. The handbook would be ready for publication in mid-1983. This handbook would include many of the research results obtained by the group and by the Modelling Sub-Group.

3.2 THE EUROPEAN PASSIVE SOLAR PRODUCTS CATALOGUE

To promote the availability of passive solar building products, the PSWG is producing a catalogue of those products available in Europe.

Manufacturers' literature would be organised in a convenient format, with periodic updates. The PSWG would also study new components or materials developed in America and provide information concerning them to potential manufacturers in the EEC.

3.3 PLANNING AND URBAN DESIGN

Factors such as urban tissue, wind sheltering, sun shading by other buildings, surrounding landforms, and vegetation can all affect a building's thermal performance. The PSWG would compile information regarding these effects for inclusion in the Passive Solar Handbook, and recommend a programme of further research.

3.4 RETROFITTING AND REHABILITATION

The potential fuel savings from effective use of passive solar retrofitting could far exceed those from new passive solar construction, projected over the next twenty-five years. The PSWG would present a method for surveying the existing building stock of EC countries, and include information on retrofitting in the Passive Solar Handbook.

3.5 EDUCATIONAL SUPPORT AND A TRAVELLING WORKSHOP

Beginning in early 1983, PSWG members would present two-day workshops in their respective countries. These workshops would acquaint design professionals with many of the economic advantages of passive solar heating, and introduce them to fundamentals of passive design and design tools.

3.6 TECHNICAL RESEARCH

Several technical subjects would receive particular attention. These include:

Air movement and heat distribution - In passive buildings the natural circulation can be an important influence on thermal performance.

Thermal Comfort - Several factors other than air temperature affect thermal comfort. Heating controls should be sensitive to some of these factors and adjust back-up heat accordingly.

Test Facilities - Many thermal processes (such as air movement) are too complex to model without proper test facilities.

The PSWG would collect information on these subjects for inclusion in the Passive Solar Handbook. It would also make recommendations for further work in these areas during the next four-year CEC R&D programme.

3.7 THE PASSIVE SOLAR MODELLING SUB-GROUP

Selecting reliable computer programmes for simulating the performance of passive solar buildings. A detailed programme for large computers would be selected and then used to calibrate a set of simpler models: those for mini/micro computers, programmable calculators, and manual calculations. The accurace of these simulation techniques will be evaluated using test data from actual buildings. Selected small models will included in the Passive Solar Handbook.

4.0 PROGRESS REPORT

The formation of the PSWG (Passive Solar Working Group) and the MSG (Modelling Sub Group) has had a few teething problems. But over the past months the momentum has been growing and the quality and quantity of output and contribution by participants has greatly improved. The work of the PSWG is very much on target according to their programme as shown in

Diagram 1. There are always weak links within any group but hopefully over the remaining contract period these can be minimised by means of example from the other members of the group.

The two greatest problems have been in the form of inconsistent participation. In the PSWG it has been the French participation, and in the MSG the Italian participation. There is no Greek participant in the MSG. The French participation has now hopefully been resolved with the appointment of M Bellanger, a private consultant in Paris. There have been a total of four meetings of both groups to date, at the following times and places:

Passive Solar Workshop, 6 May '81, Brussels
PSWG & MSG Meetings No 1, 22-25 June '81, Brussels
PSWG & MSG Meetings No 2, 26-27 August'81, Brighton
MSG Meeting No 3, 2-3 December '81, Glasgow
PSWG Meeting No 3, 17-18 December '81, Sophia Antipolis, Nice
PSWG Meeting No 4, 16-17 March '82, Liege
MSG Meeting No 4, 21-22 April '82, Liege

Future meetings:

PSWG Meeting No 5, 10-11 June '82, Venice
MSG Meeting No 5, 28-29 June '82, Brussels
PSWG & MSG Meetings No 6, 13-15 October '82, Brussels
PSWG & MSG Meetings No 7, 26-28 January '83, Paris

A separation in meeting time and place between the two groups occured in December '81 and March/April '82. For meeting No 3 the MSG went to the University of Strathclyde in Glasgow, the home of their chosen large model ESP, to see the model in operation. The MSG meeting No 4 was postponed because the participants needed more time in which to commission ESP onto their own machines. The original programme for the MSG is shown in Diagram 2.

OUTPUT OF THE GROUPS

5.0 A SURVEY OF EUROPEAN PASSIVE SOLAR WORK

Each member of the PSWG conducted a survey of the passive solar work being undertaken and proposed in their country. A standardized questionnaire form was filled in for each project. Members of the Modelling Sub-Group focussed their attention on the survey of design tools. A synthesis of these surveys was made classifying some 236 projects in Europe which are thought to be of relevance to passive solar design.

The research is classified both by country. The following categories were set up:

- Design Guidelines and Educational Support
- Planning and Urban Design
- Retrofit and rehabilitation: General
- Retrofit and rehabilitation: Technical Aspects
- Technical Aspects
- Buildings and Performance
- Design Tools

DIAGRAM 1 WORKING GROUP PROGRAMME

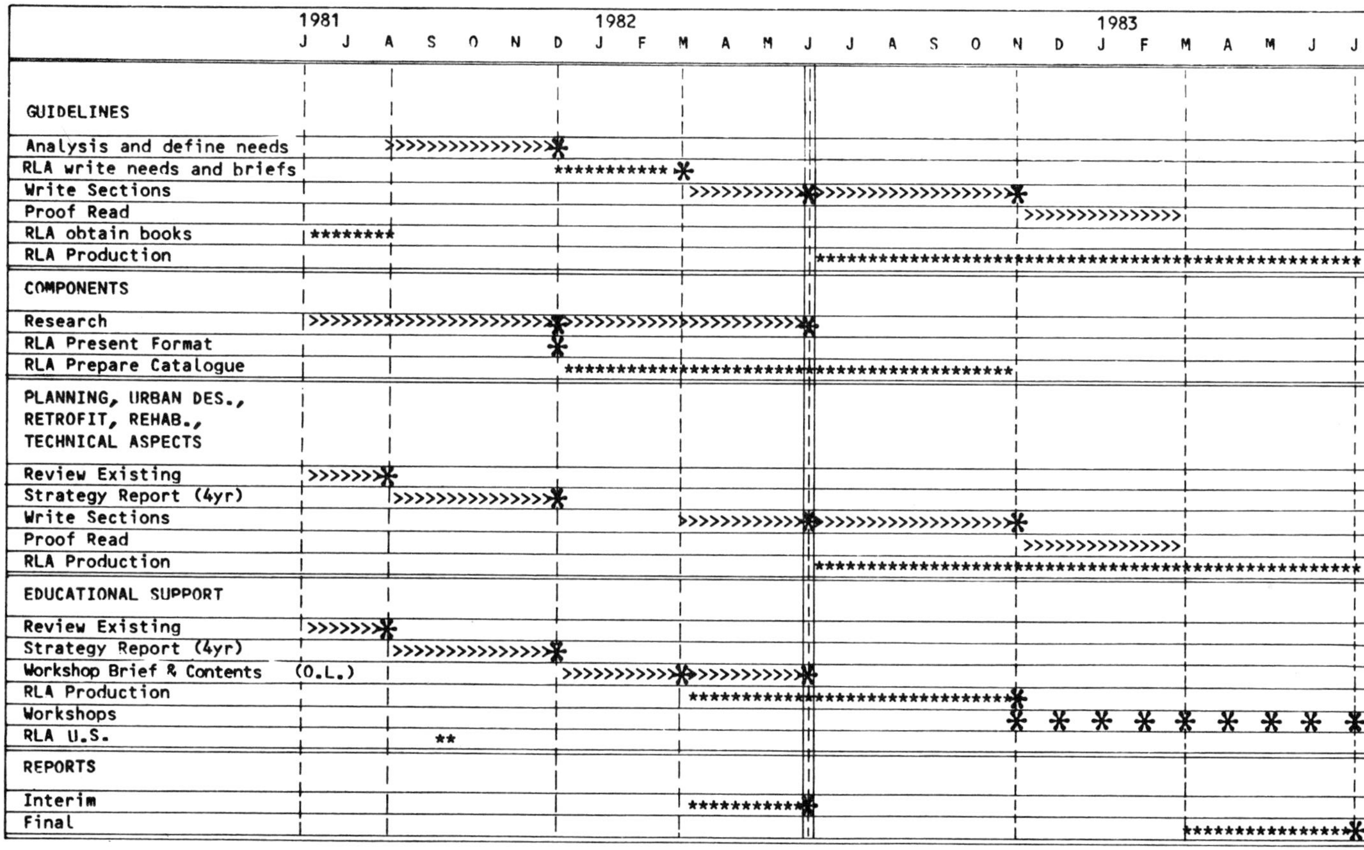

DIAGRAM 2 MODELLING SUB-GROUP PROGRAMME

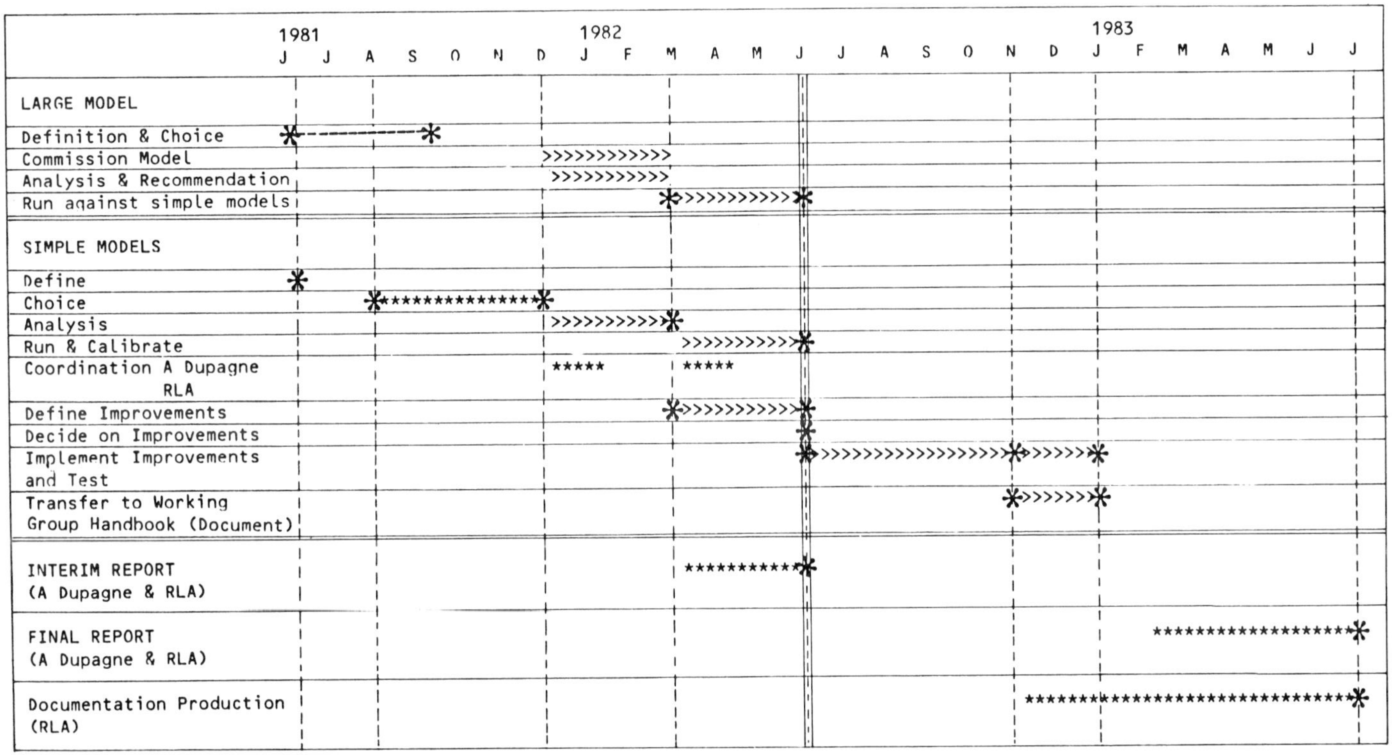

6.0 REVIEWS OF EUROPEAN PASSIVE WORK

Having completed the surveys above, the Working Group was subdivided into the following specialist teams:

- Educational Support
- Planning and Urban Design
  Retrofit and Rehabilitation
- Technical Aspects

The Technical Aspects team covered the fields defined within the aims of the group of:

Air movement and Heat Distribution
Thermal Comfort
Test Facilities

Each team reviewed the European Work which related to its specialty and as an outcome, made recommendations in the form of a strategy paper, of the work to be done in this current programme and for the next 4-year programme.

7.0 THE CEC PASSIVE SOLAR HANDBOOK

The PSWG has identified a total of 15 design handbooks and references for review. The objective was to select the good aspects of each and attempt to incorporate them in the CEC handbook. Similarly to exclude any shortcomings. Each book was reviewed in terms of presentation, structure and content. Recommendations were made about the inclusion of selected material from each reviewed book, in the CEC handbook. The group then went on to suggest a structure and contents for the CEC handbook. These suggestions were condensed into one proposal.

The proposed contents of the CEC handbook is extremely ambitious when it is considered that the intention is to produce camera-ready copy for publication by mid-1983.

The sections have been allocated to different members of the PSWG and those related to design tools assigned to the MSG. A formidable quantity of reference literature relating to each section has been collected and each member will be responsible for gleaning the best of this material for inclusion in his section of the handbook. Draft sections are currently being written. The reference literature is a combined collation of books reviewed by the group, literature known to each participant and literature collected by the co-ordinators and distributed to the group.

8.0 PASSIVE SOLAR PRODUCTS CATALOGUE

The coordinators briefly reviewed existing products catalogues from Europe and America. This lead it the proposal of a letter to manufacturers and distributors (of non EEC products), a contents list for the products investigated, a list of headings for which manufacturers to provide information, and a sample layout of the catalogue page. These proposals were accepted by the group after suggested changes were made. Each participant of the PSWG was then asked to translate the sheets and to send them out to manufacturers and distributors in his country. This action has only just been undertaken by the group and therefore no indication of magnitude of response can be given as yet.

9.0 EDUCATIONAL SUPPORT AND A TRAVELLING WORKSHOP

It is at present proposed to hold the first Workshop in early January

1983 in Dublin. It is thought that this will serve as a prototype for later workshops.

Two similar workshops have been attended and reviewed by the Co-ordinators to provide some critical input to the successful organization of such workshops:

- Passive Solar Associates Workshop, Santa-Fe, New Mexico, USA, Sept 17-18, '81
- Profs Duffie and Brinkworth Workshop, Cardiff, UK, May 17-20, 1982

Depending on the success of the first workshop in Ireland, these will be repeated throughout Europe.

10.0 DESIGN TOOLS

The Modelling Sub-Group has surveyed and reviewed desigh tools for four categories of hardware:

- Manual calculation
- Programmable Calculator
- Mini/micro computer
- Main frame computer

The first task was to select one large model to be commissioned on the computer of each participant. This has been carried out with the model ESP from the University of Strathclyde. The next task will be one of scrutinizing the simplified models, making improvements, and selecting those for inclusion into the CEC handbook. The detailed work undertaken by this group will be explained by the chairman of the group, Prof A Dupagne.

11.0 STRATEGY PAPERS

The strategy papers were undertaken as a joint effort by each team to distil the areas of weakness or lack of attention in existing European passive solar work. They were a direct outcome of the review of surveyed European passive solar projects. The strategy papers dealt with both the needs of this present programme period and the future 4-year CEC R&D programme. These lead to a combined strategy paper for the next 4-year programme which is at present in the form of a discussion paper. It is divided into the following areas of recommended work.

- Education
- Planning
- Comfort, Auxiliary Heating and Controls
- Modelling and System Optimization
- Climate
- Performance Monitoring of Passive Solar Homes
- Components
- Heat Transfer, Air Movement and Storage
- Cooling

Very little feedback from the groups has been received as yet.

EUROPEAN PASSIVE SOLAR MODELLING SUB-GROUP

Authors : A. DUPAGNE, J.M. HAUGLUSTAINE

Contract number : ESA-PM-097-B

Duration : 25 months   1 June 1981 - 30 June 1983

Total budget : £ 100,000
CEC Contribution : 100%, except for time and computer costs (50%)

Head of project : A. DUPAGNE

Organization : University of Liège
Building Physics Laboratory
Address : Avenue des Tilleuls, 15, Bât. D1
B-4000 Liège (Belgium)

Participants in the concerted action :

ALPA, G.P.	Polytechnic School of Turin, I.
CASH, J.	Dublin Institute of Technology, IRL.
KUPKE, C.	Fraunhofer Institut für Bauphysik, FRG.
LEBENS, R.	Ralph Lebens Associates, UK.
LITTLER, J.	Polytechnic of Central London, UK.
LUND, H.	Technical University of Lyngby, DK.
PERNOT, C.	Technische Hogeschool van Eindhoven, NL.
RAOUST, M.	S.C.P.A. Chaux-Pesso-Raoust, F.
DUPAGNE, A.	Building Physics Laboratory, B.

Observers : DEN OUDEN, C. Technisch Physische Dienst, Institute of Applied Physics, NL.
(project leader for the A Project : Solar Applications to Dwellings)
GODOY, R. Energy Conscious Design, UK.
(observer from the Performance Monitoring group).

SUMMARY

The task of selecting reliable computer programs to simulate the performance of passive solar buildings has been assigned to a specialized subgroup. This group has selected a detailed program for large computers, which will be used as a reference for the analysis of a set of simple models : those for mini/micro computers, pocket calculators, and manual calculations.

The accuracy of these simulation techniques is being evaluated by using test data from actual buildings and by comparing the results obtained with the large program.

The best small models will be improved and included in the Passive Solar Handbook prepared by the Passive Solar Working Group.

1. INTRODUCTION

The objectives of the Modelling Sub-Group are to unify European research in the field of computer models, to provide the best simplified models in the Passive Solar Working Group's extensive European Passive Solar Handbook and to ensure that these simple models are easy to use for design purposes.

It has defined the needs of the research program for a large model and has chosen the Scottish program ESP as a large simulation model to be used as a reference for the analysis of the simplified ones. Each participant has commissioned the chosen model on his own computer and run a very simple exercise, just to verify that the program works similarily on the different machines. The analysis of the large model is being brought on its different sections, such as : input, output, heat transfer, passive solar systems, etc. The group will run ESP using the same passive solar bulding data as inputted to simplified models to assess the sensitivity of the latter ones. It will report on the work needed for the next 4 year CEC research program.

Concerning the simplified models, the group has also defined the needs of the research program for the simplified design models. It has selected 14 simple design methods and programs within each of the following hardware groups : mini- and micro-computers, programmable desk-calculators and manual methods. A first draft analysis has been carried out on the basis of the available documentation and on first results obtained with the same very simple exercise as the one run on ESP. A more complete analysis has to be done in a near future, reporting on their abilities and limitations. The analysis will continue with the running of actual passive solar building data, the results of which will be compared with each other and with ESP results. The best simplified models will be improved -if necessary- and afterwards included in the Passive Solar Handbook, with a comprehensive documentation for their easy use by designers.

2. LARGE PROGRAMM CHOICE AND COMMISSIONING

Sophisticated computer models were reviewed with the objective of choosing one for immediate reference and for improvement during the next four year research program. After a first selection, a short list of three was drawn up :

- ESP, University of Strathclyde, United Kingdom
- LPB-1, University of Liège, Belgium
- MORE, Politecnico di Torino, Italy.

As thoroughly explained before (1), it has been decided to give option to ESP, which was already adapted to passive solar matters and would be multizone very soon.

The final multizone version was indeed ready in early December 1981 and the Passive Solar Modelling Sub-Group took the opportunity to hold a meeting in Glasgow, at the University of Strathclyde, the home of the large model ESP. Its author, J.A. Clarke, has given the group a very fruitful presentation of the program, housed on a DEC 10, but also supplied to various suscribers using other machines. The flexibility and the highly developed user interaction have impressed upon the members of the group. A partial documentation and the magnetic tape have been received by each member in mid-February 1982. According to the type of computer used and corresponding adaptations to do, the commissioning is still in progress or already finished.

3. SELECTION OF SIMPLIFIED PROGRAMS

For selecting a series of simplified models, the group has developed a questionnaire which gives a check-up on all characteristics of each program about inputs, calculation methods, passive solar matters, computer demands, outputs and ease of use for design or research purposes. According to the answers to this questionnaire, a series of 14 programs or methods have been selected and distributed into three groups, according to the hardware required by each of them : manual methods, programmable desk-calculators and mini- or micro-computers (see Table I).

Method	Contact	User-test	Machine
Manual			
1. Méthode 5000	M. Raoust	A. Dupagne	
2. CSTB's method	M. Raoust	M. Raoust	
3. 2nd European Competition	R. Lebens	G.P. Alpa	
4. Admittance method	J. Cash	G.P. Alpa	
7. Calswing (Philip Niles)	R. Lebens	C. Pernot	
Programmable desk-calculator			
5. PSP	R. Lebens	H. Lund	TI 59
6. DMT	J. Cash	C. Kupke	HP 67 or HP 97
8. Pascalc II (TEA, Los Alamos)	R. Lebens	C. Pernot	TI 59 or HP 41
9. Direct Gains	R. Lebens	J. Cash	TI 59
Mini-micro computer			
10.Spiel	J. Littler	A. Dupagne	Apple II
11.Casamo	M. Raoust	M. Raoust	Tektronix/HP 85
12.Suncode	C. Kupke	C. Kupke+J. Littler	CDC/DEC 10
13.Los Alamos	R. Lebens	J. Littler	Apple II
14.Fred 10	R. Lebens	R. Lebens	PDP 11

Table I. List of selected simplified programs with the associated hardware

The analysis of each program will be done on the basis of the available theoretical documentation and on the results obtained with preliminar runnings of passive solar building examples.

4. FIRST RUNNINGS

In order to make sure that ESP works similarily on the different machines used by the participants and to give a first assessment of the simplified models, it has been decided to do a few runnings with building and weather data as simple as possible.

Suggestions for the buildings are the Los Alamos test cells and a very simple 1 m cube (see Figure II), which has been deeply analysed and monitored by both the Belgian Building Research Institute and the Building Physics Laboratory, in the frame of a couple of research works concerning the solar gains coming into a building, the influence of the inertia on the heating energy consumption and the temperature without heating (2).

Three types of calculations are in progress with these two building examples :

1. A steady-state example in two steps. First during the minimum of days to obtain the equilibrium, the external temperature is kept constant (at 0°C) and there are no solar radiations, no wind, and consequently no air change. Then only one step in external temperature (from 0°C to 10°C) is applied to the building, while the other data remain constant.

2. The Danish climatic data are used for the calculation of the indoor air temperature obtained during a cold and a warm ten-day period ; the air change rate is 1 ac/h and there are no casual gains in the building.

3. The annual heating energy consumption is calculated with the same Danish climatic data, for an indoor consign temperature of 18.3°C.

The results will be presented in the same way on transparencies with same scales. This type of presentation is very useful to compare the results. The working time is indeed too short to give the group the possibility to do the perilous exercise of comparing outputs given by a program (with theoretical hypotheses and simplifications), with parameters measured on the field by the use of measurement tools (without any perfect accuracy). The measurements must be only used to compare the trend of the thermal parameters evolution. Then the decision is taken not to compare the absolute results given by a model to the others, but to see their trends and put in light the large differencies between them. Trends are more important than numerical values.

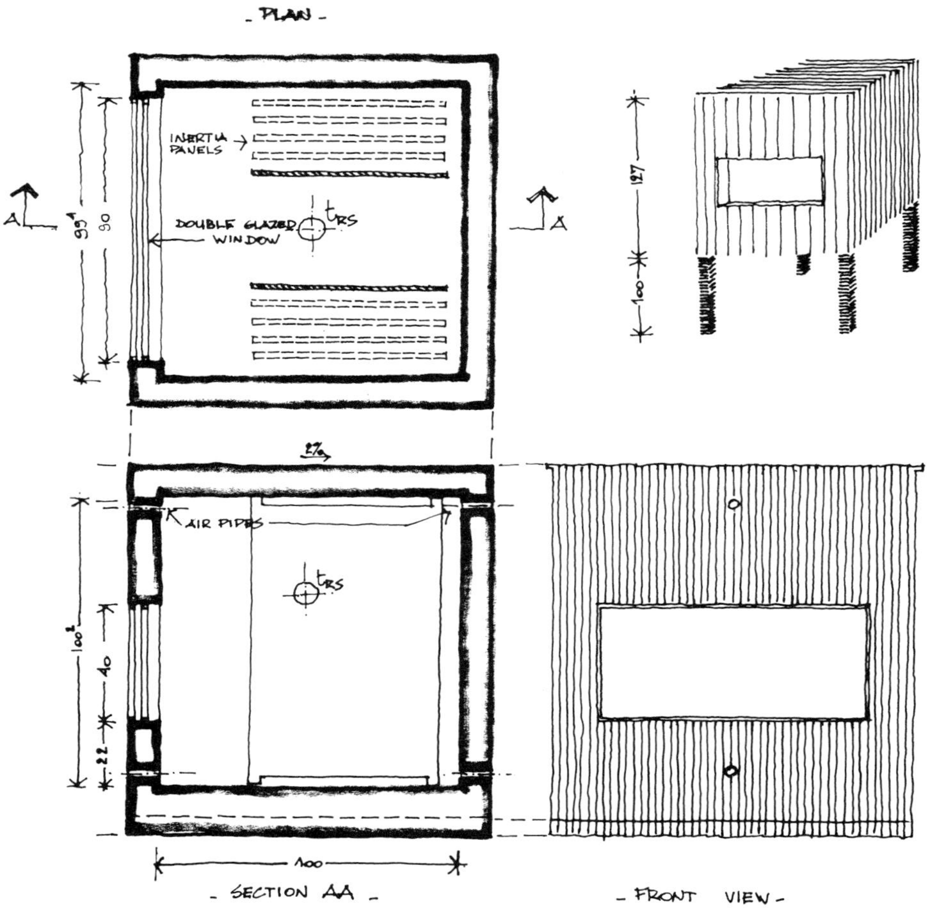

Figure II : Plans of the 1m cube cell

5. FIRST RESULTS

The first steady-state example can be simulated, for the 1 m cube, by using the two exponential formula given by L. LARET (8) : on the basis of the theory, as well as experiments, he has approximated a second order model, with the help of accurate measurement of the time constants in the climatic room at the University of Liège :

$$(t_{a,\tau} - t_i) = \left[1 - (1 - a).e^{-\frac{\tau}{\tau_1}} - a.e^{-\frac{\tau}{\tau_2}}\right].(t_f - t_i)$$

where $t_{a,\tau}$ = internal air temperature after τ hours
t_i = initial outdoor temperature (before the step)
a = 0.02 = immediate equivalent response of internal temperature
τ_1 = 8 h = equivalent largest time constant of the structure
τ_2 = 0,3 h = equivalent fast time constant of the cubic cell
t_f = outdoor temperature after the step.

This simplified formula only concerns the steps of outdoor temperature and doesn't take any other parameter into account.

Figure III gives the comparison between the simulation obtained with the above formula and ESP : both curves give nearly the same thermal behaviour of the 1 m cube. The differences between the two curves may come from a bad theoretical estimation of the thermal properties of a material, although these properties have not been experimentally measured at all.

The program ESP seems to give 9.5°C for the values of the indoor temperature, 24 hours after the outdoor temperature step. Other results have been obtained at the same time from other simplified programs, such as 8.5°C with Suncode and 8°C with Casamo. The results from other programs or other runnings are in progress.

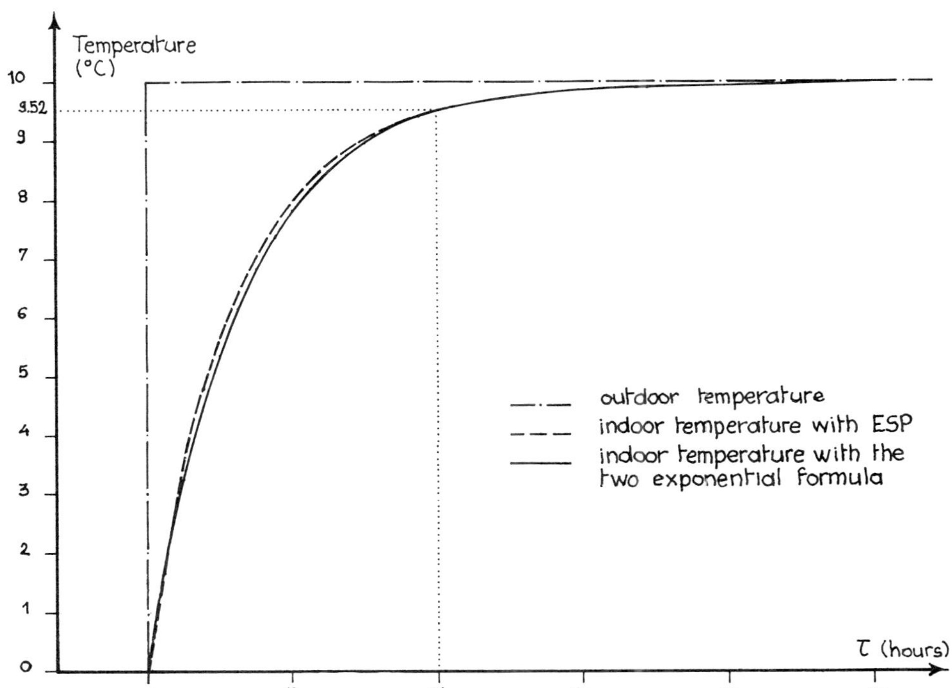

Figure III : Comparison between ESP and the two exponential formula

6. CONCLUSIONS

The analyses of the theoretical hypotheses chosen by the authors of simplified programs are being worked on ; the complete comments on each simplified method will soon be available. A few other building examples must be run, with more passive solar devices than the 1 m cube cell or the Los Alamos test cell having only one South facing window. Sunspaces, Trombe walls, etc. are to be run to see the sensitivity of the models to the different passive solar devices.

The group has devoted a lot of time to receive and commission the program ESP, but it will be very interesting to have, in the Commission of European Communities, a few scientific teams which know the same program and can run it as a common reference for the comparison of the results. The organization of the program itself seems to be so transparent that improvements, in particular routines, will be possible to develop other topics during the next 4 year research program.

Concerning the simplified models, almost one third of these seem to be sufficiently orientated to passive solar matters and to have a good scientific background. The others present lacks in passive solar orientation, in calculation hypotheses or in their adaptation of graphics or tables to European climates. The group will try to have at least one method for each type of hardware, in order to give each designer the calculation tool he can buy to take passive solar devices into account and to be accustomed to introducing these successfully in usual design. So the designer will know, for each program, the calculation hypotheses, the passive solar matters it takes into account and the ease of use for design purposes.

REFERENCES

(1) A. DUPAGNE, J.M. HAUGLUSTAINE, Passive Solar Modelling Sub-Group, paper presented at the EC Contractors' Meeting held in Athens (Greece), November 1981, in : Solar Energy Applications in Dwellings, Volume 1, D. Reidel Publishing Company, Dordrecht, Holland, 1982.

(2) H. HEIKHAUS, J. LEBRUN, Thermal monitoring of a dozen of reduced scale models of houses, paper presented at the International Colloquium "Comparative Experimentation of Low-Energy Houses", held in Liège (Belgium), May 1981.

(3) L. LARET, Contribution au développement de modèles mathématiques du comportement thermique transitoire de structures d'habitation", Thèse de Doctorat en Sciences Appliquées, Université de Liège, 1980.

EUROPEAN MODELLING GROUP FOR SPACE HEATING AND DOMESTIC HOT WATER SYSTEMS

Contract Number : ESA-M-078-B (co-ordinators' contract)

Duration : July 1, 1981 - June 30, 1983

Total Budget :

Head of project : Co-ordinator : W.L. Dutré

Organisation : Katholieke Universiteit Leuven
Laboratorium voor Warmteoverdracht en Reaktorkunde
Fakulteit der Toegepaste Wetenschappen

Adress : Instituut voor Mechanica
Celestijnenlaan 300 A
B - 3030 HEVERLEE, BELGIUM

Summary

The detailed simulation model EMGP1 has been generalised to a new programme, called EMGP2, which can be used to simulate a large variety of solar systems for space heating and/or domestic hot water production.

The proposed statistical validation procedure has been developed and is being used to validate EMGP1 against the experimental data of the SPTF-SS1 installations. The continuation of the validation work, including also the SPTF-SS2 data, will be concerned with EMGP2 only.

Introduction : Objectives and Organisation

The main objectives of this concerted action are summarised in the European Modelling Group strategy paper [2] and the EMG-paper presented at the E.C. Contractors meeting held in Athens, november 1981 [1].

The EMG-programme 1981-83 mainly consists of the following three major tasks :

- development and validation of one European model for active solar systems for application to dwellings;
- development and validation of simplified calculation methods for solar systems;
- system studies, based on the methods being developed, for various climatic zones of the European Communities.

The work related to simplified methods is guided by J. Adnot (Ecole des Mines de Paris) as a sub-co-ordinator and is presented in a separate paper by J. Adnot.

Since the contractors'meeting in November 1981, three EMG meetings have been held :

- on January 14-15, 1982 in St.-Albans (U.K.)
- on March 23-24, 1982 in Paris (France)
- on April 20-21, 1982 in Heverlee (Belgium), which was a special working meeting on the new and generalised simulation programme EMGP2.

The EMG-working documents which have been distributed during the past year are listed in the references.

Detailed Simulation Programme

Since the start of the present contracts it was decided that the EMG-work should be based on only one detailed simulation model. A first version, called EMGP1, i.e. European Modelling Group Programme 1, was developed by the coordinator and is described in [3]. The main features of this programme have been summarised in [1]. EMGP1 was succesfully implemented and tested by the EMG-participants and has been used during the period June 1981 - June 1982 for parametric studies related to the PTF-SS1 operation sequences for the '81-'82 heating season as well as for a series of reference calculations for the validation of the selected simplified methods. The validation based on the experimental data from the SPTF-SS1 installation, according to the statistical procedure which is described in [11] and briefly summarised below, has been started. Several "special tasks" have been performed by the EMG-participants in order to investigate the sensitivity of the EMGP1 results to the requested precision in the integration algorithm, the sensitivity to simplifying assumptions in the configuration of the simulated system and the use of the variable $\alpha\tau$-option in the flat plate collector model. The preliminary conclusions from these studies, in which the PTF-SS1 system was considered, can be summarised as follows :

- the results are not very sensitive to the requested precision in the Runge-Kutta-Gill integration algorithm, while the required computer time for an annual system performance calculation can be reduced considerably (up to 30 %) when this precision request is reduced [14]. Based on this study it was decided, as a compromise between precision and cost of the computer calculations, to perform the validation calculations as well as the system studies with a one degree precision request. This feature of EMGP1 is mainly due to the use of a self adaptive procedure of the integration time step and the high precision of the selected integration algorithm.

- a further reduction in required computer time may result from the use of less sophisticated integration algorithms. The Runge-Kutta-Merson algorithm was implemented and used succesfully [15], and the second order Runge-Kutta procedure was proposed as well. These two integration algorithms are now implemented in the new version of the programme and may further reduce the required computer time by about 30 to 40 %.
- for the PTF-SS1 system it was shown that several simplifications in the simulated configuration of the system can be allowed, such as using a one element approximation for the complete collector array and modelling of the piping without thermal capacity. The thermal capacity of each complete loop should however be such that the effective time constants are preserved [16]. From this study it was concluded that calculations will be performed with such simplified system configurations.
- the use of the variable $\alpha\tau$-option in the flat plate collector model results in a decrease of the calculated system performance [17]. Because in such calculations more detailed optical properties of the collector components are required, some of which are not sufficiently accurately known, such as the angular dependance of the absorption coefficient, extinction coefficient of the glass cover..., and because some approximate assumptions are used for the effect of the angular distribution of the diffuse radiation, it is not clear for the time being how the differences between the constant $\alpha\tau$-model and the variable $\alpha\tau$-model should be interpreted. These differences are however restricted to a few percent and will not be further investigated in the present work of the EMG.

Several tasks have also been devoted to the further development of the simulation programme, resulting in several proposals for programme modifications and for additional features and subroutines. Based on the group experience with EMGP1, the various requests and proposals for modifications and the proposed physical models for additional components, the coordinator has developed EMGP2, which is the generalised version of EMGP1 and which will be used in the future work of the European Modelling Group. The physical models, the calculation methods, the EMGP2-user guide and several examples of solar system simulations are described in detail in [13]. The present version of EMGP2 includes twenty different system elements which can be combined according to the system configuration to be considered : flat plate collectors, conduit-pipes with or without thermal capacity, stratified and unstratified sensible heat liquid storages, flat plate collector structure, collector and regenerative storage heat transport fluid, rock bed storage, phase change material storage, dynamic building heat balance as part of a heat extraction algorithm, two-fluid flow heat exchanger, submerged heat exchanger, mixing point, branching point, heat pump evaporator, heat pump condensor, heat meter, heat loss meter, integrator of an external load function, temperature integrator and operation time integrator. Available controllers include differential thermostats, thermostats, safety thermostats, time switches with daily, weekly or annual cycle as well as any combination of such elementary control devices. For the simulation of the heat withdrawal from the simulated system different space heating and domestic hot water draw off algorithms are available. EMGP2 offers a choice of three different integration algorithms : Runge-Kutta-Gill method with compensation for accumulated round-off errors (which is also the EMGP1-integration algorithm), second order Runge-Kutta method and the Runge-Kutta-Merson method. The user has almost complete freedom in selecting the variables and time period of the output. This generalised programme version is now distributed to the EMG-participants and will shortly be implemented on the participants' computer installations.

For the system studies to be performed with EMGP2 and which are being prepared, the meteorological data of seven different test reference years are available : Copenhagen, Valentia, London, Ukkel, Hamburg, Trappes and Carpentras. A detailed examination of these test reference years revealed several physical incompatibilities in the radiation data. Corrections have been introduced and the data are now available for system studies and have been dispatched to the participants.

Validation of EMGP1 and EMGP2.

The validation procedure being used by the EMG is a statistical method, accounting for the error margins of the experimental data as well as for the uncertainties to be associated with the theoretically calculated values as a result of errors on measured input data and the uncertainty interval of the various parameters which describe the solar system. The procedure is described in detail in [8] and [11] . A first version of a computer programme, called EMSTA, performing the statistical calculations on the experimental PTF-data and theoretical results of EMGP1/2 as well as on the difference between theory and experiment, has been developed by the co-ordinator. This programme has been dispatched to the participants together with a first series of 29 periods of uninterrupted PTF-SS1-measurements, ranging from 7 up to 44 days.
For this first statistical validation exercise, according to an agreed work sharing scheme, EMGP1 is used (EMGP2 was not yet available) and starts from the house heat demand and a heat extraction algorithm as being used in the actual PTF-installations. In this type of validation the regulation errors of the PTF-interfaces may strongly influence the results of the validation calculations. Therefore, only those PTF-measurement periods for which the house heat demand is available and the interface was functionning properly, could be considered.

In the validation procedure, integrated energy input and output values of the main system components are considered, rather than instantaneous values of temperatures. In the calculation of the uncertainty margin of the theoretical values, a linear system error analysis is assumed to be applicable. This assumption has been verified for one PTF-SS1 sequence and it was shown [18] that the resulting error on the estimated uncertainty interval of the calculated output quantity is smaller than five percent.

The results of the first series of validation calculations are only partially available at this time, mainly because some unexpected difficulties were encountered, such as erroneous formatting of some of the data files and an error in EMGP1 which was detected from the results of validation calculations related to measurement sequences with a reduced collector surface area. From the few correct validation calculations which are already available, it is not yet possible to make reliable conclusions.
It appears that the scattering of the discrepancy between experimental and theoretical results is very large and a detailed and critical analysis of the results of every validation calculation will be required.

From the preliminary results it appears that in the case of PTF-SS1, the combined effect of the various uncertainties gives rise to a relative uncertainty of the discrepancy between the measured and theoretically calculated system performance, i.e. the useful energy output as a percentage of the heat demand, decreases from about 10 % at system performance values below 10 %, to about 3 % for performance values above 50 %. For the most successful validation calculations performed thusfar, some overall results are given in the following table, which is only meant to illustrate the type of results being obtained from the validations calculations but which cannot yet be generalised.

1	PTF-SS1 Data file number		2	5 (first half)	7	8
2	Number of days		44	20	30	23
3	Storage Energy Input (MJ)	m c	2834 ±38 2854 ±195	2103 ±29 2006 ±69	3964 ±47 3964 ±168	640 ±12 649 ±43
4	Storage Energy Output (MJ)	m c	2479 ± 53 2561 ±206	1082 ±10 1104 ±30	4026 ±36 4068 ±149	829 ±17 831 ±44
5	% Solar	m c	17.3 ± 0.4 17.9 ± 1.4	63.5 ±0.6 61.0 ±1.7	47.7 ±0.4 48.2 ±1.8	9.1 ±0.2 9.1 ±0.5
6	Relative discrepancy of % solar (% of measured value)	value σ	3.3 8.6	-4.0 2.8	1.0 3.8	0.3 5.7

m = measured; c = calculated.

Conclusions.

The generalised European Modelling Group Program (EMGP2) has been developed and can be used to simulate a large variety of different solar systems, including the SPTF-SS2 systems.

The validation of the programme is progressing, but the scatter of the agreement of calculated and measured values does not yet allow to make reliable conclusions about the degree of accuracy of the simulation programme. The validation work is to be continued intensively and will from now on also include the PTF-SS2's. Simultaneously, parametric system studies will be performed with EMGP2 in view of system evaluation for different climatic zones in the European Communities.

References and EMG-working documents

[1] W.L.Dutré : page 14-18 of "Proceedings of the EC Contractors Meeting held in Athens (Greece), 11-13 November 1981" - Project A : Solar Energy Applications to Dwellings. D. Reidel Publ. Co., Dordrecht, Holland

[2] EMG81-83/WD01, W.L.Dutré : Proposed Research Programme of the European Modelling Group for Solar Systems (June 1981).

[3] EMG81-83/WD02, W.L.Dutré : Description of EMG81 (June 1981).

[4] EMG81-83/WD03, W.L.Dutré, P.De Ceuninck : EMGP1-Test Runs (August 1981).

[5] EMG81-83/WD04, W.L.Dutré, P.De Ceuninck : EMGP1-Modifications and Short Term PTF-SS1-Parameter Studies (August 1981).

[6] EMG81-83/WD05 : J.Adnot, B.Bourges : Synthesis of Simplified Methods for Active Solar Systems (September 1981).

[7] EMG81-83/WD06 : J.Adnot, B.Bourges : Standardised Versions of Simplified Methods (September 1981).

[8] EMG81-83/WD07 : W.L.Dutré : Statistical Analysis of PTF-validation Calculations (September 1981).

[9] EMG81-83/WD08 : W.L.Dutré, P.De Ceuninck : Analysis of PTF-cassettes (September 1981).

[10] EMG81-83/WD09 : W.L.Dutré : Proposals for Additional EMGP1-subroutines (September 1981, January 1982).

[11] EMG81-83/WD10 : W.L.Dutré, P.De Ceuninck : Statistical Validation Calculations for PTF-SS1 - Part 1 : Procedure and Programme for Statistical Calculations (January 1982).

[12] EMG81-83/WD11 : J.Adnot, B.Bourges, F.Lasnier : First set of runs for the validation of simplified methods.

[13] EMG81-83/WD12 : W.L.Dutré : EMGP2 - A solar system simulation program (June 1982).

[14] E.Van Galen : EMG-Progress Report (March 1982).

[15] B.Rogers : EMG-Progress Report (February 1982).

[16] M.Frank : EMG-Progress Report (15 November 1981).

[17] J.Reichert : EMG-Progress Report (January 1982).

[18] M.Frank : EMG-Progress Report (15 March 1982).

SIMPLIFIED METHODS FOR THE SIZING OF SOLAR THERMAL PLANTS

Authors : J. ADNOT, B. BOURGES

Contract number : ESA-M-086 F(G)(and others for the participants)

Duration : 24 months 1 july 1981 - 30 june 1983

Total budget : 348 000 FF CEC contribution : 174 000 FF

Head of project : B. BOURGES (Ecole des Mines de Paris)

Contractor : ARMINES - Centre d'Energétique

Address : 60, bld St Michel - 75272 PARIS CEDEX 06 - FRANCE

Participants in the concerted action : W. Dutré (coordinator of the European Modelling Group), A. Debosscher (Katholieke Universiteit Leuven, BE), O. Balslev-Olesen (Technical University of Denmark, DK), B.Verdier (Commissariat à l'Energie Atomique, FR), J. Reichert (Fraunhofer Institut für system technik und Innovations-forschung, GE), M. Frank (Trinity College, IR), A. Biondo (Phoebus SpA, IT), E. Van Galen (TNO-TH, NE), R. La Fontaine (Faber Computer Operations, UK) , J. Adnot (Ecole des Mines de Paris, FR), D. BOYD (Polytechnic of Central London, UK), S. Grove (Plymouth Polytechnic, UK), J. Lee (J. Lee Computing, UK), B.Rogers (University college, UK).

Summary

Among the activities of the European Modelling Group (coordinated by Prof. Dutré, Katholieke Universiteit Leuven), one of the most important and fruitful for the European users of Solar Energy is the design and validation of Simplified Sizing Methods. This activity of the group is beeing coordinated by Ecole des Mines de Paris since june 1981 and our working plan and some preliminary results are presented here:

1 - Existing Simplified Methods : qualities, shortcomings, typology
2 - Working Plan for the comparison of the methods with the European Reference Model (EMGP1) and European Experimental Data
3 - Preliminary results for six existing Simplified Methods.

Résumé

Parmi les activités du Groupe Européen de Modélisation (coordonné par le Prof. Dutré de l'Université Catholique de Louvain) l'une des plus importantes et des plus fructueuses pour les utilisateurs Européens de l'Energie Solaire est la conception et la validation de Méthodes Simplifiées de Dimensionnement. Cette activité du groupe est coordonnée par l'Ecole des Mines de Paris depuis juin 1981 et nous présentons ici notre plan de travail et quelques résultats préliminaires :

1 - Les Méthodes Simplifiées existantes: qualités, défauts, typologie
2 - Plan de travail pour la comparaison des méthodes avec le modèle de référence Européen (EMGP1) et les Données Expérimentales Européennes
3 - Résultats préliminaires pour six méthodes simplifiées existantes.

1 - EXISTING SIMPLIFIED METHODS : QUALITIES, SHORTCOMINGS, TYPOLOGY

1.1 - Bibliographical review

The number of Simplified Methods published is regularly increasing. This is related to the fact that Solar Heating is going from Childhood to Adolescence : Architects and Heating Consultants have to design Solar Houses in common practice and they cannot use the huge Simulation Programs available in the laboratories. Public bodies and Audit Offices have also to check the validity of the expected Energy saving, or, as it is the case in France and in other countries, to define Solar Regulations. Existing Methods aim at these objectives, but their accuracy is sometimes dubious and one rarely knows exactly their application field (thermal parameters range, geographical area). Our activity for the present contract (july 1981/june 1983) is to give an assesment of the validity of the Existing Methods for (1st generation) active systems, to select one or a few methods well fitted to the European Climates and the European Heating Practice, and possibly to do the same with 2nd generation systems like some of the systems 2 of the European PTF.

We hope that our activity will give confidence in the simplified computations of the Designers and harmonize the results in the Community, to avoid that one may use an "adequate" method to obtain a given "result" even if completely wrong when compared to reference programs and experimental data.

Within the present two-years contract a first bibliography had been made : six teen Methods were considered among which six were kept for validation and improvements, concerning Classic Active Space Heating. The review and selection had been made in Working Document 5, with the agreement of all participants ; 21 papers have been used at this stage of the work but later we produced standardized versions of the selected methods (working document 6) and computer translations of them (BASIC, FORTRAN). The process of selection will be now described.

1.2 - First selection among Existing Methods

Among the sixteen methods, two were only proposed for DHW systems. We rejected them until we begin with the study of the water solar heater in the group. We rejected also the methods with restriction of use to only one country of the Community (like two methods with specific data needed not allowing the variation of the collector parameters) were put apart.

Among the remaining eight methods, six were directly applicable and two needed large adaptations : their principle was kept to be introduced later in our activity but we began with standardized versions of the six "ready to use" methods. The fact of wanting to arrive to one or a few "ready to use" methods before june 1983 (and possibly december 1982) is not contradictory with a bettering of methods and a definition of new methods after this date.

Consequently, the methods considered in our present activity are :

(1) f-charts
(2) φ-f-charts
(3) Lunde 2 (version C)
(4) Lunde 2 (version C1)
(5) CFC 2
(6) CFC 3

Three categories can be found in the sixteen available methods: they are based on three different types of approaches : analytical, empirical and hybrid approaches.

1.3 - Analytical methods

Seven methods are related to this class, generally old methods like utilisability.

Their common principle is to compute a simplified thermal balance of the collection loop, based on the notion of critical radiation intensity (or operating threshold radiation). Let us consider Hottel-

Whillier relationship, giving the power supplied by a flat-plate solar collector in steady-state conditions :

$$Q_u = AF_R \ (\overline{\tau\alpha}) \quad I - AF_R \ U_L(T_i - T_a)$$

This power is positive (and the collector operates) only if the solar radiation intensity on the collector surface is greater than a critical radiation sometimes called "threshold radiation" :

$$I_c = U_L \ (T_i - T_a) \ / \ (\overline{\tau\alpha})$$

The available power is thus written:

$$Q_u = AF_R \ (\overline{\tau\alpha})(I - I_c)^+ \text{ where } (I-I_c)^+ = \begin{cases} 0 & \text{if} \quad I < I_c \\ I-I_c & \text{if} \quad I > I_c \end{cases}$$

If hourly measurements of solar radiation are available, one can compute the available energy over a specified period of time. Assuming the ambient temperature T_a and the collector inlet temperature T_i to be constant, it leads to :

$$E = A \ F_R \ (\overline{\tau\alpha}) \ \sum^{n} \ (I - I_c)^+$$

For a given period this energy E is a function of the collector inlet temperature Ti, because the critical radiation, I_c depends on T_i ; The available energy can be plotted as a function of the temperature T_i (fig. 1).

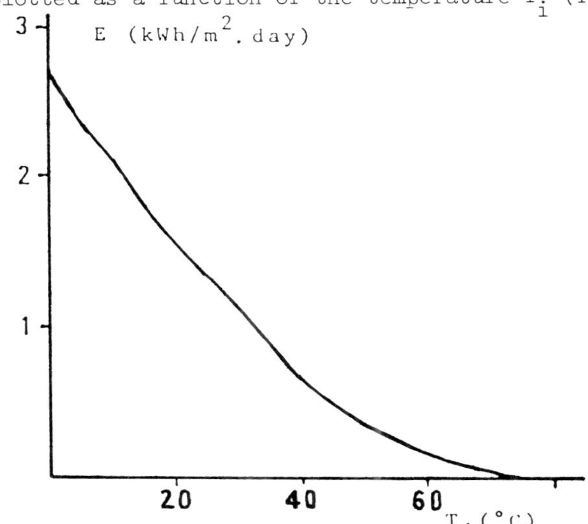

Fig. 1 - Available energy vs fluid collector inlet temperature

The dimensionless form presented under the heading of "Utilisability" has to be noticed

$$\phi = \frac{E}{N \ AF_R \ (\overline{\tau\alpha}) \ \overline{H}} = \frac{1}{N} \ \sum^{n} \ \left(\frac{I}{\overline{H}} - \frac{I_c}{\overline{H}}\right)^+$$

The utilisability curves are only influenced by the distribution of solar radiation and not by the parameters of the specified system.

The best representation of this distribution is given to our mind by the Cumulative Frequency Curves, studied systematically by B. BOURGES and coworkers in the frame of project F of the Community Research Program:

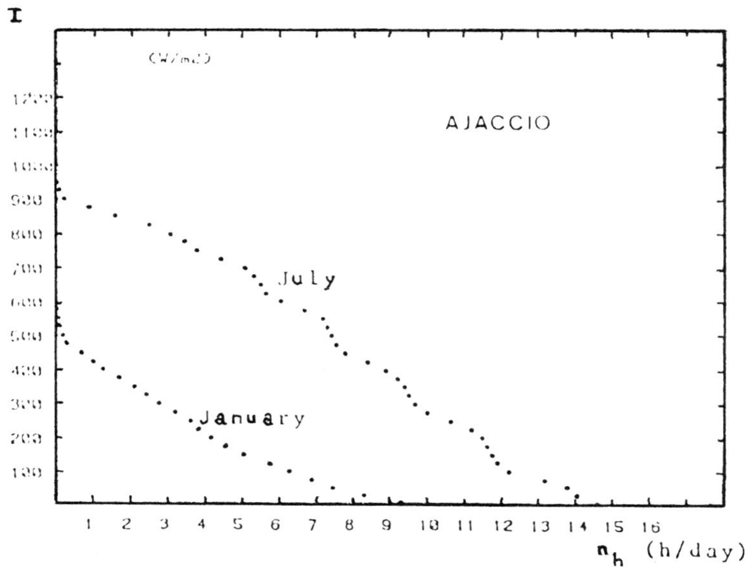

Fig. 2 - Example of Cumulative Frequency Curve of Global Solar Radiation (here example for the horizontal plane)

The CFC give on the x-axis n_h, the number of hours per day for the given month when solar irradiance has been over the value I on the y-axis. For instance, at Ajaccio, in january the threshold radiation of 300 W/m^2 is overstepped during the average of 2,75 h/day (see fig.2).

The intercept with the x-axis gives the average number of hours between sunrise and sunset for the given month, t_d.

The area between the curve and the axis is the average daily sum of the solar energy H. These curves can be drawn for different orientations and tilting of the collecting surface.

These curves are calculated by sorting hourly values of total solar radiation data in classes of 25 W/m^2.

These curves are fitted by a polynomial formula :

$$x = \frac{n_h}{t_d} = (1-y)(1+B_1y+B_2y^2+B_3y^3+B_4y^4)$$

where $y = I/I_{max}$

The dimensionless parameter ν is an important characteristic of the curve

$$\nu = \overline{H}/(I_{max} \cdot t_d) = 1/2 + B_1/6 + B_2/12 + B_3/20 + B_4/30$$

The B_i parameters are known simply for any place where there are meteorological measurement. In other places they can be obtained by interpolation or extrapolation rules based on the knoledge of ν (see B. BOURGES thesis).

The knowledge of the B_i parameters allows very simple computations of the utilisability φ, the collector operating time t_c, etc ... The analytical methods are not generally used presently but they served as a basis for the hybrid methods (see here under) which use the same meteorological data.

However a more sophisticated presentation of the same computation, based on the search for an average operating point of the system, is the basis of one of the methods considered in our activity (CFC3).

1.4 - Empirical methods : fitting of simulation results

A completely different approanh is to start from a computer simulation programme of the system. This programe isrun for a lot of system parameters and climatic hourly data files.

The great number of monthly results are then fitted as a function of a few groups of characteristic parameters. The fitting is presented on easy to use charts like f-charts (considered in our present activity) and four other methods reviewed.

These methods, when the simulation programme is precise enough, give good results and are easy to use for designers. They need, as meteorological inputs, only mean monthly values of temperature and solar radiation intensity. However their validity is restricted to a limited number of systems and to a limited variation range for the parameters. Extrapolation could not be warranted.

1.5 - Hybrid methods

Among the sixteen methods, four methods were hybrid methods. These methods are a synthesis of the two preceding approaches. They are established in three steps :
- an estimation of available energy at a minimum temperature level is made
- a computer simulation of the system is run
- a statistical corrective factor is defined from the simulation results and applied to the "available energy".

Thus the systematic bias of simplified calculations as "available energy at a given temperature level" is avoided ; the "energy really used by the system" is computed taking into account the size of storage and heat distribution.

As for the firstkind of design methods, statistical frequency distribution of solar radiation (on a monthly basis) is the principal climatical input.

Four of the six methods considered in our present activity are of this kind and as soon as CFC3 needs corrective terms it will also enter it.

2 - WORKING PLAN FOR THE COMPARISON OF THE METHODS WITH THE EUROPEAN REFERENCE MODEL (EMGP1) AND EUROPEAN EXPERIMENTAL DATA

2.1 - Systematic comparison with EMGP1

This systematic comparison is the basis of our whole activity. Only the comparison of a simplified model to a detailed one allows for variations of the parameters in a large range, and for a logical search of the differences encountered. The European Reference Model, EMGP1, written by Katholieke Universiteit Leuven is run for a certain nomber of input data and for the same input data the six available Simplified Methods are also run.

Test Reference Years available for six European Meteorological Stations are used : Copenhagen (DK), Hamburg (GE), Valentia (IR), Ukkel (BE), Trappes (FR), Carpentras (FR). The diffuse and beam components of solar radiation being on the tapes, EMGP1 performs the meteorological computations for the tilted plane with the classic assumptions. The results are edited under the form of :

- yearly results (Solar Energy used E_a and Solar Fraction f_a)
- monthly results (Solar Energy used E_m and Solar Fraction f_m).

The monthly time-step seems to be the best for the study of simplified methods : with it we take into account part of the meteorological variations of the resource and of the load and we are able to compute the simultaneous effect of various equipments (active, passive, Heat pump ...).

The system considered in the first stage of the activity of the group is the following :

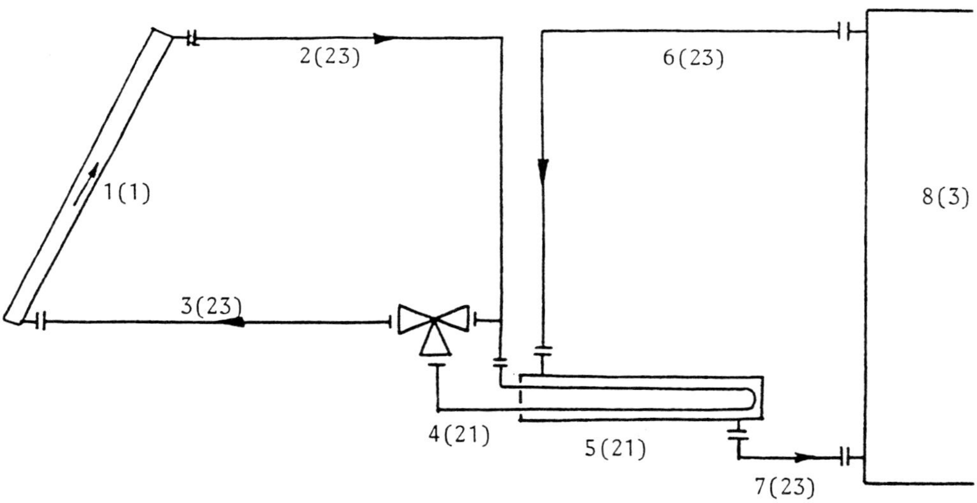

Figure 3 : description of the system for simulation by EMGP1

Figure 3 shows the lay out of the systems ; the figures indicate the number of the elements in the list one has to give to the programme ; number into brackets refer to elements'types for simulation purposes. The system is PTF.like but the sizings have been varied largely in the course of the exercise :

- area of collectors,
- storage capacity,
- distribution temperatures,
- type of collectors.

However at the first stage some parameters that are not taken clearly into account in most Simplified Methods were set to zero : they are now studied one by one to see their effect and at the same time we derive formulae to introduce them in Simplified Methods. Among them let us mention : the tank losses, stratification of the store, pump losses, pump power, etc ...

2.2 - <u>First analysis of the results</u>

For each simplified method and for each sizing of the system, the use of the six TRYs gives more than fifty pairs of results for the comparison (if done on a monthly basis). The points may be plotted as follows :

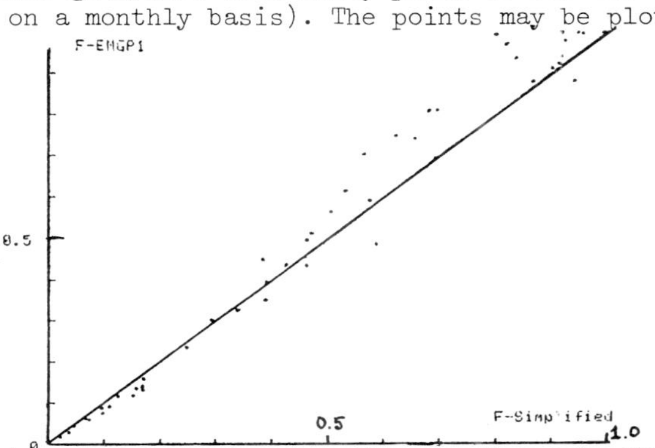

figure 4 - Monthly solar fraction predicted by EMGP1 compared with the CFC2 prediction

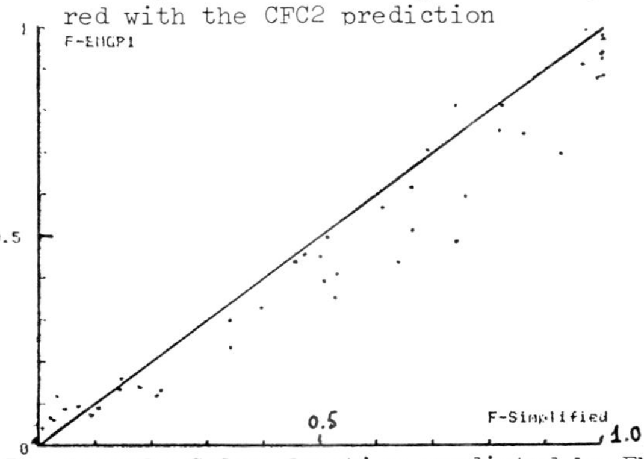

figure 5 - Monthly Solar fraction predicted by EMGP1 compared with the f-chart method

The visual assesment indicates here, for specified conditions, a large scattering for f-chart and a smaller for CFC2. Some methods seem to have a systematic bias which may be corrected by corrective terms like Lunde 2-C1 in this case .

The statistical analysis described in part 3.1 aims at giving figures for the phenomena shown up by the graphs : bias, scattering, differences between methods, differences between locations, correct sensivity of the methods to some parameters or not, etc ...

2.3 - <u>Utilisation of Test Facilitites data (PTF group)</u>

The direct comparison between PTF-experimental results and Simplified Methods results is not unnecessary even if we consider the comparison with EMGP1 as completed. It will give practical examples of the

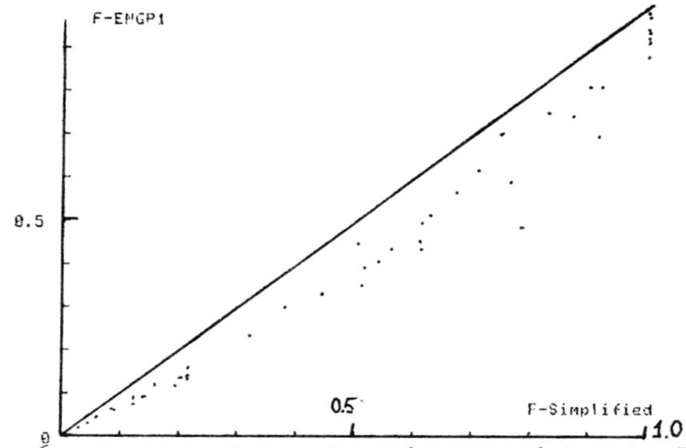

figure 6 : Monthly Solar fraction predicted by EMGP1 compared with the Lunde 2-C1 method

accuracy of Simplified Methods for the designers and, the number of PTF cassettes being larger and larger, an estimate of the statistical characteristics of the errors.

The work includes : compression of the PTF data (partially done already for the validation of EMGP1), preparation of meteorological data needed for Simplified Methods, comparison between results and statistical analysis of the agreement. This work is shared among the participants in the same way that the validation of EMGP1 itself is shared (see the companion paper by Prof. Dutré).

The definition of the process of comparison between PTF results and simplified methods and the statistical analysis of the agreement brings new problems, differing from the previous validation process :

- the parameters to be used in the Simplified Methods are not perfectly known ; they have to be defined for each sequence and they may differ from one moment to another within the sequence ;
- there is an uncertainty on all the input quantities to the Simplified Methods ;
- there is an uncertainty on the reference results (PTF data instead of EMGP1 outputs).

So the direct experimental validation gives new dimensions to our work : closer to the field problems.

2.4 - <u>Utilisation of Performance Data of Houses monitored by P.M. Group</u>

Eight Active Space Heating Systems are presented in the documents of the PMG : among them six are liquid-based, one is air-based (Dourdan) and another one is a Solar + Heat Pump System (Carcassone). The use of these last two data-sheets is delayed. For most of the houses one year of data is available.

Simplified Methods are only available presently within the group for the liquid-based active systems. As soon as the air-based and the Heat Pump Assisted Systems are considered in the normal activity of the Group the Simplified Methods activity will be extended to those systems. The nature of the Methods may be completely different for such systems, and the number of methods is very small.

Another problem is related to DHW production : as all systems monitored include it, it is possible to check only simplified methods taking DHW into account.

The following table gives some indications about the data

which will be used rapidly in the activity of the group. In one case (Milton Keynes House) some results are already given in part 3.3 of the present paper

Project	Country	Lat.	Floor Area (m^2)	Collector Area (m^2)	Storage Volume (m^2)
Bourgoin	France	45.6	573	152	10
Fiume Veneto	Italy	46.2	1011	129	3
Eindhoven	Netherlands	51.5	220	52	4
Zoetermer	Netherlands	52.1	130	35	2
Milton Keynes	United Kingdom	52.0	85	34	4

Table 1 : Results provided by the PMG to be used in the EMG activity

3 - PRELIMINARY RESULTS FOR SIX EXISTING SIMPLIFIED METHODS

3.1 - Methodology of statistical analysis

The analysis can be divided into three sub-questions :

- Is there a systematic bias between the detailed model and the simplified method (for f or E) ?
- Is there a large scaterring of the results ?
- Is the considered method equally correct for different system parameters or climates ? Has it a limited range of validity ? (i.e. does the bias change according to various parameters ?).

Plots as figures 4, 5 and 6 already give the first indications : on monthly solar fraction, the f-charts method presents a reasonable bias but with a large scattering. Methods starting from an "Utilizability" approach show a smaller scattering but with important bias (except CFC 2).

These empirical consideration have to be completed with a numerical approach:it is then necessary to introduce a model to analyse relationships between both sets of results.

From the existing results it seems that a linear model is an acceptable assumption under the form

$$Y = a + b X + \varepsilon$$

Y is a characteristic of EMGP1 results (monthly or annual solar fraction or solar absolute contribution)

X is the same characteristic for the simplified method result

ε is a random-residual, with a standard-deviation σ_ε and a zero-average

If we consider a set of n results (x_i, y_i), we can estimate values of a and b with a least-square procedure.

If there is no bias between the simplified method and the detailed model, a and b do not differ significant by from 0 and 1.

The standard-deviation σ_ε of ε may be estimated also.

It seems a better index of scattering than the regression coefficient (ρ^2) because it gives directly an idea of the error of the linear model.

An interesting result for the linear model is the confidence interval relationship : we have formulae giving confidence intervals at a given confidence level.

3.2 - Preliminary results with use of EMGP1

Eighteen sets of simulation results were defined :

(S1) basic set of runs with normal distribution temperatures and medium storage and the selective collector used in the PTF
(S2) lower distribution temperature and medium storage
(S3) small storage and normal distribution temperature
(S4) small storage and lower distribution temperature
(S5) large storage and normal distribution temperatures
(S6) large storage and lower distribution temperatures
(S7) to (S12) = (S1) to (S6) with a non selective collector (lower performances)
(S13) to (S18) = (S1) to (S6) with a higher performance collector (reduced losses).

Each set of runs or agregate of sets will be characterized by the computation of parameters of the linear model will be performed and particularly of the confidence interval (confidence level : $1 - \alpha = 0,95$). Results could be analysed on the following way :

- if the regression line of basic set of runs (S1) differs significantly from the straight line $y = x$, there is a bias between the detailed model and the simplified model ;
- the value of σ_ε for the basic set of runs will give an information on scattering ;
- the regression line of each other set of runs will be compared (from the point of view of confidence interval) with the basic set regression line
- the regression line of the basic set of runs for every climatic place will be compared with the overal basic regression line to investigate a possible geographical bias of the method.

The results will be available in september 1982 for the existing methods and modified methods will be probably ready at the end of the year.

The parameters a, b, σ_ε are given in table 2 as a preliminary result for the basic set of runs S1 and only for the f-chart method. The six locations are mixed and the collector used is the selective collector of the PTF

	a	b	σ_ε
Basic set (S1) f-chart (f_m)	- 0.0085	0.897	0.058

Table 2 : Preliminary results for the set of runs (S1)

3.3 - Preliminary results with use of data collected by the PMG

Two methods have been compared with the results obtained by the Performance Monitoring Group for the Milton Keynes House (UK) : CFC2 and the f-charts. Table 3 gives the results for CFC2 and table 4 for f-charts. The load (8870 kWh) had been measured. From the meteorological point of view only monthly means were available : the B_i needed for CFC2 method were predicted from $\bar{H}$ using results of project F (action 3.3) of the European Research Program.

Month	φ	Solar fraction (computed)	Solar fraction (measured)
1	0,27	0.13	0.15
2	0,35	0.21	0.23
3	0,54	0.55	0.61
4	0,62	0.81	0.84
5	0,77	1	0.91
6	0,78	1	0.97
7	0,80	1	0.94
8	0,79	1	0.99
9	0,79	1	0.99
10	0,68	1	0.86
11	0,53	0.74	0.64
12	0,25	0.13	0.19

Yearly solar fraction : 0,475 (Measured : 0,483)

Table 3 : Comparison of CFC2 with experimental results

The agreement between predicted and measured values is very good at monthly level both for winter months (space heating) and for summer months (Domestic Hoc Water Heating). On a yearly basis the prediction by CFC2 agrees perfectly with the measurements, in this case.

Month	Load (kWh/day)	$\bar{H}$ ($kWh/m^2/day$)	Solar fraction (computed)	Solar fraction (measured)
1	53.9	1.16	0.077	0.15
2	56.9	1.63	0.231	0.23
3	39.9	2.55	0.631	0.61
4	30.4	3.39	0.885	0.84
5	16.1	4.77	1.000	0.91
6	6.7	4.56	1.000	0.97
7	6.7	4.28	1.000	0.94
8	5.5	4.86	1.000	0.99
9	6.7	3.89	1.000	0.99
10	11.6	2.32	1.000	0.86
11	21.9	1.56	0.618	0.64
12	37.6	0.85	0.087	0.19

yearly solar fraction : 0.466 (measured 0.483)

Table 4 : Comparison of f-chart with experimental results

On a yearly basis the result given by f-charts is in this case very close to the result predicted by the CFC2 method and to the experimental one. However the discrepancies at the monthly level are larger.

Obviously no conclusions can be drawn from only one case of validation, but the feeling that both the referred methods are close to an achievment.

4 - CONCLUSIONS

A common exercise of design and validation of Simplified Methods is being performed within the European Modelling Group with use of :

- Reference Runs with the European Simulation Program (EMGP1) ;
- Experimental Results of the European Test Facilities (PTF) ;
- Experimental Results of the Houses monitored by the PMG.

The results of the Statistical Analysis for the existing methods will be available in september and "ready to use" new methods a few months later. The activity of Simplified Modelling within the Modelling Group will then move from the "classic" 1 st generation active systems to 2nd generation systems : air based, heat-pump assisted, hybrid (active + passive), etc ...

NOMENCLATURE

A : collector area (m^2)
E : Collected Energy over a given period (Wh)

F_R - collector heat removal efficiency factor
$\overline{H}$ - average daily total radiation on the collector surface in a specified period (Wh/m^2 day)
I_c - critical radiation level (W/m^2)
I - intensity (or mean hourly intensity) of total solar radiation on the collector surface (W/m^2)
n - number of hours in a specified period
N - number of days in a specified period
Qu - available collected power (W)
T_a - air temperature (K)
T_i - collector inlet fluid temperature (K)
U_L - collector overall energy loss coefficient ($W/m^2.K$)
α - solar absorbtance of the collector plate
τ - solar transmittance of the collector transparent covers.

WORKING DOCUMENTS USED BY THE GROUP

EMG/WD2 : Description of EMGP1 (W. Dutré)
EMG/WD3 : Test Runs of EMGP1 (W. Dutré)
EMG/WD5 : Synthesis of Simplified Methods for Active Solar Systems (B. BOURGES)
EMG/WD6 : Standardised Versions of Simplified Methods (B. BOURGES)
EMG/WD11: First Set of Runs for the Validation of Simplified Methods (J. ADNOT)
EMG/WD12: Utilisation of PMG-data for Simplified Models Validation (B. BOURGES)
EMG/WD13: Utilisation of compressed PTF-data for the validation of Simplified Methods (J. ADNOT)
EMG/WD15: Methodology of Statistical Validation of Simplified Methods (B. BOURGES)
PMG : Performance Monitoring of Solar Heating Systems in Dwellings-Part II

THE FIRST TWO YEARS OF OPERATION OF THE SOLAR PILOT TEST FACILITIES

Author : G. OLIVE

Contract number : ESA-T-118-F

Duration : 21 months 01.10.81 to 30.06.83.

Total budget : 670 400 FF CEC contribution : 100 %

Head of project : G. OLIVE

Contractor : G. OLIVE, Consulting Engineer

Address : 16 rue Nansouty, 75014 PARIS

Participants in the concerted action : W.L. DUTRE, Catholic University of Leuven, Belgium
K. ELLEHAUGE, Technical University of Denmark, Denmark
R. PLOYART, Nuclear Agency, France
R. HANITSCH, Technical University of Berlin, Germany
I.J. COWAN, Institute for Industrial Research & Standards, Ireland
L. GRASSI, ENEL-PHOEBUS, Italy
C. den OUDEN, TPD-TNO, The Netherlands
G.J. BAKER, BSRIA, United Kingdom

Summary

The SPTF group supplies the Modelling Group with the maximum amount of data for the improvement of simulation models for domestic solar heating systems. The SPTF group has completed 4 stages of its work : from 01.78 to 06.79 : construction of the SPTF. From 07.79 to 06.80 : SPTF trails and validation; mastering the use of the SPTF. From 07.80 to 06.81 : first year of operation. From 07.81 to 06.82 : second year of operation.
The main results and conclusions to be drawn from these two years of work are described schematically in order to define precisely the work which remains to be done for the third year of operation (07.82 to 06.83).

1. Use of test facilities with a physical simulator

In order to validate simulation models for the thermal behaviour of domestic solar heating systems, the Commission of the European Communities (CEC) chose to use what have become known as Solar Pilot Test Facilities (SPTF).

The facility consists of a real solar system from the collection to the storage area. The only part which has been eliminated is the heat distribution section. It has been replaced by a physical simulator, which is automatically controlled and which reconstitutes the behaviour of this section in transient conditions in a realistic way. In order to do this, the house's heating demand (referring to a house equipped with this system) is numerically simulated while taking into consideration the climatic environment in which the house would be. The way in which the replaced section of the system would respond to the heating demand is also simulated.

In comparison with a complete solar system for a house which would be the subject of experimental study, the test facility previously described has many advantages. First of all each physical simulator, in so far as it simulates the house load realistically, ensures that the performance testing of the solar system installed on the test facility is similar to that of a particular solar system for a house designed and used in a particular way. A fairly easy simulation change permits the economy of a section of the system, and above all that of the experimental house. Secondly, within the limits of one's knowledge concerning how to simulate the house's heating requirements and the response of the system in transient conditions, the only advantage of an experimental house is the reality of the occupation factor. However, since correlation between these effects and their cause is far from being obvious, one may say that in an experimental house one is governed by the living habits of the occupants. On the other hand, it is quite simple to simulate the effects of occupation in a realistic way. Finally, metrology adapted to experimental investigation conditions in situ is far more difficult to attain than metrology adapted to laboratory conditions.

For these reasons the CEC has chosen this test facility technique. So that a wider field of investigation may be created, the CEC has promoted the construction of eight test facilities, each one being situated in a different European country.

Figure number I indicates how the real closed loop "weather conditions -occupied house-system" is replaced by another closed loop including a real section and a physically simulated section, the latter being automatically controlled by a numerical simulator. The "realism" of physical simulation has been proved in reference (1).

As for the practical configuration of the Solar Pilot Test Facilities (SPTF) it has been described in reference (2). It is sufficient to note that each SPTF has been built to accommodate two solar systems: a reference solar system (SS1), as far as possible identical in all countries to supply validation data for the same system over a wide range of climatic conditions, and a system (SS2) reflecting the national research-development interests and practices of the participating countries.

After having built and verified the behaviour of the SPTF (01.78 to 06.80) the group began operating the SPTF for a period of three years (07.80 to 06.83).

2. The various activities of the SPTF group

The group's fundamental task is to produce sound data for the European Modelling Group which is needed for model validation. However, as early as April 1979, the co-ordinator indicated the necessity to carry out system behaviour studies so that the SPTF would be used to ensure the good quality of experimental data, and on the other hand to acquire reliable knowledge of different types of systems. Figure n° II illustrates the various activities of the group to carry out this double research programme "model improvement - system behaviour studies".

It is immediately evident that the SS1 is the tool which is preferred for supplying useful experimental data for the purposes of model improvement and for establishing work procedures for the concerted action of the eight teams involved.

The SS2 are tools which are well adapted for system behaviour studies due to their varied designs.

As for the system development activity, it must be stated that the test facilities have been built for experimental research, but that it is rather difficult to organize this area of research in a concerted way. One may hope to draw conclusions from the teams' efforts which will be of benefit for all concerned.

3. Work procedures of the SPTF Group

The first year of SPTF operation (07.80 to 06.81) enabled, among other things, the elaboration of work procedures, concerning operation of the SPTF and data exchanges, which have been proved to be most efficient.

Concerning the operation procedures, they are essentially a good definition of the work sequences. A year of operation for a SPTF is a consecutive series of operation sequences and intermediary sequences.

An operation sequence (OS) is a run of an unmodified SPTF. A particular configuration of a solar system is characterized by its system characteristics which are measured during an intermediary sequence. They are written down in an installation descriptor which is sent together with the experimental data supplied to the Modelling group for model improvement purposes.

An operation sequence is a series of validation sequences (VS) which correspond to continuous runs of a SPTF.

This continuity is necessary so that the time series of experimental data may be used to validate a model. Each validation sequence is itself a series of data storage sequences (DS), not necessarily uninterrupted, during which experimental data are acquired and stored.

An intermediary sequence (IS) corresponds to the necessity of carrying out one or several of the following tasks: installation modifications, software modification of the central control system of the SPTF, measurement of a solar system's characteristics.

The data exchange procedures mainly concern the nominal data package, a set of data necessary for model validation.

This package consists of experimental data corresponding to a validation sequence of at least fifteen days, as well as a series of documents which describe these experimental data: a log sheet (giving a detailed account of the various data storage sequences), a data descriptor giving the data storage mode and an installation descriptor giving a description of the solar system being studied.

These nominal data packages are systematically the subject of central data storage in Heverlee in a data bank for security reasons, and to facilitate transmission of data to the Modelling group. Certain significant nominal data packages have been selected for a package catalogue for researchers interested in the field of solar energy applications to dwellings.

From experience, normal operation of a SPTF permits realising at least six operation sequences and ten nominal data packages per year and per system. This corresponds to operation sequences and intermediary sequences of an average duration of five and two weeks respectively.

4. Production of validation data

It is a question of producing time series of experimental data which permit real model validation. Five questions must be answered.

First of all, which part of a time series must be rejected, since a test facility reachs its normal operation mode after a certain period of time, at the beginning of a validation sequence? The answer would seem to be: the first twenty four hours.

Secondly, what are the quantities of data necessary for model validation? Apart from weather data and system behaviour data (about 15) and the system characteristics which constitute the experimental quantities which obviously seem necessary, the outputs of the model must be decided on and with which significant SPTF data they could be confronted (see figure n° I . Reference (3) contains a reliable answer to this question.

Thirdly, what is the time step for data storage which is of use for validation? Reference (3) tends to show that one may be satisfied with a time step of one hour on most occasions.

Fourthly, what is the effect of measurement error propagation on the utilisability of experimental data? According to the work of the SPTF and Modelling group co-ordinators, it would seem that the present accuracy of measurements is adequate for weather and system behaviour data. Requirements have been voiced for values of the system characteristics.

Fifthly, what are the requirements concerning the length and period of time for the time series? Research undertaken by the Danish, Dutch and co-ordinators of the SPTF and Modelling group give one to believe the necessity of using consecutive time series for winter and for half a season of four or two weeks respectively.

Figure III indicates the frequency and location of the production of data packages which constitute an exceptional information source in the world.

5. System behaviour studies

Reference (2) gives a résumé of the areas of research and the necessary multiplicity of procedures to be adopted. Concerning the areas of research, one may say that the SPTF deal with nearly all solar system families:

- space heating water systems
- space heating air systems
- space cooling systems
- space heating or cooling and domestic hot water production systems.

Ten aspects of each system are investigated:

- primary circuit
  1. on/off control
  2. flow rates
  3. pump control
- solar storage
  4. Storage volume/collector area ratio
  5. stratification
- secondary circuit
  6. requested distribution temperature
  7. heating demand
  8. control system strategies
  9. combination of functions
- others
  10. components and arrangements

Figures IV and V show the aspects of the SS1 dealt with by the participants over the first two years of SPTF operation. A document is being prepared which tries to draw the main conclusions from these studies for SS1 as well as SS2.

As for the multiplicity of research procedures, (system investigation, parametric studies, reliability studies), one may say that they have all been investigated.

The third year of operation will essentially consist of ensuring that time series of experimental data of adequate quality (with respect to length and period) for all areas of research will be acquired, which will permit reliable extrapolation of the behaviour of solar systems by the use of correctly validated models. Obviously all practical data helping in the evolution of design and use of solar systems will be the subject of a document to be distributed.

REFERENCES

(1) G. OLIVE. "Pilotage d'un simulateur physique pour une plate-forme d'essais thermiques." Solaire 1. 10/11.81.

(2) G. OLIVE. "Solar Pilot Test Facility Programme of the Commission of the European Communities". "Solar Energy Applications to Dwellings" Vol. 1. D. Reidel Publishing Company. 82.

(3) O. JORGENSEN. "Common solar simulation and validation in Europe". CEC Report. 2-82.

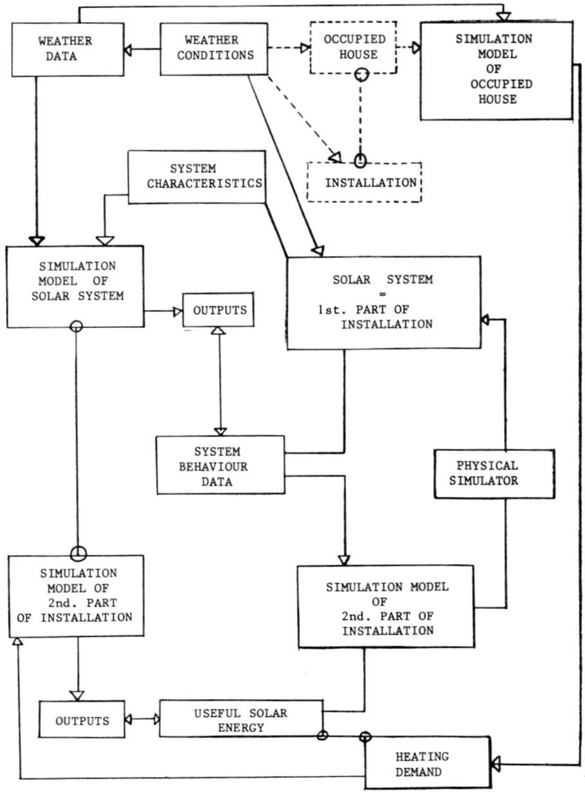

Figure I - Practical conditions of model validation

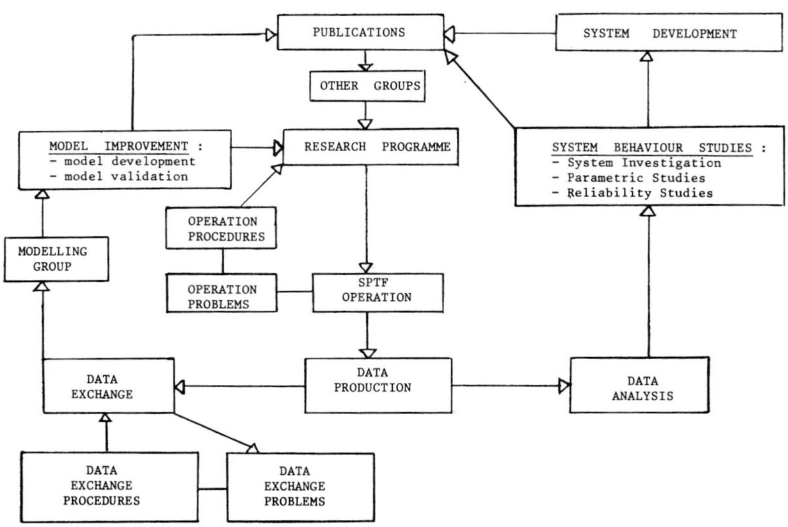

Figure II - Different activities of the SPTF Group

NUMBER OF CASSETTES

SYSTEM	COUNTRY / §	BELGIUM	DENMARK	FRANCE	GERMANY	IRELAND	ITALY	NETHER-LANDS	UNITED KINGDOM
SS1	80 - 81	16	19	7	3	9	2	16	11
	81 - 82	0	16	6	8	8	6	9	7
SS2	80 - 81	0	11	0	10	6	0	15	11
	81 - 82	0	10	0	1	9	11	8	86

§ year of operation

Figure III - Data packages produced

SS1 1980 - 1981

DATES OF ACTIONS CARRIED OUT

§ \ §§	I	II	III	IV	V	VI	VII	VIII	IX	X	RC
BELGIUM	3/4/81 12/5/81			19/11/80 9/3/81		8/10/80 6/11/80					9/7/80 6/11/80
DENMARK					6/4/81 30/6/81	30/10/80 5/1/81			4/6/81 30/6/81		23/6/80 16/7/80
FRANCE			31/7/80 29/6/81								
GERMANY	20/2/81 12/4/81										1/7/80 21/7/80
IRELAND				12/6/81 29/6/81							25/6/80 8/6/81
ITALY											2/2/81 1/5/81
NETHER-LANDS		9/4/81 15/6/81		2/2/81 6/4/81							28/7/80 1/2/81
UNITED KINGDOM											3/2/81 18/3/81

§ = country §§ areas of research

Figure IV - 1st year of operation of SS1
RC = reference configuration

SS1 1981 - 1982 DATES OF ACTIONS CARRIED OUT

§§ / §	I	II	III	IV	V	VI	VII	VIII	IX	X	RC
BELGIUM											
DENMARK	4/9/81 13/12/81 3/2/82 27/4/82				4/7/81 20/7/81 3/2/82 27/4/82	4/9/81 13/12/81 3/2/82 27/4/82	4/9/81 13/12/81 3/2/82 27/4/82		4/7/81 20/7/81	3/2/82 27/4/82	
FRANCE			27/7/81 8/9/81	17/3/82 1/5/82						17/3/82 1/5/82	
GERMANY	26/10/81 14/12/81 15/12/81 15/1/82	5/4/82 7/5/82	10/9/81 19/9/81 15/12/81 15/1/82	5/4/82 7/5/82				30/1/82 18/3/82 5/4/82 7/5/82			
IRELAND	15/10/81 22/11/81			19/3/82 5/4/82	1/3/82 11/3/82	23/11/81 7/1/82	23/4/82 13/5/82	13/5/82 28/5/82			10/2/82 28/2/82
ITALY		11/2/82 15/3/82				6/4/82 29/4/82					10/11/81 30/12/81
NETHER-LANDS		15/2/82 19/3/82	9/12/81 26/4/81			15/1/82 15/2/82	8/4/82 26/4/82	9/12/81 15/1/82			
UNITED KINGDOM					28/3/82 19/4/82 23/4/82 4/5/82			23/4/82 4/5/82			

§ = country §§ = areas of research

Figure V - 2nd year of operation of SS1
RC = reference configuration

PERFORMANCE MONITORING OF SOLAR HEATING SYSTEMS IN DWELLINGS : PHASE 3

Authors : R. FERRARO, R. GODOY, D. TURRENT.

Contract No : ESA-P-116-UK(H)

Duration : 18 months 1st January 1982 - 30th June 1983

Total Budget : £209,250 CEC Contribution: 100%

Head of Project : R. Ferraro

Contractor : Energy Conscious Design

Address : 44 Earlham Street, London WC2H 9LA, UK.

PARTICIPANTS IN THE CONCERTED ACTION

A. Debosscher - Katholieke Universiteit Leuven, Belgium.

P. Kristensen - Thermal Insulation Laboratory, Lyngby, Denmark.

G. Kuhn - AGEDES, Grenoble, France.

U. Luboschik - IST Energie Technik, Kandern Wollbach, Germany.

J.O. Lewis - University College Dublin, Ireland.

F. Cecchi Paone - Fidimi Consulting, Rome, Italy.

D. Brethouwer - Institute of Applied Physics (TNO), Delft, Netherlands.

Co-ordinators - Energy Conscious Design, London, UK.

Summary

The Performance Monitoring Group (PMG) has been receiving performance data from monitored solar heating systems designed 4-8 years ago. These systems are generally over-complicated, oversized, economically unattractive and have poor performance. Based on experience, the PMG believes that it is possible to design solar heating systems with higher outputs per unit collector area and/or lower capital costs. Such systems will provide better cost-benefit balances and a greater potential for market penetration.

This paper describes the progress of the PMG from the conclusions of its second phase of work (1981/82) into the third phase (1982/83), currently underway. A programme of Performance Optimisation Studies has been initiated, in which a series of better optimised solar heating system designs are being developed. These are based on existing knowledge and experience, simulation modelling, detailed cost analysis and commercial practice. The results from the work should represent the most promising developments in domestic scale solar heating system designs for Europe.

The exercise will illustrate the process by which initial designs are optimised to achieve higher performance at minimum capital cost, under a set of practical constraints. The output document will consist of a series of case studies, which could form the basis for a programme of field trials that might be built and monitored starting in 1983.

Finally, other continuing PMG activities are also described.

1.0 General Introduction

The work of the Performance Monitoring Group is a concerted action under Project A of the CEC programme. The second phase of the Group's work was carried out in the period 1981/82 and the third phase continues now through to the end of the current programme.

This paper sets out the relationship between past, present and future work, and to do this it describes briefly the objectives and main conclusions of the previous phase as well as covering in more detail the work of the current phase up to the middle of 1983.

2.0 Phase 2: Objectives

The main objectives of the last phase were:

- To produce Reporting Formats to enable uniform recording of system parameters, cost and performance results from different types of solar system.
- To report on the performance of domestic solar heating systems in the EEC.
- To produce guidelines for monitoring solar heating systems to ensure uniformity of approach, reliability and accuracy of results.
- To draw conclusions on the feasibility of selecting solar system types suitable for optimisation, with a view to achieving lower cost and/or better system performance in the future.

2.1 Reporting Formats

Three Reporting Formats are now available ior recording design and performance data from different solar system types as follows:

- SOLAR WATER HEATING SYSTEMS
- ACTIVE SPACE HEATING SYSTEMS AND DOMESTIC WATER HEATING SYSTEMS (combined)
- PASSIVE SPACE HEATING SYSTEMS

The Formats are particularly suitable for the inter-comparison of different installations, and are also suitable for recording data to be used for the validation of simplified models. Each is accompanied by a comprehensive User Guide.

2.2 Phase 2: Final Reports

The main body of work from the last phase, which included the compilation and analysis of detailed performance data from over 60 monitored installations throughout the Community, has resulted in the production of four final reports, as follows:

- MONITORING SOLAR HEATING SYSTEMS: A Practical Handbook.
- SOLAR WATER HEATING: An Analysis of Design & Performance Data from 28 Systems.
- SOLAR SPACE HEATING: An Analysis of Design & Performance Data from 33 Systems.
- PERFORMANCE MONITORING OF SOLAR HEATING SYSTEMS IN DWELLINGS: Executive Summary & Conclusions

(NB: The last three reports above are available direct from the Co-Ordinator. The first report is to be published.)

The Practical Handbook on Monitoring Solar Heating Systems covers both active and passive systems and consists of a sequence of chapters dealing with monitoring programmes generally, and detailed requirements. It covers detailed measuring requirements, instrumentation (including heat meters and flux meters), data acquisition systems, installation of equipment, data storage and processing, check procedures and maintenance. Additionally there are appendices on one-time measurements and post monitoring procedures, together with catalogues of monitoring organisations, measuring instruments and data acquisition and storage devices, found in the monitored projects covered by the work of the Group. Altogether, it is a useful, practical document which represents the latest European experience in the field of performance monitoring in practice.

The reports on Solar Water Heating and Solar Space Heating contain copies of all the completed Reporting Formats from the projects in the data collection exercise, together with analysis sections covering most aspects. Conclusions are drawn on performance, system design and construction, system costs and the prospects for optimisation. Additionally, recommendations are made for future work.

The main conclusions and recommendations from these reports are reproduced in the Executive Summary where, in addition, there are appendices on cost optimisation techniques and performance optimisation procedures. Also a methodology for deriving optimum sample sizes for field trials of occupied housing with solar heating systems is presented.

2.3 Phase 2: Conclusions

The detailed conclusions are not set down in this paper as they are too numerous but, in general terms, the following points should be noted:

- Solar heating systems are generally found to be reliable, and improvements could be made to increase this reliability.
- Average performance of domestic solar water heaters in the data collection exercise was found to be 820 $\pm$ 400 MJ/m^2 per annum (225 $\pm$ 110kWh/m^2 per annum). Detailed modelling exercises carried out at the Thermal Insulation Laboratory in Denmark suggest that it is possible to achieve 1080 to 1440 MJ/m^2 per annum (300 to 400 kWh/m^2) in Northern Europe (1). Recent measurements from a test rig at the same laboratory for the 1980/81 heating season confirm this (2).
- Average performance of active space heating systems (with domestic water heating combined) was found to be 560 $\pm$ 240 MJ/m^2 per annum (154 $\pm$ 66 kWh/m^2 per annum). Detailed modelling exercises carried out at the Institute of Applied Physics (TNO) in The Netherlands suggest that, for central and northern Europe, a small combined system with short term storage could deliver approximately 1000 MJ/m^2 per annum (275kWh/m^2 per annum). It was also found by modelling at the same institution that this figure could be raised to 1400 to 1800 MJ/m^2 per annum (385 to 495kWh/m^2 per annum) with interseasonal storage (3).
- Performance data for passive space heating systems is very sparse, and that which does exist is believed in most cases to be unrepresentative.

(NB: More passive projects are included in the current exercise - see 3.7 below. Also see * Footnote at bottom of following page.)

- As far as solar water heating systems and active space heating systems are concerned, there are areas where cost reductions can be made to improve cost effectiveness. Principally these are in the collector, thermal storage and installation costs.

It has come to the notice of the Performance Monitoring Group that most installations available for study have been designed rather extravagently, and that few, if any, have been designed with the benefit of hindsight to earlier system types. This is, of course, not universally true, but sometimes more recent installations perform less well than others designed earlier. In general, the installations so far studied must in any case be described as "first generation systems".

Most active space heating systems, for example, are over-sized for their load and, in most of these cases, a high solar fraction has been sought. The penalty incurred in such an instance is, inevitably, a low system output per m^2 of collector, with consequently lower system cost effectiveness.

2.4 Performance Criteria

The amount of solar energy used per m^2 of collector is, in the opinion of the Performance Monitoring Group, the most important performance figure relating to any type of solar installation, together with the overall system efficiency.

The solar fraction, defined as the percentage of the total load satisfied by solar energy, is a potentially misleading indicator of performance, for it gives no indication at all of the solar contribution made by the system. (Further details on this topic are given in the reports.)

3.0 Phase 3: Introduction

Referring again to the conclusions from the previous phase of work, the Performance Monitoring Group has become aware that there is a specific need to undertake system optimisation studies in areas where detailed experience of system design and monitored performance exists and where better performance and/or reduced costs are possible. Indications of these possibilities come either from modelling exercises, or from evidence that cost savings might be made by modification.

Additionally, it has always been an objective of the Performance Monitoring Group to get to a position where it can make recommendations to the Commission for "second generation" solar heating system types that might be designed, built and monitored in the future. "Second generation" in this context means systems whose design, specification, predicted performance and cost represent significant improvements over those demonstrated and monitored to date. Figure I illustrates in general terms where it is we want to go.

It is broadly on the basis of these various points that the Performance Monitoring Group has embarked on a programme of Performance Optimisation Studies, which form part of the work of our current third phase programme. This work is described below, and is followed afterwards by a brief resume of other current tasks.

Footnote (refer previous page) re. Passive Space Heating Systems.

Performance estimates have been made, however, based on various modelling predictions for the UK. These vary for different system types and from researcher to researcher. Predictions range from 70 to 435 MJ/m^2 per annum (20 to 120kWh/m^2 per annum) for simple direct gain systems; from 220 to 545 MJ/m^2 per annum (60 to 150kWh/m^2 per annum) for sunspaces; and as high as 900 MJ/m^2 per annum (250kWh/m^2 per annum) for specialised hybrid systems (4).

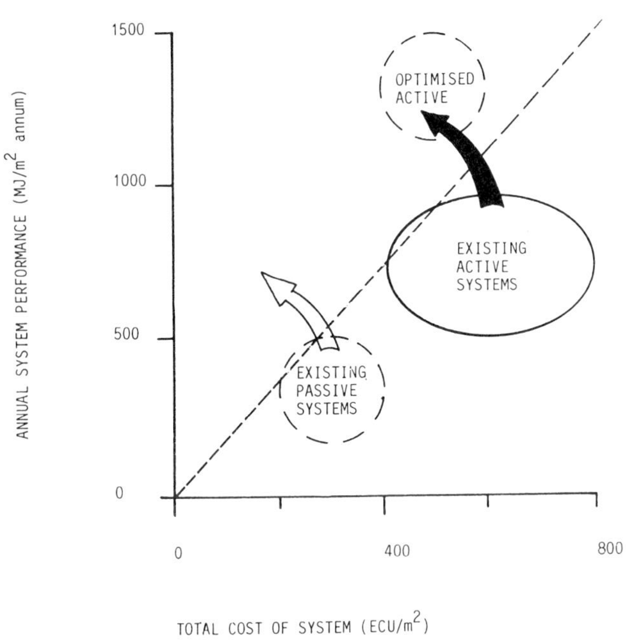

Note: Due to lack of sufficient performance data, 'Existing Passive Systems' are based on estimates.

Fig. I - Approximate cost-performance of solar heating systems

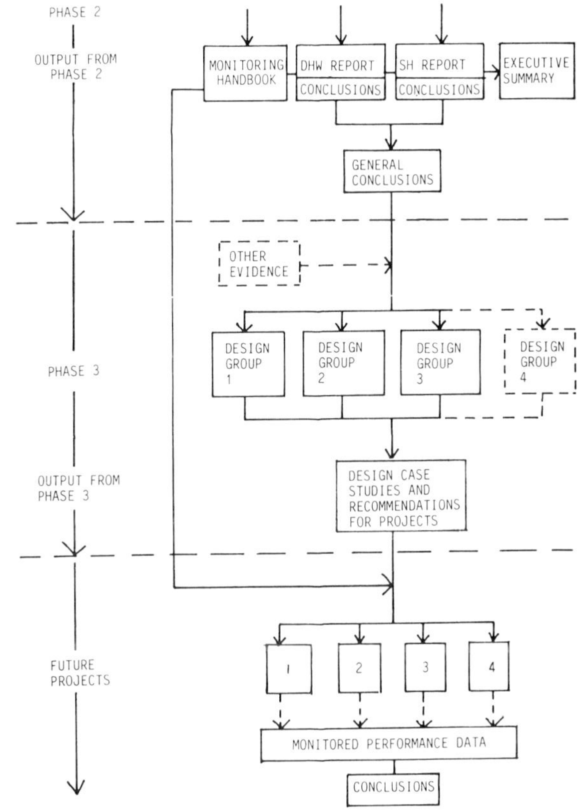

Fig. II - PMG: Design groups

3.1 Phase 3: Objectives

The main objective of the current phase is to produce a series of solar system case study designs based on a set of realistic practical constraints, industrial expertise and existing PMG experience. The case studies will be based on low to medium cost housing, and will be presented in such a way that they could act as the basis for detailed proposals for future monitored field trials and demonstration projects.

In more detail the objectives are:

- To examine and illustrate the process of design optimisation needed to achieve lower system costs and higher system outputs.
- To illustrate design configurations for selected system types and the general implications on house design.
- To produce case study designs based on existing experience, sensible engineering, industrial practice and cost restraints.
- To produce a document suitable for use by those involved in system design. The document would contain the case studies and would:
  a) illustrate the process of design development and system optimisation;
  b) illustrate some of the most promising system types in domestic solar heating applications;
  c) act as the basis for detailed briefs for actual field trial/ demonstration proposals;
  d) help prevent the repetition of non-optimised designs, by example.

Figure II illustrates schematically the process whereby the work from Phase 2 is linked to Phase 3, and indicates that the output document in 1983 could be used as the basis for future field trial and demonstration project specifications.

3.2 System Types to be Studied

It was decided that emphasis should initially be placed on domestic hot water systems and active space heating systems (with domestic water heating combined) due to the fact that design and performance data for these system types has been studied by the PMG over the recent periods. It was also decided that a series of climate sensitive solutions should be produced, one for each of these system types for north, central and southern Europe.

Additionally, a more investigative optimisation case study is to be carried out on a passive system for central Europe, this receiving less emphasis at this stage due to the relative lack of good performance data from first generation passive installations so far.

To summarise, therefore, the PMG Design Groups will produce optimised solar system case studies as follows:

- DOMESTIC SOLAR WEATER HEATERS: One for north, one for central and one for southern European climates.

- ACTIVE SPACE HEATING SYSTEMS (WITH DOMESTIC WATER HEATING COMBINED) One for north, one for central and one for southern European climates.

- PASSIVE SPACE HEATING SYSTEMS: One investigative study for a central European climate.

It is, of course, realised that the Design Groups undertaking this work will be located in countries with different local regulations, building and design practices, and it is not intended to deviate from those constraints. Where such variations do occur, they will be noted and comparisons drawn between the different local situations under several headings, including Architectural and Construction practice, Cost and Engineering.

3.3 PMG Design Groups

Under the direction of the Co-ordinators, all the members of the Performance Monitoring Group are participating in this work, either directly in the Design Groups or in Associated Tasks (see 3.4 below). The tasks are all of similar weight and, in all cases, consideration has been given to special expertise or experience. This has resulted in a practical framework for activity, represented schematically in Figure III.

There are three main Design Groups, represented by vertical lines in the diagram. In the case of the first and second, each will undertake case study designs for both domestic solar water heaters and active space heating systems (with domestic water heating combined). In the case of the third, responsibility for each of the two system types is split between two participants. The fourth deals on an investigative basis with a passive space heating system.

Design Group 1, for northern Europe, is based around Poul Kristensen at the Thermal Insulation Laboratory at Lyngby in Denmark.

Design Group 2, for central Europe, is based around Dick Brethouwer at TNO in Delft, Netherlands.

Design Group 3, for southern Europe, is based around Gerard Kuhn from A.G.E.D.E.S. at Grenoble, France for the space heating system; and around Fabio Cecchi Paone at Fidimi Consulting in Rome, Italy for the solar water heating system.

Design Group 4, dealing with the passive space heating system for central Europe, is based around the Co-ordinator, Energy Conscious Design in the UK.

In each case, local expertise will be used as necessary in each of the following areas, represented by the thinner vertical lines in Figure III:

- Architectural input
- Cost expertise
- Industrial input
- Performance modelling

I would stress that the input from industry is seen as a most important factor in this activity. This is intended to create a useful dialogue and to achieve sensible, commercially orientated results.

3.4 Associated Tasks

Other PMG participants, as mentioned previously, will undertake tasks of comparison and liaison between the Design Groups, with responsibility for different areas. These are represented by horizontal lines in Fig III. Their principle objectives are either to service aspects of the work or to extract information to be used in assessment of the product.

These tasks have been allocated as follows, and each task will result in a chapter in the Final Report:

- Architectural and Construction Aspects: Owen Lewis at University College Dublin, Ireland
- Cost Control and Engineering Design: Ulrich Luboschik at IST Energie Technik, Germany (with cost support from a UK based consultant as necessary)

- Comparative Modelling: Arnold Debosscher at Katholieke Universiteit Leuven, Belgium.

3.5 Method of Work

In general terms, the method of working adopted reflects the process of design optimisation. Initially, designs will be produced based on current experience, then costed, and their performance predicted by local modelling. The designs will then be reviewed and modifications undertaken. Re-costing and re-assessment of performance predictions will then take place, followed by a further review and so on. This iterative process is intended to result in better optimised designs.

In the case of each Design Group (or sub-Group in the case of France and Italy), the responsibility for organising local effort will rest with the PMG member concerned.

In the case of the "horizontal" comparative and liaison Tasks (see Fig. III), input and effort is being organised to take place there in step with the work of the Design Groups. In some cases, e.g. Architectural and Construction aspects, input begins at the outset of the work. This also applies to the Modelling task, where modelling requirements at local level will be specified by the PMG participant. Later on, Cost Control and Engineering Design input will be brought into the design process. Finally, all the "horizontal" roles will be exercised at the review stages.

A detailed programme for the work has been produced.

3.6 Relationship to Other Groups

Quite obviously, a link is being forged with the European Modelling Group through this work, which the PMG sees as an important step. The nature of this link derives from the fact that a correlation will be carried out between the local modelling by the different Design Groups. This work will be undertaken by the Belgian PMG participant in his "horizontal" PMG Task, as he is also closely involved in the work of the European Modelling Group.

Some of the work also touches on areas of common interest shared with the Passive Solar Working Group, and the PMG views its work here as a complimentary effort to the ongoing work of that Group.

3.7 Other PMG Tasks

In addition to the work mentioned above, the Performance Monitoring Group will continue to carry out other tasks as before, although in this third phase it has been agreed that these should have less emphasis than before.

Principally these are as follows:

- The maintenance of monitoring equipment catalogues;
- The production of study reports on specialist areas of monitoring, e.g. the special requirements of large projects involving large numbers of houses, and passive monitoring techniques;
- The production of reduced versions of the Reporting Formats, for use in larger scale projects with lower levels of monitoring;
- The continued collection of design and performance data from existing monitored solar installations in buildings.

In the case of the last item, the PMG has agreed that more stringent selection criteria be employed, so that only those projects with good performance or interesting features are included. The Group is keen to obtain more data from passive projects in this phase as, in the previous phases, performance data in this area has been sparse.

After an initial survey the Group expects that approximately 30 Reporting Formats from European projects will be completed during the current Phase 3: 10 solar water heating projects, 10 active space heating projects and 10 passive space heating projects. It is therefore expected that the PMG will be able to continue its role of providing useful analysis of measured performance data, for use by all involved in Project A, well into the future.

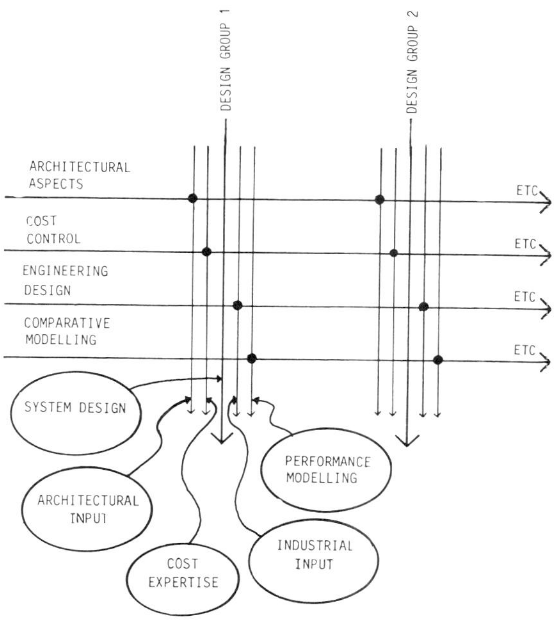

Fig. III - Organisation and structure

N.B. Only two design groups are illustrated in the diagram, whereas there will be four begun at this stage

References:

(1) Solar Water Heating: An analysis of design and performance data from 28 systems.
Appendix D.
CEC Performance Monitoring Group
Edited by D. Turrent, R. Godoy, R. Ferraro.

(2) Reporting Format, completed for Lyngby DHW Test Rig by Thermal Insulation Laboratory, Denmark.
Reproduced in (1) above.

(3) Figures supplied by the Institute of Applied Physics (TNO), Delft, The Netherlands.

(4) Passive Solar Housing in the UK.
Energy Conscious Design. 1980
ISBN: 0 9507409 0 X.

SOLAR COLLECTOR TESTING

(The Co-ordinators' Report on a Concerted Action)

Authors	:	W. B. GILLETT, J. E. MOON	
Contract Number	:	ES-A-G-047-UK(N)	
Duration	:	36 months	1 July 1980 - 30 June 1983
Total Budget	:	£72 400	CEC contribution : 100%
Head of Project	:	Dr. W. B. Gillett, Solar Energy Unit, University College, Cardiff.	
Contractor	:	University College, Cardiff.	
Address	:	University College, Cardiff, P.O. Box 78, GB-Cardiff, Wales CF1 1XL	

Participants in the Concerted Action:

Belgium	:	CRES, Faculte Polytechnique de Mons; Katholieke Universiteit, Leuven.
Denmark	:	Technical University of Denmark, Copenhagen.
E.C.	:	Joint Research Centre, Ispra, Italy.
France	:	CEA, Saclay & Cadarache; CSTB, Sophia Antipolis; Ecole Nationale Superieure des Mines, Sophia Antipolis; CETIAT, Lyon; Electricite de France (until 1981); Ecole Superieure d'Ingenieurs de Marseilles (until 1979);
Germany	:	KFA, Julich; Ludwig-Maximilians-Universitat Munchen; BLL Technische Universitat Munchen; Universitat Stuttgart; TUV ev Bayern, Munich; DFVLR, Cologne (until 1982); Brown Boveri et cie, Heidelberg (until 1982).
Greece	:	University of Athens.
Ireland	:	Institute for Industrial Research & Standards and University College, Dublin.
Italy	:	CRAIES, Verona; Politecnico di Torino; Phoebus, Catania, Sicily; Fiat, Torino (until 1979); Zanussi, Pordenone (until 1981).
Luxembourg	:	Gradel S.A., Steinfort.
Netherlands	:	TPD-TNO-TH, Delft; Philips, Eindhoven.
United Kingdom	:	Building Research Establishment, Garston, England; University College, Cardiff, Wales.

Summary

The collector Testing Group which contains approximately 20 laboratories began work in 1976, and has continued its development of test methods for the performance, durability and reliability of thermal collectors. Round Robin tests are nearing completion on selectively coated flat plate and evacuated tubular collectors. These have been performed both in solar simulators and in outdoor conditions. Test facilities for air heating collectors are under construction and the first round robin air collector will be distributed in June 1982. A joint workshop was held with the International Energy Agency collector testing experts in February 1982 on the subject of solar simulators, and proceedings of this are soon to be published. Work on durability and reliability aspects of collectors is gathering momentum and a specialist meeting was held in March 1982 on the measurement of optical properties of solar collector components.

The Group remains very active and is continuing to develop new and better test methods for characterising collectors. Its next major tasks are to update its recommendations for collector test methods and test facility design and to draft some guidelines for collector design. Planning has begun for a future work programme in which the experimental expertise of the Group will be used to develop short term tests for domestic solar water heating systems.

1. INTRODUCTION

The Collector Testing Group (CTG) contains experts from about 20 European laboratories, and has been working since 1976 in close collaboration with the Joint Research Centre at Ispra. The work includes the development of collector thermal performance test methods for which recommendations were published in 1980 (1), and studies of collector durability and reliability.

The participating laboratories change from time to time in line with national research and development priorities, and a current list is given at the front of this report. A recent addition to the group is Dr. M. Tsamparlis from the University of Athens, Greece.

The work of the group has been co-ordinated by University College since 1978, and the most recent half yearly meeting was held in the Hague on 24-26 May 1982. The work of the Group was last summarised for the CEC Contractors' Meeting in Athens in November 1981 (2). This report therefore concentrates on developments made in the period November 1981 to June 1982.

2. ROUND ROBIN COLLECTOR TESTING

Round robin tests, where collectors from a single batch of production are distributed to all the laboratories in the Group provide the main data on which test recommendations can be based. Results from the first four round robin tests were summarised in a paper to the Solar World Forum in August 1981 (3). Current work includes an extension of the round robin programme using the DRU collector (CEC 4) to study test methods in solar simulators, and a Round Robin CEC 5 employing the Philips' evacuated tubular collectors. A Round Robin (CEC 6) employing a Belgian air collector will begin later in 1982.

2.1 Solar Simulator Round Robin Series. Eight laboratories with solar simulators submitted results for the DRU collector in time for the final group analysis. All laboratories used the "Quasi Steady State" method as a basis to determine the collector's instantaneous efficiency curve. As; far as possible, tests were performed in accordance with the "CEC Recommendations for European Solar Collector Test Methods, 1981" (1). At present these recommendations only give outline guidance on testing in solar simulators, stating that "the simulator and indoor environment should be representative of the outdoor environment", and that the test methods developed for outdoor testing may be generally applied.

Analysis of Results. The results have been analysed to produce thermal performance graphs with both a linear and a second order fit calculated by use of a 'least squares' curve fitting routine (Fig. 1). In order to allow better comparison of group results the curve fit values of efficiency (η) for selected values of T* (0,0.02,0.04,0.06,0.08) have been extracted for the second order fit, and are presented instead of the usual equation coefficients a_1 and a_2G.. This new presentation overcomes the problem of the inter-relationship of the second order coefficients that has made previous comparisons difficult.

The efficiency values of the raw data have also been'adjusted' using a theoretical computer model in order to compensate for differences in the environmental conditions found between laboratories. The adjustment takes into account variations in solar irradiance, percentage of diffuse radiation, local wind speed, ambient air temperature, sky temperature (excess thermal radiation in the simulator room) and collector tilt. The resulting adjusted values (η') have been used to produce the $\eta' \sim T^*$ graphs in the same manner as for the raw data (Fig. 2). The adjusted results have been compared with the theoretical performance derived from the computer model (Figs. 3 and 4).

Discussion of Results. The figures clearly show the reduction in data scatter after adjustment of the raw data to standard environmental conditions. The tight grouping of data points at low T* is particularly good giving a very close agreement for η'_o. The slope of the curve is rather less well defined, but the all data adjusted fit gives good agreement when compared with the theoretical model except at high T* values. The majority of laboratories have adjusted efficiency values within around 0.03 of the theoretical model throughout the chosen T* range.

The linear fit always gives higher η_o values than the GT^{*2} fit. After adjustment, the U values from individual laboratories lie within 0.6 $Wm^{-2}K^{-1}$ of the All Data value. The Raw Data U values have greater variation due mainly to the low local wind speed values used by some participants.

The overall result compares favourably with the outdoor data (adjusted) result of the previous (CEC4) Round Robin series using the 'DRU' collector.

Conclusions. The scatter of data between laboratories can probably be attributed in some extent to differences in test conditions, but some may also be caused by differences in test procedure. The tests at low local wind speed appear to have less stability and give inaccuracies that cannot be readily corrected. The point of measurement of ambient temperature is not consistent between laboratories and may lead to errors or at least to a shift of data points along the T* axis.

The theoretical and measured effects of collector tilt are not in agreement and so the non-standard tilt angles may have a greater effect on the collector efficiency than supposed.

The number of test points and the period between them varied considerably between laboratories. This leads to the 'weighting' of data and also to potential inaccuracies.

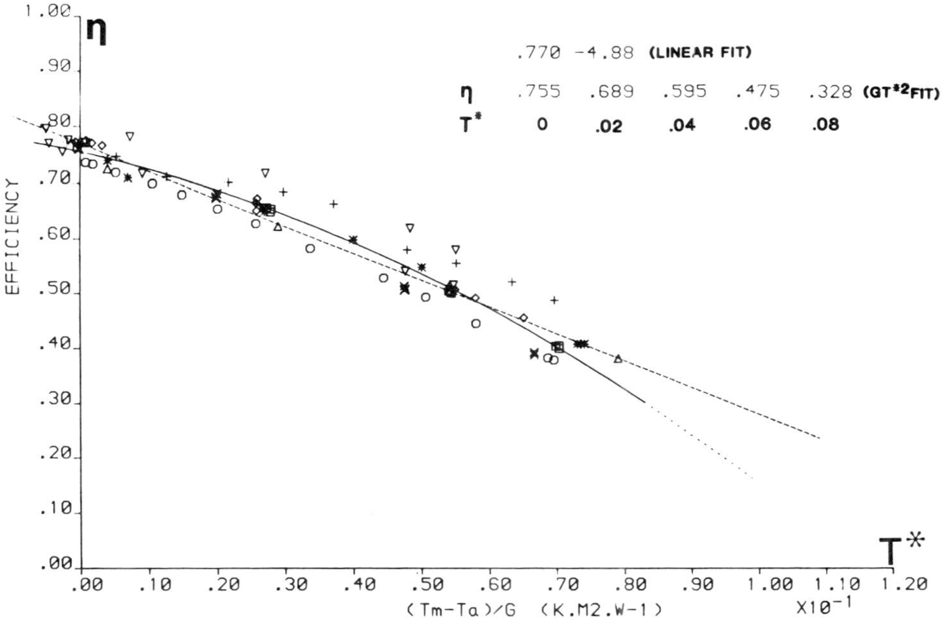

Fig. 1 Simulator Round Robin (CEC 4) *All Data*

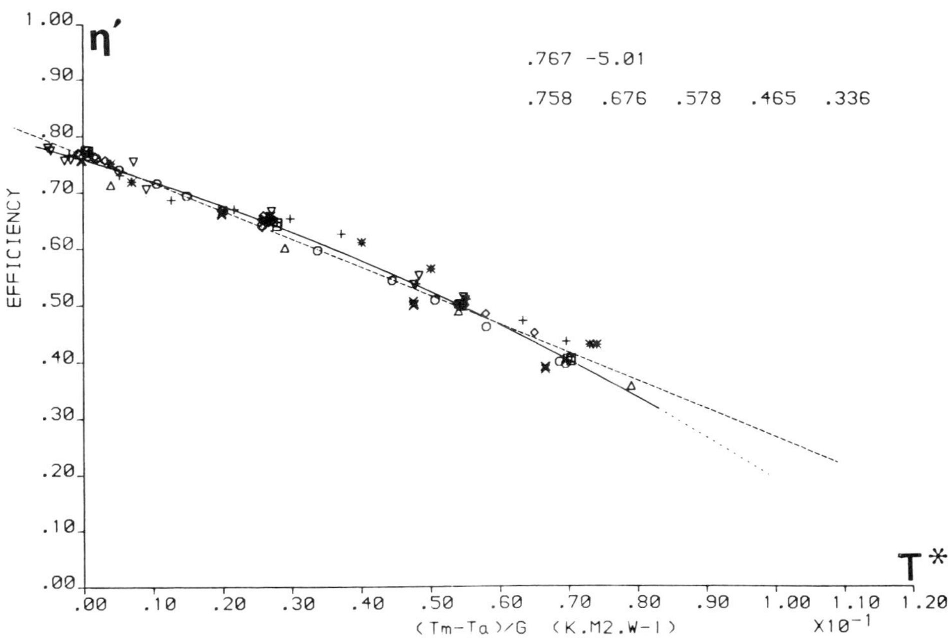

Fig. 2 Simulator Round Robin (CEC 4) *All Data 'adjusted'*

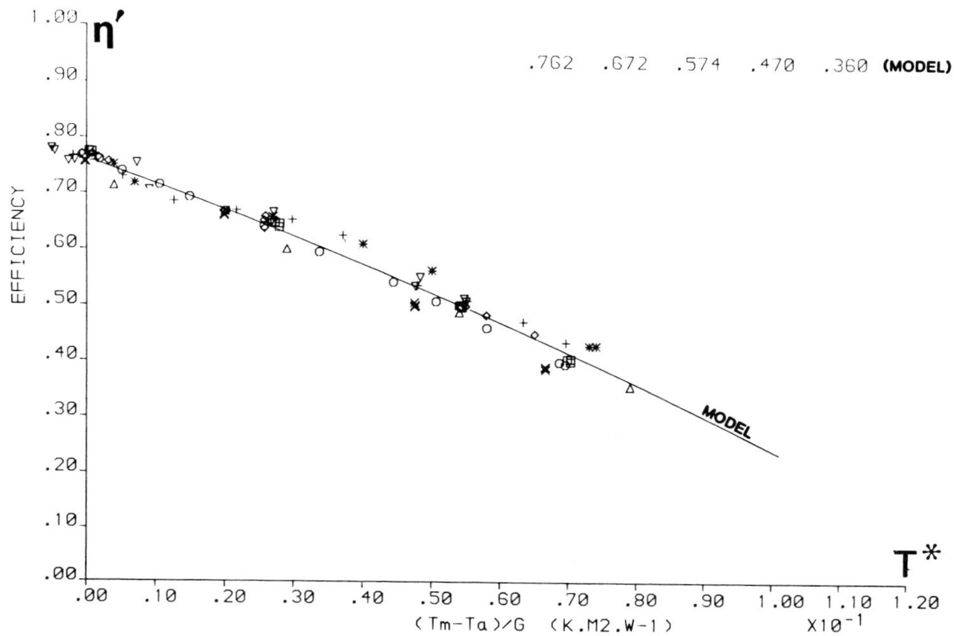

Fig. 3 Simulator Round Robin (CEC 4) *All data 'adjusted'*

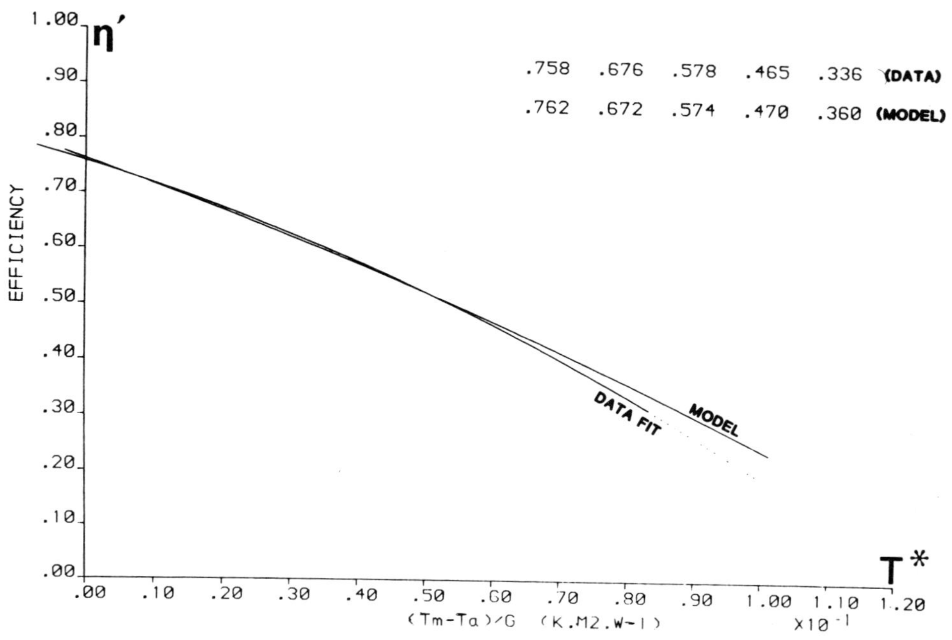

Fig. 4 Simulator Round Robin (CEC 4) *All data 'adjusted'*

These inconsistencies need to be overcome and the test procedures clarified as the Group continues to use and develop solar simulators for collector testing.

2.2 CEC 5 Round Robin Series (Philips Evacuated Tube Collector). As a result of an extension of the test period for this round robin until the end of May 1982, several additional sets of test results have been received since this work was reported in November 1981 (2). These include some 'retests' and make a total of 43 sets of test results from 13 laboratories. The new tests have been undertaken in accordance with revised test recommendations that include a higher test flowrate and the monitoring of individual tube performance.

Results. A full analysis of this round robin has not yet been completed. Each data set has been provisionally analysed by U.C.C. using a 'least squares' curve fitting routine to produce both linear and second order (GT^{*2}) curve fits.

The provisional Combined Group Results are shown graphically in Figures 5 and 6.

All data sets have been included at this stage even if 'Retests' have been performed. Later, it may be appropriate to delete test results that are known to have been affected by damaged tubes and insufficient flowrates.

No attempt has yet been made to'adjust' the data to standard reference environmental conditions.

COMPARISON OF CEC5 COMBINED GROUP RESULTS

	η_o	U	η_o	$\eta_{0.02}$	$\eta_{0.04}$	$\eta_{0.06}$	$\eta_{0.08}$
WHITE REFL.							
All Data	0.629	-2.21	0.622	0.588	0.546	0.498	0.444
Outdoor	0.639	-2.61	0.638	0.588	0.536	0.483	0.428
Simulators	0.624	-2.07	0.615	0.588	0.552	0.505	0.449
BSE I/O	0.619	-1.92	0.610	0.583	0.549	0.509	0.461
BLACK REFL.							
All Data	0.543	-2.05	0.537	0.505	0.466	0.421	0.371
Outdoor	0.544	-2.37	0.537	0.500	0.455	0.402	0.340
Simulators	0.545	-1.98	0.538	0.510	0.474	0.429	0.375
BSE I/O	0.547	-1.78	0.551	0.510	0.473	0.439	0.409

Discussion. Both η_o and loss coefficient values still vary between laboratories, more than might be hoped, but variations may be reduced by normalising results to reference conditions.

In addition to the well known instrument errors the following factors are expected to affect the performance and lead to data scatter with this collector:

1) Individual tube performance variations (including failure)
2) Flow rate variations (including low T* instability)
3) High irradiance (affects heat pipe performance)
4) Extremes of tilt
5) Incidence angle
6) Dirt on tubes and reflectors
7) Percentage of diffuse radiation
8) Wind speed around collector

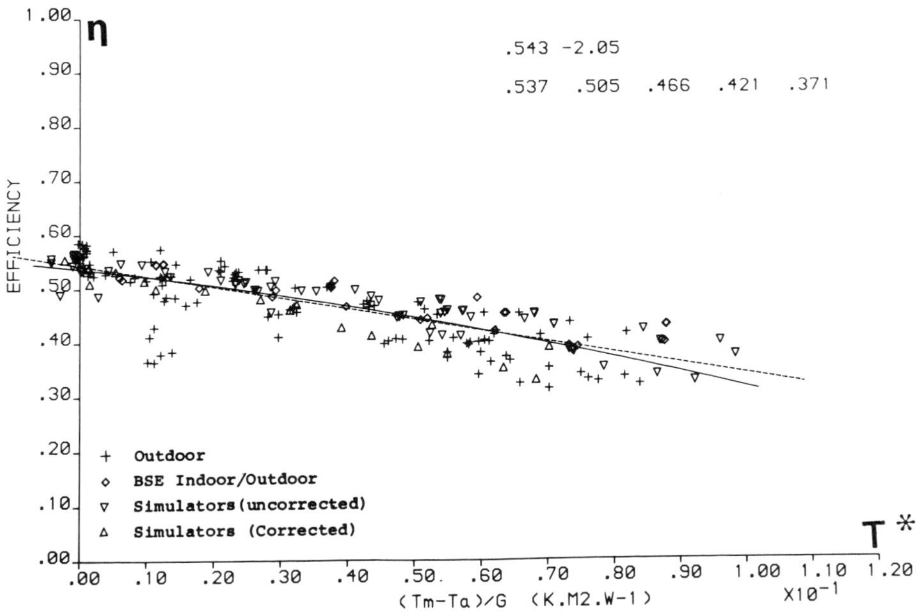

Fig. 5 Philips Round Robin (CEC 5) *All data (black reflector)*

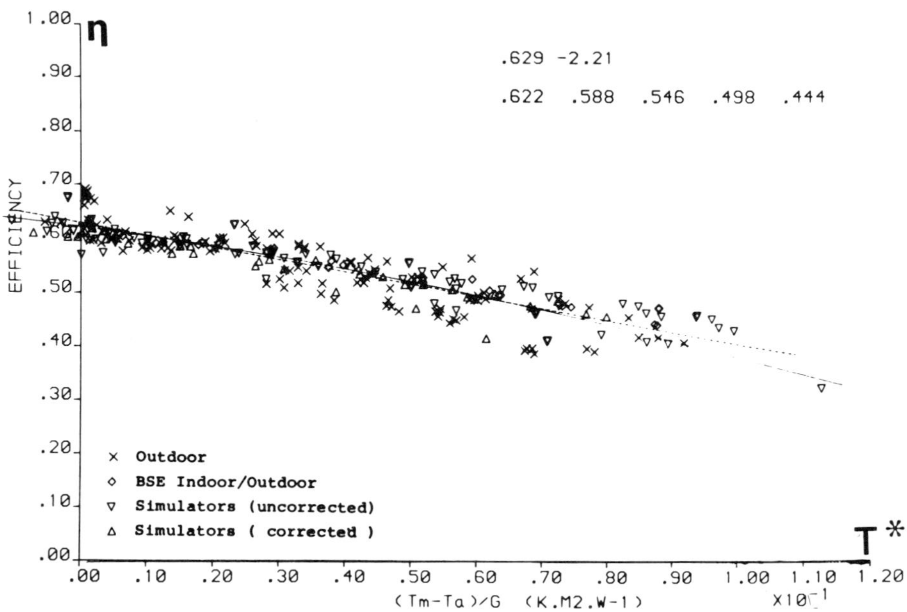

Fig. 6 Philips Round Robin (CEC 5) *All data (white reflector)*

The rather complex heat transfer behaviour of this collector may be expected to make it difficult to normalise test results to reference conditions, but work on a computer model for the collector is in progress.

A draft final report on this round robin will be distributed later in 1982, but much has already been learned from this round robin series that should enable useful recommendations to be included in the new edition of the "CEC Recommendations for European Solar Collector Test Methods".

2.3 'Air' Collector Testing. Fifteen laboratories are now actively engaged in the design and construction of test loops suitable for the testing of air heating collectors. This work is being supported financially by the CEC and participants will report on the design of their facilities by the end of 1982. Some of the test loops will be for use in solar simulators, and others for outdoor collector testing.

A Belgian 'air' collector has been selected for the next (CEC 6) Round Robin series and will be distributed to participants around the end of June 1982.

Recent experience from laboratories already active in air collector testing has been compiled as an "addendum" to the earlier "Draft Recommendations: The Testing of Air Heating Solar Collectors" (4). A draft test procedure for the round robin series was formulated at the May Group meeting.

There appear to be two major problem areas to be overcome in air collector testing, even before attempting to deal with the complex flow rate dependent performance of collectors themselves. These problems are (i) the measurement of air mass flow rates to better than ±2%, and (ii) the measurement of mean air temperatures in a duct to ±0.01°C.

3. HIGH TEMPERATURE TESTING

Five participants have developed test loops using oils or pressurised water for the testing of high temperature collectors.

The testing of concentrating collectors is also being undertaken by at least three participating laboratories and will be continuously reviewed by the Group.

At this time there is no plan to include a major collaborative task on high temperature testing in the current programme. This subject may be considered in more detail at some time in the future.

4. REPRODUCIBILITY AND TEST ACCURACY

Participants have agreed to undertake reproducibility tests for all existing and new facilities that are to be used in the collaborative programme.

Other tasks are also being encouraged such as the exchange of collectors between laboratories and the testing of several collectors at individual laboratories.

This is considered as ongoing work which will be analysed and documented by the co-ordinators for use in Round Robin analysis, and reported on at suitable intervals.

Pyranometers remain a subject of concern to the CTG, since the accuracy of these instruments directly influences collector test results. The CTG was represented at a meeting in the World Radiation Centre (WRC), Davos in February 1982, where the progress of pyranometer experiments was reviewed. This meeting was called by the International Energy Agency (IEA) Tasks III and V, but was also attended by EC experts. The meeting concluded that WRC and manufacturers' calibrations for good Kipp CM11 and Eppley PSP pyranometers could agree to within ±0.5%, but that calibration requirements were not the same for collector testing as for long term meteorological network recording. Further work is planned by the IEA, and the CTG is continuing its encouragement of EC work on pyranometers, particularly on the production of a European Pyranometer Handbook.

5. SOLAR SIMULATOR DEVELOPMENT

A joint workshop was hosted by JRC, Ispra in February 1982 with participants from IEA Task III, the CEC Collector Testing Group and lamp manufacturers. The attendance was 33 and the proceedings which will be published soon will include all papers presented, a comprehensive summary and some comments by each chairman on his session.

It was shown during the workshop that only about 30 simulators were in use for thermal collector testing around the world, and co-operation on the drafting of standard test methods should therefore be possible. About 7 lamp types were in use including Thorn CSI and CID, Osram HQI, Vortek, Philips, Xenon and Heraeus, but over half of the simulators in regular use employed CSI lamps. Simulators at Munich, CSIRO (Australia) and DSET (USA) had special facilities for measuring collector incidence angle modifiers, but as yet experience with these measurements was rather limited. Interest in methods for normalising results to reference conditions such as that reported by the CTG co-ordinators (5) was growing, but more work was needed on the measurement of simulated solar irradiance and thermal irradiance.

The proceedings of this workshop will provide a comprehensive "state of the art document" in themselves, but it is also envisaged that an enhanced section will be written on solar simulator testing in the revised "Recommendations for European Solar Collector Test Methods". A solar simulator design guide is under consideration.

6. NEW TEST METHODS

During the current and the previous CEC solar energy programmes, a number of new test methods have been proposed. A report summarising most of these was tabled at the May 1982 CTG meeting (6). Following some discussion it was concluded that several of these methods need not be pursued further but that others warranted further consideration.

The method recommended in reference (1) for measuring the effective thermal capacity of a collector was still felt to be the most appropriate, and it was agreed not to pursue test methods which had been developed only to allow testing with varying fluid inlet temperatures. However, transient test methods which allowed testing under variable irradiance conditions were found attractive by participants who do not possess solar simulators. In particular, the simple integral method proposed by KFA, Julich, and the response function method proposed as a British Standard Draft for Development were identified for further study. New tests proposed by CRES and Ecole des Mines will also be investigated.

7. NEW STUDIES

The CTG expects to include the development of short term test procedures for domestic solar water heating systems in its next work programme. The main emphasis of this work will be experimental, and will be collaborative with regard to the methodology employed. Experimental measurements will be used to validate some kind of DHW system model, and the work of the CTG in this field will therefore be complementary to that of the European Modelling Group and of the Performance Monitoring Group, with whom discussions have already begun. The resulting test procedure is foreseen as a system qualification test or commissioning test, and it is intended that it should be of practical use to commercial installers.

8. COLLECTOR DURABILITY AND RELIABILITY

This subject was discussed in detail at the May 1982 group meeting, and a well founded basis for a future programme of work began to emerge. The initial development work carried out by the different participants has identified a number of tasks in which collaborative work could be pursued, and these will be included in the next work programme.

a) *The Inspection Reporting Format* developed jointly with IEA Task III has now been used to provide information on the collectors from more than 90 systems. The quality of the reports varies widely, and analysis will not be easy. However, from this size of data sample, bearing in mind that several systems contain more than 50 m^2 of collector, it should be possible to confirm a positive set of guidelines for collector design. The first draft of the guidelines will be distributed later in 1982. The inspection reports will also be used to assist with the identification of a recommended set of collector durability tests.

b) *Natural Ageing of Collectors* by dry stagnation and by stagnation with a small fluid reservoir and natural circulation are both being investigated. The diagnostic tools used to assess the results of ageing are:

(i) Periodic thermal performance testing (important but insensitive)
(ii) Visual inspection (important but limited)
(iii) Periodic weighing (indicates ingress of moisture)
(iv) Periodic optical property measurement (samples need to be removed).

c) *Qualification Tests* of many types have been investigated by specialist laboratories, but following discussions at the meeting in May 1982 it was agreed that very few of these tests could be economically justified for widespread use. The most important are (i) high temperature stagnation, (ii) rain/wind penetration, (iii) absorber internal pressure and (iv) external thermal shock. Optical property measurements and surface adhesion (tear) tests are important for absorbers and covers.

9. OPTICAL PROPERTY MEASUREMENT

In support of the work on collector durability and reliability it was found necessary to compare measurements of solar absorptance and thermal emittance for solar absorbers. A small round robin was carried out in 1981, and a meeting of optical property specialists was held to discuss the results at TNO, Delft in March 1982 (7).

Three main methods of solar absorptance measurement were identified and 5 main methods of thermal emittance measurement. It was concluded that the most important part of the measurement equipment was the standard surface used for calibration. When standard surfaces which have values of α_s and ε close to those of the specimen are used in well controlled instruments, then the results agree very closely between laboratories.

It was agreed that:

1. Participants would investigate the availability of an appropriate range of standard surfaces for α_s and ε measurements.
2. The Co-ordinators would produce draft recommendations for α_s and ε measurements, and participants would help to refine these.
3. A round robin of collector covers would be performed to measure τ_s, α_s and τ thermal for Teflon, glass and polyethylene.

A second meeting of the optical property specialists is anticipated in Autumn 1982, possibly in association with the Autumn meeting of the Collector Testing Group. This meeting will review the results of the Collector Cover Round Robin and the draft recommendations on optical property measurement methods.

10. CO-ORDINATION AND DOCUMENTATION

In addition to analysing the round robin programmes and preparing the reports discussed above, the co-ordinators have distributed CTG Newsletters in December 1981 and April 1982. A colour brochure describing CEC work on Solar Collector Testing has been prepared and detailed minutes have been written for each group meeting.

The main CTG document "Recommendations for European Solar Collector Test Methods" will be revised in close consultation with the participants during the remainder of 1982, such that a detailed discussion can be held at the next Group meeting. In parallel with this, work will begin on the analysis of the system inspection reports with a view to drafting a collector design guide. Final reports for the current round robin series will also be prepared.

Two further group meetings are planned within the current programme before June 1983, and three more Newsletters will be distributed.

11. CONCLUSIONS

The enthusiasm of the participants in the collector testing group is resulting in a very ambitious work programme. The co-ordinators have attempted to accommodate this by arranging for specialist groups to work together, for example the optical property measurement specialists, and the solar simulator specialists. Before embarking on new studies, however, it will be important to document thoroughly the work which is currently in progress.

Future work on durability and reliability shows every sign of expanding and may also need to be tackled by a specialists meeting.

Tests for the efficiency of simple flat plate liquid heating collectors may not require very much more work, but the problems of incidence angle modifier measurements and irradiance measurement in solar simulators have yet to be solved. Those laboratories without simulators may also continue to develop test methods for use in variable irradiance outdoor conditions. New problems have been identified with the more advanced 'second generation' collectors and these require further study.

The main technical effort in the thermal performance field will now be concentrated on air collector testing, and in the next programme on the development of short term tests for domestic solar water heaters.

12. REFERENCES

1. Derrick, A. and Gillett, W.B. "Recommendations for European Solar Collector Test Methods", DGXII, Brussels. (Jan. 1980).
2. Gillett, W.B. and Moon, J.E. "The Collector Testing Group", Solar Energy Applications to Dwellings. Vol. 1 Series A pp 31-35, Reidel Pub. Co. (1982).
3. Moon, J.E. and Gillett, W.B. "Collector Testing in the European Communities", Proc. ISES Solar World Forum, Brighton. (Aug. 1981).
4. Moon, J.E. and Gillett, W.B. "Draft Recommendations: The Testing of Air Heating Solar Collectors", DGXII, Brussels (1981).
5. Gillett, W.B. and Moon, J.E. "Results from Solar Simulators in the CEC Collaborative Collector Testing Programme", Proc. CEC/IEA Solar Simulator Workshop (Feb. 1982) (in press).
6. Gillett, W.B. "The Development of CEC Transient Collector Test Methods" UCC Departmental Report 878 (SEU 322). (May 1982).
7. Gillett, W.B. and Moon, J.E. "CEC Collector Testing Group, Optical Properties Measurement Meeting, 30 March 1982", UCC Report 873 (SEU 319), (April 1982).

CONCERTED ACTION EUROPEAN SOLAR STORAGE TESTING GROUP

Author : E. van Galen

Contract number :

Duration : 1982-01-01 until 1983-06-30

Total budget :

Head of project : E. van Galen

Contractor : Technisch Physische Dienst TNO-TH

Address : P.O. Box 155, 2600 AD DELFT, The Netherlands

Summary

The European Solar Storage Testing Group has been established by the Commission of the European Communities to draw up recommendations for test-procedures for solar storage systems. The working group programme is discussed. Background of and demands made upon the testprocedures are elucidated.
The activities consist mainly of a joint action in which tests shall be carried out according to a provisional testprocedure. Each individual group member tests a complete different storage system. The global contents of the testprocedures are formulated. A selection of storage systems is made. The need for simulation models to diminish the number of tests required is expressed.
Subtasks to solve problems identified already in the early discussions are being formulated and are allocated to the individual group members. The objectives for the period until June 1983 are stated.

1.1 Introduction

During the last couple of years the main point of the research on thermal storage of solar energy was related to the development of advanced storage systems. Indeed a large variety of solar storage systems arised from this research with respect to the storage principle, such as sensible, latent, hybrid and thermo-chemical heat stores and all of them in various designs and constructions.

For a better understanding of all these new developments 'Recommendations for a European Reporting Format on the Performance of Solar Heat Stores' were drawn up in 1981 [1]. There was no problem in answering questions concerning design and construction, but the thermal performance of the storage systems had to be expressed in terms as laid down in the only official testprocedure known the ASHRAE Standard 94-77 and this procedure turned out to be inadequate. On the last meeting of the ASHRAE Standards Project Committee SPC 94.1-77 R on January 1982 in Houston it was concluded that the ASHRAE test procedure does not account for realistic cycling of the storage system and therefore does not provide adequate comparisons of different systems under actual operating conditions.

About this time the European Solar Storage Testing Group has been established by the Commission. The ultimate objective of their research programme is: to draw up recommendations for testprocedures for solar storage systems, which must provide useful tools for modellers, system designers and industry. These testprocedures must be as simple as possible, applicable over a wide range of storage systems and the validity must be proven. The objectives of the work in the current four year programme of the Commission till June 1983 are: to draw up draft recommendations for testprocedures for solar storage systems and to draw up guidelines for the design of testfacilities.

1.2 Organisation and working methods

Apart from the coordinator the Solar Storage Testing Group has six active members:

Denmark : S. Furbo, Technical University of Denmark
France : P. Achard, Ecole des Mines de Paris
Germany : J. Sohns, Universität Stuttgart
Italy : R. Vellone, E.N.E.A.
Netherlands : G.J. van den Brink, Institute of Applied Physics TNO-TH
United Kingdom: : R. Marshall, University College Cardiff.

At the meetings, Belgium and Ireland are represented by an observer.

There were a few qualifications for participation: Participants must have disposal of:

- testfacilities to carry out the experimental programme of the group
- a prototype or a commercial solar storage system
- computerfacilities to carry out the required modelling work.

The coordinator is responsible for the presentation of draft testprocedures. These initiative proposals can be amended by the group members and identified problem areas shall lead to the formulation of a programme of subtasks, which shall be allocated to individual group members. The group members then are responsible for the execution of these subtasks and the testprocedure agreed upon for a storage system at their disposal. The results shall be reported according to a reporting format. Apart from the test results, participants have to submit a description of their testfacilities and measuring equipment to judge their results. The coordinator then carries out an analysis of the data and in close collaboration with the group members the final reports of the contract will be prepared giving respectively guidelines for the design of testfacilities and draft recommendations for testprocedures for solar storage systems.

The detailed information necessary for the programme of action is given in working documents, prepared by the coordinator. The participants will report their results in progress reports. After each working meeting a newsletter will be brought out, in which the minutes of the foregoing meeting and the updated action programme will always be part of the contents. The group meets three to four times per annum. The Working Group programme is given in figure 1.

1.3 Demands made upon the testprocedure

It has been recognized that the testresults should be given in the form of storage efficiency curves, usable for system designers and modellers. Each single influence of the relevant parameters on the storage efficiency must be determined by a limited experimental programme. This is in fact the same approach as followed in collector testing and in establishing efficiency curves for heat exchanger and cooling machines.

If a solar installation is divided into three parts, the collector system, the storage system and the distribution system, then the following efficiencies can be defined (see figure 2).

collector efficiency $\eta_{col}(t) = \frac{Q_2\ (t)}{Q_1\ (t)}$

storage efficiency $\eta_{sto}(t) = \frac{Q_3\ (t)}{Q_2\ (t)}$

system efficiency $\eta_{sys}(t) = \frac{Q_3\ (t)}{Q_1\ (t)}$

The results of demonstration projects show that a realistic value for both the yearly collector efficiency and storage efficiency is 0.4 for these first generation solar systems, resulting in a yearly system efficiency of:

$$\eta^{y}_{sys} = \int_{o}^{1\ year} \eta_{sys}(t)\ dt = \int_{o}^{1\ year} \eta_{col}\ (t)\ \ \eta_{sto}\ (t)\ dt = 0.16$$

Improvement of the yearly storage efficiency by a factor 2 will consequently result in a two time higher system efficiency. With a good design such an improvement in efficiency is technically feasible. It is therefore expected that a large variety in storage efficiencies for prototype and commercial solar heat storage systems will occur. Because of the large thermal capacity effects, instantaneous efficiencies are meaningless and mean values over a finite time period are to be considered. Consequently there is a special interest in describing a storage system in terms of a mean storage efficiency for charging and one for discharging. The heat loss rate shall be presented as a seperate result.

Additional results are required to judge the design and construction e.g. with respect to the choice of the storage material, the type of heat exchanger and the rate of thermal stratification under operating conditions. There are three main aspects in solar storage system testing:

- Testing of the storage material (properties and ageing phenomena).
- Testing of the storage system under specific input and output conditions. Two main lines can be distinguished: firstly the temperature step response and secondly the constant power charge/discharge operation.

- Testing of the storage system under dynamic heat input and output conditions, derived from computer based model calculations, simulating the actual collector and system behaviour under actual weather conditions.

In view of the existing testfacilities, no tests under dynamic input and output conditions are yet required in the joint action of the group. To make the most of the available facilities, these types of tests shall be carried out in the form of subtasks in appropriate cases. They are required to prove the validity of the storage efficiency curves under operating conditions.

The testprocedures must be applicable over a wide range of storage systems. For practical reasons the work has been restricted for the period till June 1983 to some short term storage principles, which have already reached a grown-up stage.

1.4 Present status of the work

1. Joint action.

Two documents describing background and provisional test procedures for material tests and component tests, according to a black box approach have been prepared and were presented to the group.

In the document on material tests, testprocedures are proposed so far for measurements of heat effects associated with a temperature change for phase change materials according to the 'differential thermal analysis' method and for measuring the thermal conductivity according to the 'hot-wire method'. It was shown that both measuring methods and instruments have to be adapted to the special characteristics of some phase change materials, namely inhomogeneity, evapouration of water, expansion and electric conduction.

The document on component tests gives a testprocedure which requires the following tests:

a. tests to determine the heat loss rate at finite flow rate
b. tests to determine the heat loss rate at zero flow rate
c. charge tests (temperature step) around ambient air temperature to determine the heat exchanger efficiency and the rate of stratification in the flow direction.
d. charge tests (constant heating power) to determine the storage capacity.
e. charge tests (temperature step) to determine the storage system efficiency for charging and the storage capacity.
f. discharge tests (temperature step) to determine the storage system efficiency for discharging and the storage capacity.

A relatively large number of tests is required to ensure that there will be no lack of experimental data during the phase of interpretation. Reducing the number of tests required for the final recommendations to be drafted by this group shall be one of the aims of that phase.

In view of this each individual member has to develop a detailed model of his storage system to be tested as part of the joint action. Once validated by means of the output temperature response for some typical charge and discharge tests, more storage efficiencies can be calculated for any set of operating conditions using the validated model. The development of these models has started.

Within the restrictions made a wide variety of solar systems have been selected out of the available systems:

Denmark: A hybrid storage system with water in a hot water tank situated inside a container with an inorganic salt hydrate based on the extra water principle. The storage material is partly water, partly a salt water mixture. Thermal stratification is aimed at. The heat exchanger is an embedded spiral. The heat transfer fluid is a liquid.

France: A latent heat storage system. The storage material is $Ca(NO_3)_4 \cdot 4H_2O$. A plastic modular heat exchanger is used. The heat transfer fluid is water.
Germany: A cold hybrid heat storage system. The storage material is magnesium silicate packed in equally sized ceramic spheres (packed bed) and water. Heat transfer by means of an embedded heat exchanger. Stratification is possible. The heat transfer fluid is ethanol.
Italy: A sensible heat storage system. The heat storage material is water. Bottles make up the heat exchanger/container. Stratification is aimed at. The heat transfer fluid is air.
Netherlands: A latent heat storage system using an organic phase change material with a metal matrix for increased thermal conductivity. The storage material will be Shell paraffin wax 60/63. Thermal stratification is aimed at. The heat exchanger is of the heat exchanger/container type. The heat transfer fluid is water.
United Kingdom: A hybrid storage system. A water store is enveloped by a vertical shell with paraffin wax. So the storage materials are water and paraffin wax. There is no stratification in the flow direction. The storage system has an embedded heat exchanger. The heat transfer fluid is water.

A description of the testfacilities is being made. Figure 3 gives a scheme of the Dutch testfacility.

2. Subtasks.

With reference to the discussions resulting from the presentation of the provisional testprocedures, the global contents of ten subtasks have been formulated. A detailed formulation is being made by the individual group members for the tasks allocated to them. The subject-matters of these subtasks cover questions concerning the definition of testconditions, the interpretation of results, the development and validation of simulation models and the actual execution of tests under dynamic input - output conditions.

1.5 Future work

According to the workplan the group plans to carry out the testprocedures and the subtasks mentioned above and after a careful analysis of all results two documents shall be drawn up:

- Draft recommendations for testprocedures for solar storage systems
- Guidelines for the design of testfacilities.

Finally the coordinator plans to update the Recommendations for a European Reporting Format on the Performance of Solar Heat Stores [1] according to the latest insights in the presentation of the results resulting from this concerted action.

References

[1] E. van Galen, J.K.M. Verdonschot, A.J.T.M. Wijsman, C. den Ouden. 'Recommendations for a European Format on the Performance of Solar Heat Stores'.

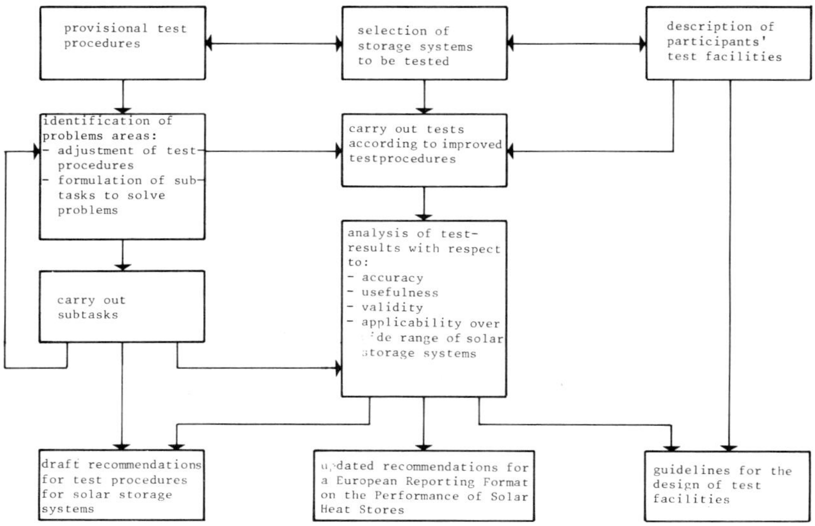

Figure 1 - Working Group Program

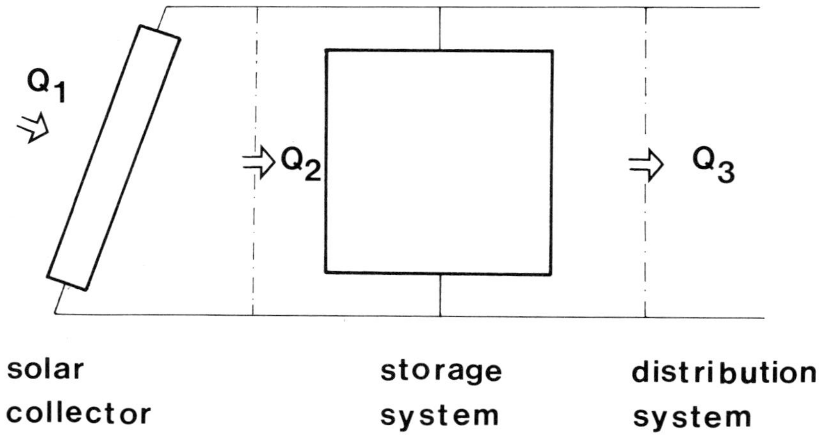

Figure 2 - Definition of energy flows

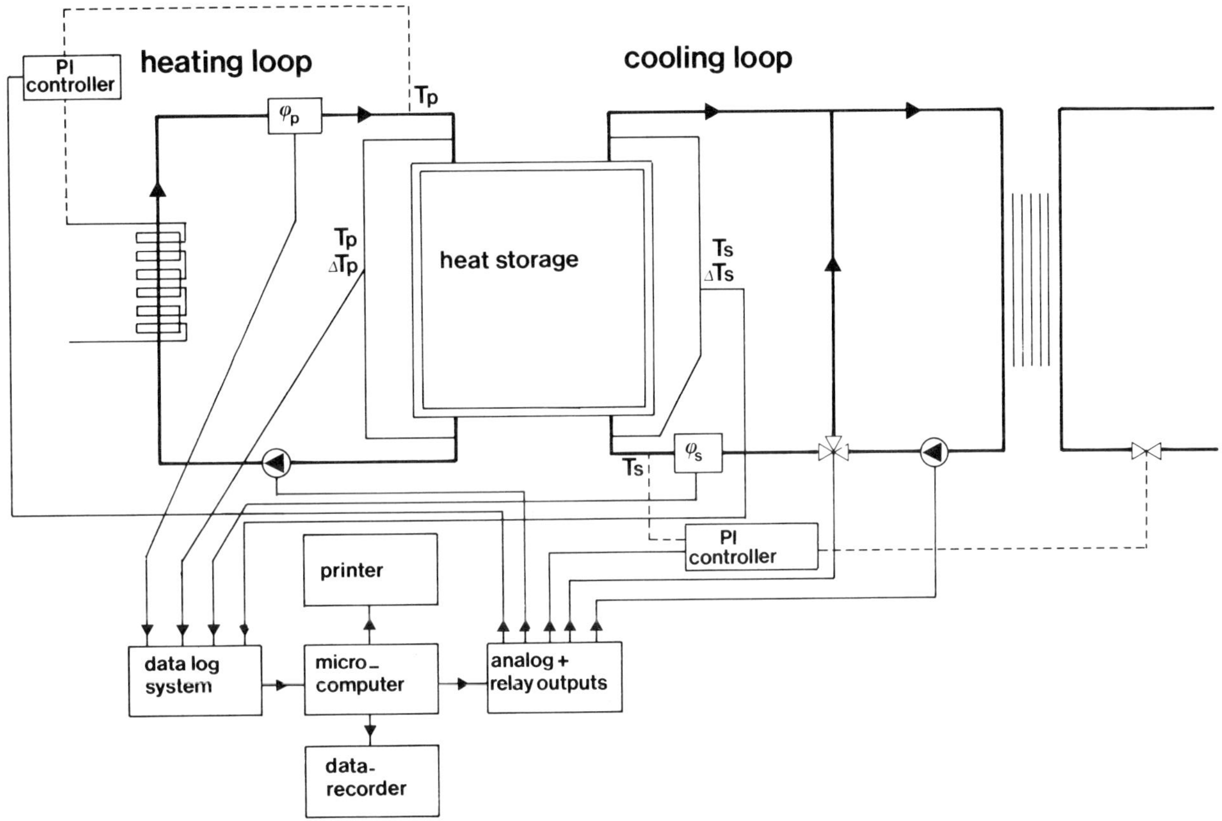

Figure 3 - Scheme of the Dutch test facility

DIGITAL CONTROL SYSTEM FOR A PASSIVE SOLAR HOUSE SPACE HEATING

Authors	: Pierre DUBOIS, Dusan SINOBAD
Contract number	: ESA PS 144 F
Duration	: 12 months
Head of project	: Pierre DUBOIS, directeur de la Division Energie
Contractor	: Laboratoires de Marcoussis, Centre de Recherches de la Compagnie Générale d'Electricité
Address	: Route de Nozay -91460 Marcoussis (France)
Total budget	: 500 000 FF CEC contribution : 250 000 FF

Summary

The reported work relates to a digital control system for a passive solar house space heating. The conceptual design of this control system and the design study of its components has been completed. An experimental model, now beeing completed, will be installed at Marcoussis in an experimental house and fully tested from next september.

O. INTRODUCTION

L'étude porte sur un système numérique de régulation de chauffage pour un logement solaire passif. La définition des éléments constituant le système de régulation et la définition de la stratégie de régulation sont maintenant terminées. Nous avons par ailleurs commencé à réaliser et à assembler la maquette complète du système de régulation qui sera mise en place à Marcoussis sur une maison expérimentale et essayée à partir du mois de septembre 1982.

Nous allons présenter successivement :

- les systèmes de chauffage étudiés,
- l'étude de définition de la régulation :
  - . principe de fonctionnement,
  - . principaux composants utilisés.
  - . dialogue entre l'utilisation et la régulation : choix donnés à l'utilisateur et renseignements fournis par la régulation,
  - . variantes des systèmes de régulation,
- les travaux de réalisation de composants et de sous-ensembles.

1. LES SYSTEMES DE CHAUFFAGE ETUDIES

Les systèmes de chauffage étudiés font largement appel à la captation pasive d'énergie solaire dans une serre et dans des murs captant et stockant de l'énergie solaire qui est ensuite utilisée pour réchauffer l'air neuf fourni à la maison. L'appoint électrique est fourni, soit par des convecteurs (cf. fig. 1), soit par un stock thermique à accumulation nocturne avec résistance d'appoint (cf.fig. 2).

2. L'ETUDE DES REGULATIONS

21. Principe de fonctionnement

211. Chauffage solaire passif avec appoint par convecteurs électriques (cf. fig.1)

L'air neuf aspiré à l'extérieur passe d'abord dans la serre, si elle existe, puis dans le mur captant et stockant l'énergie solaire. Si la température intérieure de consigne choisie par l'utilisateur est de 20°C :

- lorsque la températzture intérieure est inférieure à 19°C, les convecteurs électriques sont alimentés. L'air neuf est réchauffé par l'énergie solaire. Le volet V1 est en position a et le ventilateur F1 est en marche,
- lorsque la température intérieure est comprise entre 19 et 21°C, les convecteurs électriques ne sont plus alimentés. On agit sur la position du volet V1 pour maintenir la température dans la plage 19/21°C aussi longtemps que possible,
- lorsque la température intérieure dépasse 21°C, on prend l'air neuf directement à l'extérieur ; le volet V1 est en position (b). Lorsque la température du mur dépasse une valeur fixée ou lorsque la température intérieure du logement a dépassé 23°C pendant 24 heures, on ouvre le volet de sécurité V2 de façon à assurer un refroidissement du mur par circulation naturelle.

212. Chauffage solaire passif avec appoint électrique à stockage thermique

Le principe de fonctionnement est analogue à celui qui vient d'être décrit pour l'appoint électrique par convecteur. Si la température intérieure de consigne choisie par l'utilisateur est de 20°C, on fait appel au déstockage (mise en route du ventilateur F3) quand la température intérieure devient inférieure à 19°C ; si cette température devient inférieure à 18°C, l'appoint est complété par la mise en service d'une résistance électrique.

22. Principaux composants utilisés

La régulation sera conduite par un microprocesseur utilisant les données fournies par des capteurs de température. Nous indiquerons les principaux éléments sur lesquels agiront les ordres élaborés par le microprocesseur d'abord pour le chauffage solaire et ensuite pour le chauffage électrique de complément:

a) Chauffage solaire

Les principaux composants de la régulation de la partie solaire sont :

- un volet modulant V1 comportant deux entrées et une sortie d'air : la modulation de la position du volet permet de doser la température d'air passant par chacune des entrées. Ce volet sera équipé d'un moteur de commande "pas à pas",
- un volet de sécurité V2 qui aura une commande "tout ou rien",
- un ventilateur de soufflage d'air neuf F1 ayant un débit de un volume/heure,
- un ventilateur d'extraction d'air vicié F2 ayant un débit de un volume/heure.

b) Chauffage électrique de complément

-Pour le chauffage électrique d'appoint à convecteurs, on trouve :

. les convecteurs C qui peuvent être munis d'un thermostat individuel, ce qui ajoute aux possibilités offertes par la régulation centralisée des convecteurs une possibilité de moindre chauffage pièce par pièce,

. un relais statique "S" permettant de mettre l'ensemble des convecteurs en service ou hors service. Sa présence permet notamment de donner priorité au chauffage solaire en coupant le chauffage électrique si la température intérieure dépasse 19°C.

-Pour le chauffage électrique d'appoint avec stockage thermique, on trouve :

. un accumulateur thermique rempli la nuit,
. des relais statiques pour commander le stockage, le déstockage et la mise en service de la résistance électrique d'appoint.

23. Variantes de la régulation du chauffage

Au cours des essais de la maquette de régulation, nous comparerons l'intérêt de plusieurs variantes de la régulation du chauffage différentes par le mode de fonctionnement du volet V1 (cf. figures 1 et 2) :

- volet V1 régulé par tout ou rien,
- volet V1 régulé de façon progressive selon la température constatée dans le logement,
- volet V1 mettant le mur captant hors circuit (position b) pendant les heures de bas tarif électrique.

24. Dialogue entre l'utilisateur et la régulation : choix donnés à l'utilisateur et renseignements fournis par la régulation

241. Choix donnés à l'utilisateur

La régulation gère le chauffage solaire passif et l'appoint électrique en donnant à l'utilisateur des possibilités de choix portant sur les programmes de chauffage et sur les températures de consigne.

a) Programme à une seule consigne de température choisie par l'utilisateur.

b) Programme journalier à deux consignes de température choisies par l'utilisateur.
L'utilisateur peut choisir deux températures de consigne pour deux périodes durée variable au cours d'une journée. Par exemple, pour obtenir une programme jour-nuit :

- de 7 heures à 23 heures : 20°C
- de 23 heures à 7 heures : 16°C.

c) Programme "absence"
Pour une période pouvant aller de un à quinze jours, choix d'un premier niveau de température, puis de l'heure et du jour du changement de ce niveau. Exemple :

. 12°C jusqu'à 15 heures au 3ème jour,
. 20°C à partir de ce moment.

d) Programme "absence prolongée"
En cas d'absence prolongée, l'utilisateur peut afficher l'une des deux consignes suivantes :

- consigne "hors humidité". La régulation assure alors une température de 14°C,
- consigne "hors gel". La régulation assure alors une température de 6°C.

e) Possibilités supplémentaires données à l'utilisateur

- correction temporaire de la température de consigne sans toucher à la programmation par déplacement d'un curseur "plus chaud - moins chaud",
- possibilité d'obtenir, dans une certaine mesure, un réglage de température pièce par pièce si les convecteurs électriques individuels sont munis d'un thermostat qui peut laisser un convecteur particulier "hors service" même lorsque la régulation centralisée prévoit de les alimenter.

242. Informations fournies à l'utilisateur

- En dehors des périodes de programmation, le système de régulation affiche l'heure en permanence,
- l'opérateur dispose de deux boutons pour demander :
  . la température du logement,
  . la consommation du chauffage électrique.

Ces informations restent affichées pendant 30 secondes.

3 LES TRAVAUX DE REALISATION DES COMPOSANTS ET DES SOUS-ENSEMBLES

On voit sur la figure 3 le schéma de la maquette du système de régulation que l'on mettra en place sur une maison expérimentale. Ses principaux composants sont :

a) Le micro-ordinateur (2) de conduite utilisé pour la maquette est un MINC MNC 11 BD de Digital Equipement, comprenant 64 kilooctets de mémoire vive, 2 disques souples, une entrée numérique de 16 lignes, une sortie numérique de 16 lignes, deux horloges programmables et une console opérateur VT 105.

b) Les capteurs numériques (1) sont réalisés à base du composant Analog Devices 537 qui est un convertisseur tension-fréquence avec une jonction semi-conducteur intégrée sur le substrat. Ils sont équipés pour délivrer une fréquence variable proportionnelle à la température ou à l'intensité du courant de chauffage.

c) L'ensemble d'interface (3) assure la connexion du micro-ordinateur avec les actuateurs (cf. fig. 1 et 2). Un dispositif de décodage permet d'adresser l'actuateur désiré :

. le moteur "pas à pas" pour la commande des volets V1 et V3,
. le moteur "tout ou rien" pour la commande du volet de sécurité V2,
. les ventilateurs F1 et F3,
. les convecteurs C (cf. fig. 1), le stockage et la résistance électrique d'appoint (cf. fig. 2).

L'ensemble d'interface permet la commande modulante des moteurs et les commandes par "tout ou rien" des autres organes :

. les commandes modulantes s'obtiennent par l'ordre de positionnement du moteur à la position voulue. Une électronique associée dans l'ensemble (3) décompte le nombre d'impulsions nécessaires et produit les impulsions de polarités et de phases successives.

. les commandes "tout ou rien" sont obtenues par des relais statiques de puissances convenables.

d) Les actuateurs sont achetés dans le commerce et utilisés tels quels : volets, convecteur, ventilateur, moteur "tout ou rien". Seul le moteur de commande de volet modulant a été modifié par remplacement du moteur alternatif d'entraînement pour un moteur "pas à pas".

L'ensemble des éléments constitutifs de la maquette ont été achetés ou réalisés et interconnectés en laboratoire. Nous avons écrit le logiciel d'acquisition pour les capteurs et le logiciel de commande pour les actuateurs et ces logiciels ont été essayés en laboratoire.

La maquette de régulation, qui doit être complétée par l'écriture du logiciel de la régulation, sera ensuite montée sur une maison expérimentale à Marcoussis et essayée.

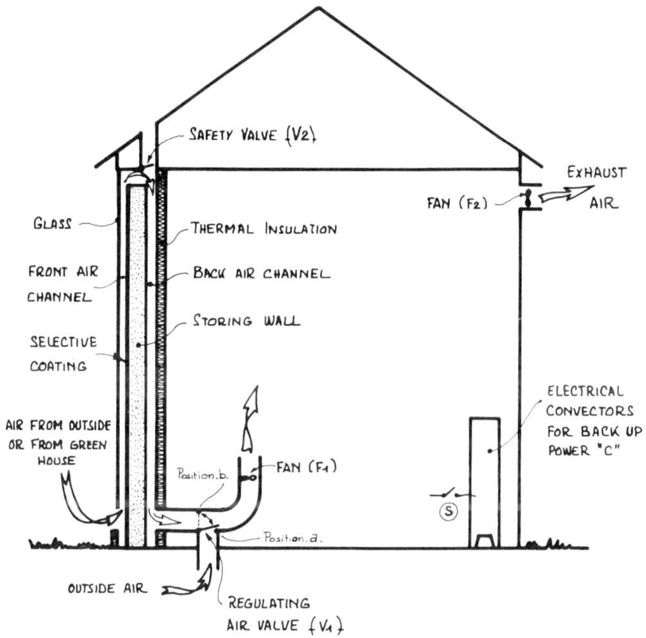

Fig. 1 - Passive solar space heating :
Electrical back up power supplied by convectors

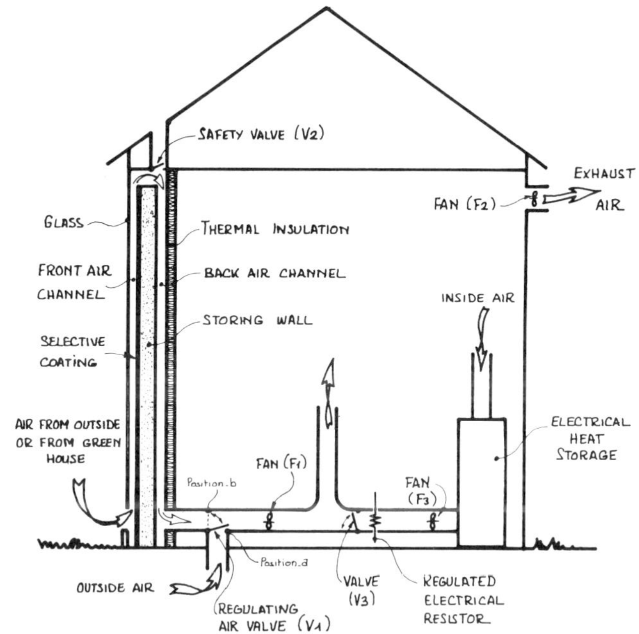

Fig. 2 - Passive solar space heating :
Electrical back up power with heat storage

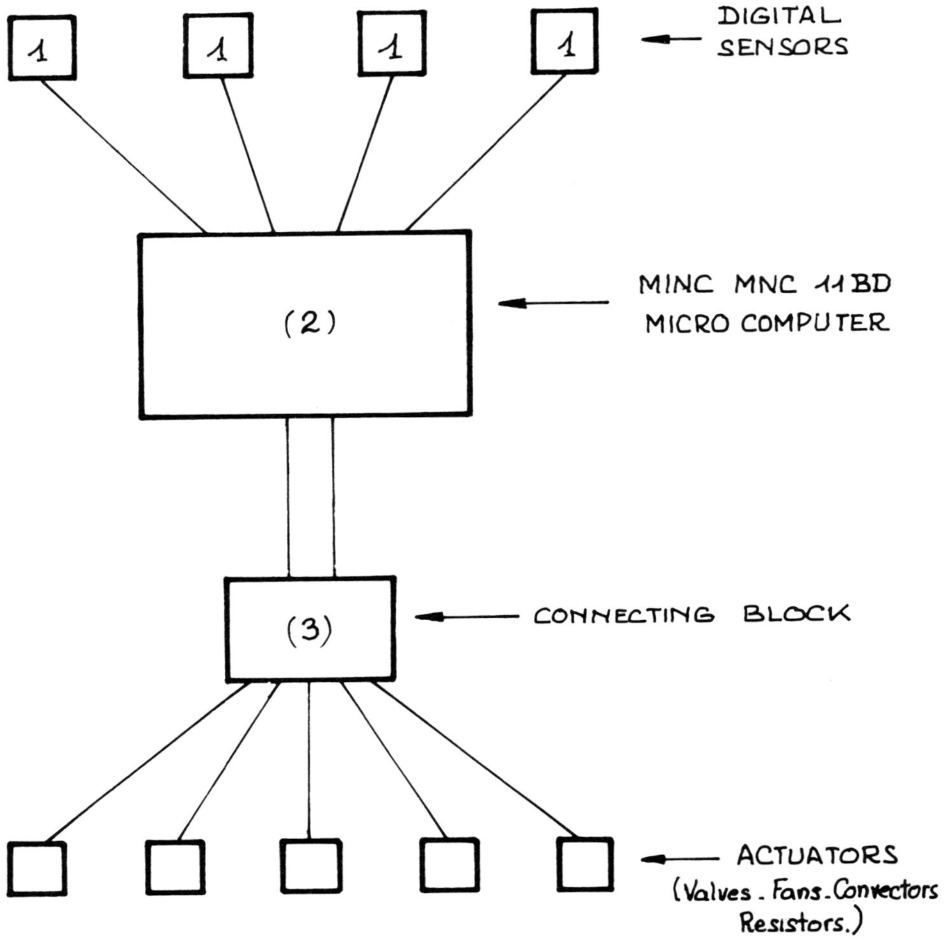

Fig. 3 - Regulation model

SECTION 2 - PASSIVE SOLAR COMPONENTS

Summary on passive solar components

A. Subsection : walls, façades, etc.

Hollow wall : architectonic structural element alternative to trombe wall

Research in and development of a passive solar energy façade in the "council housing" sector

High performance passive solar heating system with heat pipe energy transfer and latent heat storage modules

Transparent insulation for thermal storage walls

A ventilated trombe wall unit

Study to design and manufacture an insulating translucent glass panel made with evacuated pyrex tubes

Research and development of a passive solar façade industrial component

B. Subsection : Shutters, window systems, etc

Temporary thermal insulation in the passive solar utilisation

Development of an independent shutter system for passive temperature control in buildings

Development of a thermal store and a solar blind for use in conservatories and glazed roof spaces

Development and testing of architectural components for passive solar energy use in transparent building façades

Development of a range of double windows including a solar control blind and internal shutters for use with passive solar houses

Use in passive solar architecture of solar dousers with a seasonal effect

C. Subsection : Control devices, etc.

Development of an automatic temperature sensitive air control device for use with passive solar gain collectors

Self-acting power cylinders for controls in passive solar buildings

Unit for local storage of solar gain

Design and development of an easy made convective type solar air collector

Solar wall with tiles covered by a layer of spectral selective enamel

System for warming the new air using the solar energy in an individual house

SUMMARY ON PASSIVE SOLAR COMPONENTS

Authors: R. Lebens and D. Turrent

1. Introduction

The morning was devoted to two subsections of passive solar component contracts: a. walls, façades, etc.; b. shutters, window systems, etc. There are six contracts within each of these two subsections.
The afternoon session on passive solar components was split into two sections; the first dealing with control devices for passive solar systems and the second covering miscellaneous items such as heat storage and integrated systems designed to preheat ventilation air. Because the contracts were so recently awarded, few, if any, had major results to report.

2. Review of the presentations

Contract ESA-P-152-I: Hollow wall Mr Erbaggi from CTIP presented the proposals of this research which were to optimise, build, and test a prototype transparent Trombe Wall using hollow blocks. The wall will be low cost and operable in several ways: insulating panels, insulating venetian-type blinds, and possible openable double glazing to allow through ventilation. Studies will be made to optimise the shape and spacing of the voids in the hollow concrete blocks.

Contract ESA-PS-148-NL: Development of a passive solar energy façade in the 'Council housing' sector. Mr Lammers from ERA Bouw BV presented a south façade as a double acting solar collector. In one part of the façade solar energy is gained directly, in the other part the solar heat is partially stored in concrete slabs. This façade will be built in a low cost housing project, in which the individual houses will have excellent air tightness and part of the collector will be used to preheat the air of an adapted air heating system.

Contract ESA-PS-140-NL: High performance passive solar heating system with heatpipe energy transfer and latent heat storage modules. Mr Van Dijk from TNO (Delft) presented his paper which describes the proposed research to develop a passive solar wall which collects and stores energy with increased efficiency. It attempts to do this by means of heat pipes to create a one-way valve for the transport of heat to storage, and by means of phase change material storage. The wall will be constructed in modular elements in order to reduce manufacturing costs and to increase application flexibility. The work will include detailed analysis of heat pipe and phase change storage and the design of a prototype.

Contract ESA-PS-146-DK: Transparent Insulation for Thermal Storage Walls. Mr Olsen of the Technical University of Denmark (Thermal Insulation Laboratory) described various possibilities for transparent insulating structures between thermal storage walls and the glazing:
- horizontal reflective ridges, triangular in cross-section and insulated,
- slatted and v-corrugated honeycomb,
- selective absorbers and transparent films.

The options are analysed theoretically. A ray-tracing model has been developed to assess transmittance. The most promising options will be built into a thermal storage wall in a test house and compared with a reference flat glass cover.

Contract ESA-PS-158-F: A ventilated Trombe Wall unit. Mr Bourdin presented work which is being conducted at Saint Gobain vitrage with the aim of developing components which will be used instead of the normal Trombe Wall. The units are to be modular and incorporate double glazing (openable to the outside), an aqueous gel encapsulated by glass, the overall framing, and various options for insulating devices both in front of and behind the storage wall. An aqueous gel is used to increase collection efficiency. Much emphasis is being placed on the durability of the product.

Contract ESA-PS-139-F: Study to design and manufacture insulating translucent glass panels made with evacuated pyrex tubes. Mr Courtier of Corning France described the research at his company which has the aim of developing an insulating glazing system using evacuated tubes of circular cross-section. Their proposals are for a K-coefficient of between 0.6 and 1.3 W/m^{2o} C. The work will include investigating optimum jointing material, measurement of thermal and solar transmittance. The system has great physical strength with minimum hazard. Corning have conducted much of the testing in relation to active systems using the same construction. One novel possibility is the use of reflecting shutters within the evacuated space which can be operated by means of a magnet.

Contract ESA-PS-143-D: Temporary thermal insulation for passive solar utilization. Dr. Fuhse presented his firm's research into external insulating roller shutters giving an insulating value (K-coefficient) of 0.5 W/m^2 ^{o}C. Initial work has been into the design of the horizontal strips. Their shape has been thermally analysed by means of computer simulation. The work includes serious consideration of architectural integration of such bulky devices and their durability under working conditions. Also development of control gear. A prototype will be constructed at any early date.

Contract ESA-PS-15-D: Development of an independent shutter system for passive temperature control in buildings. Dr. Schmid of the Fraunhofer Gesellschaft presented this ingenious system under research. It consists of a scrolling device which has various types of film, attached to each other, and rolled between top and bottom rollers, above and below the window. The films are different for winter day, winter night, summer day and summer night, depending on how they transmit or reflect visible, near infrared and far infrared radiation. It is estimated that the system will cost 500 DM/window, which includes the cost of the DC drive motor (100 DM) and will give a payback of about 6 years. There is strong industrial interest in this project.

Contract ESA-P-147-UK: Development of a thermal store and a solar blind for use in conservatories and glazed roof spaces. Mr Stuart presented the work being undertaken by the Fulmer Research Institute in conjunction with Calor Group Limited and Perma Blinds Limited. The Calor test house is being retrofitted with phase change material storage tubes (stabilized Glauber's salt) designed to change phase at 32 ^{o} C. The development of the blind will be of interest to all passive solar applications. Future work will be data acquisition and analysis from the test house.

Contract ESA-PS-149-D: Entwicklung und Test architektonischer Komponenten passiver Solarenergie - nutzung im Bereich transparenter Gebaudeoberfachen. Dr. Bauman of the Gesamthochschule Kassel presented his paper on research into window control devices. The controls are to be housed in the inter-pane space and will control both thermal and solar transmittance. The presentation did not clarify the details of this work.

Contract ESA-PS-142-UK: Development of a range of double windows including a solar control blind and internal shutters. Dr. Houghton-Evans of Stephen George and Partners presented the aims of this research. It is being undertaken in conjunction with a window manufacturer and the University of Leeds. The project will assist a window manufacturer in putting well-thought-out passive solar windows and associated components (blinds and shutters) onto the market. Prototypes will be tested.

Contract ESA-PS-138-I: Development of solar dousers with a seasonal effect. Dr. Flisi described the work of Montepolimeri CSI in their aims to produce a profiled acrylic glazing material in which the angles incorporated into the profile allow transmittance of solar radiation at winter solar altitudes but total solar reflection at summer solar altitudes. This is the exploitation of well known physical principles. A computer model has been developed which describes the light passage and a comparison is made between this system and fixed overhang shading devices. A test cell with Trombe wall will be built before December 1982 and data acquisition and analysis will commence.

Contract ESA-PS-155-UK: Un régulateur d'air automatique sensible aux variations de température, à utiliser avec des collecteurs de gain solaire passifs. This paper, presented by the Waterloo Grille Company, addressed the problem of how to control air movement from Trombe Walls and conservatories. The aim of the research is to develop a low cost, reliable, temperature sensitive air control device which is based on current technology but requires no external power supply or manual assistance. A prototype is to be developed and tested before proceeding with commercial development.

Contract ESA-PS-144-F: Système numérique de régulation du chauffage d'un logement solaire passif. This paper reported on work being carried out at the Laboratoires de Marcoussis on the design of a digital control system for passive solar space heating systems. An experimental model, including digital sensors, micocomputer, connecting block and actuators is now being built, for installation in an experimental house later this year.

Contract ESA-PS-154-UK: Self-acting power cylinders for use in controls in passive solar buildings. This paper, by Stephen George & Partners, describes a proposal to study currently available self acting power cylinders for use as control devices in passive solar buildings.

Contract ESA-PS-150-DK: Unit for local storage and solar gain. Mr Svensson of the Technical University of Denmark described the development of a unit for local storage of heat, using latent heat storage. The storage medium consists of a salt-water mixture based on the extra water principle, with a melting temperature close to room temperature. The crucial point is to charge and discharge the store.

Contract ESA-PS-51-F: Recherche et développement d'un composant fenêtre comme element de captation solaire. The aim of this project, by Rossignol S.A., is to design a passive solar component as a part of a building façade, with several functions such as:
- closure of the building, from floor to ceiling, over the width of a window; - lighting and solar gain collection through glazing materials, - solar control with shading devices, - good insulation for the night. Improved reliability can be obtained by the transfer into the workshop of assembly operations previously taking place on the building site.

Contract ESA-PS-157-F: Tiles with a layer of spectral selective enamel. This paper, again from the Laboratoires de Marcoussis reported on the development of steel tiles covered by a layer of spectral selective enamel which can be fixed to solar walls. The whole element is to be tested on a rig before being installed in an experimental house.

Contract ESA-PS-153-F: Systèmes de préchauffage d'air neuf de renouvellement utilisant l'énergie solaire. The final paper of the session was presented by the Societé les maisons Bruno-Petit. This described a system for pre-heating ventilation air using passive solar energy from a sun-space. Two variants were presented; one using a heat recovery device and the other, using solar air collectors coupled to a heat storage device. The systems are now installed in several houses and there are plans to build several hundred in France in the next few years. The total additional cost of the system is estimated to be about 20% of the construction cost.

Conclusions

It is encouraging to see some very real possibilities of new passive solar components being developed and the necessary attention to the questions of durability, cost effectiveness and architectural integration that some contractors are placing on their work.

In other contracts it is observed that there are the following deficiencies:
- some systems are still being developed without enough attention to cost
- some research is not giving sufficient regard to previous work in the same or similar fields.

Some research contracts do not include the building and testing of components being developed. This makes it extremely difficult to assess the value of the component.

HOLLOW WALL : ARCHITECTONIC STRUCTURAL ELEMENT ALTERNATIVE TO TROMBE WALL

Authors	: C. ERBAGGI, G. FUNARO
Contract number	: ESA-PS-152-I (S)
Duration	: 18 months
Total budget	: Lit. 103.200.000
CEC Contribution	: Lit. 46.440.000
Contractor	: CTIP Solar S.p.A.
Address	: P. le G. Douhet, 31 I - 00144 - ROMA

Summary

The aim of this research project consists of the realization of an architectonic structural element that allows light to pass while acting as a curtain wall and as a system for heat storage and transfer.
This "Hollow Wall" is an alternative to the traditional Trombe wall which is not trasparent.
Low cost, the ease with which the materials making up the system can be found on the market, its pleasing aesthetic aspects and its efficient thermal performances, could make the "Hollow Wall" an attractive construction element in passive architecture for the housing industry.

1. <u>Introduction</u>

The "Hollow Wall" is made of prefabricated modular blocks in cement (or other good inert material) in various forms that, when assembled, can give a structure with various performances :

a) curtain wall
b) heat storage mass
c) heat transfer element

The parts that make up the wall, in fact, have holes (whose shapes will be optimized) allowing light to pass into the area behind and providing for the transfer of heat by natural convection through the holes themselves and, at the same time, also forming a storage mass since they have good thermal inertia. The removable panels in transparent material to be fixed next to the wall, on the outside, will control the passage of heat by conduction and convection from the air space to the adjacent room.

2. <u>Description of the "Hollow Wall"</u>

The "Hollow Wall", integrated into the south side of the building, consists of the following (see fig. 1) :

A. <u>Double glazing</u>: a simple system of sliding glass panels will be designed so that the amount of solar energy to be stored or transferred to the internal part of the building can be regulated. Furthermore, the "Hollow Wall" can be useful for summer ventilation as the sliding double glass panel will make it easy to change the air during the hot months of the year.

B. <u>Air space</u>: this will be designed to the optimum size to give natural circulation of the hot air coming into the building.

C. <u>Hollow block in cement</u> (or other material): the blocks will be painted in a dark colour on the south side of the building to increase their solar energy absorbing capacity and assembled together to form the wall, (the size of which will be optimized during the design phase). The hole shapes of the blocks will be optimized to obtain the right ratio hollow/filled for the better heat storage mass and the better inside lighting.

D. <u>Removable panels</u>: the control system for solar radiation, heat passage and heat loss consists of both transparent material panels and removable insulation system. The removable panels in transparent material will control the passage of heat from the air space to the room, while the removable insulation system can be a "venetian blind" type, made, eventually, with "sandwich" element, which will be made in light coloured plastic on one side and expanded polystyrene on the other to increase the insulation effect. The blind will thus serve a dual purpose since it can be used as a sun-shade in summer and as an insulation panel in winter or during the night.

Fig. 1 - SKETCH OF "HOLLOW WALL"

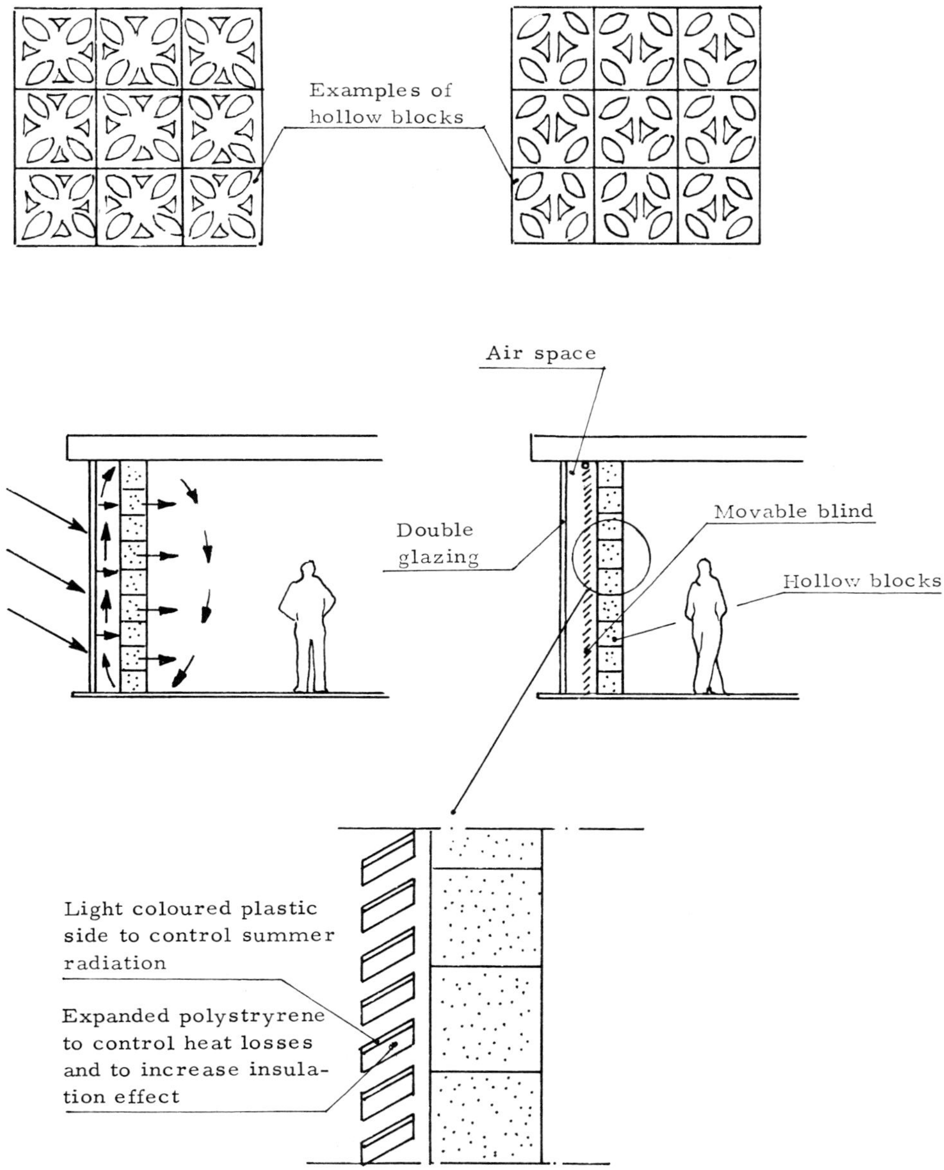

Detail of shade system to control solar radiation and heat losses (Venetian blind type)

3. Performance and Advantages

The "Hollow Wall" operates in a simple manner very similar to the traditional Trombe wall: the system is based exclusively on air circulation by convection and on the conduction of heat through the storage mass.

A rough evaluation of performances, based on theoretical calculations on the K coefficient of heat transmission, is that the "Hollow Wall" could provide a thermal contribution of about 35÷40%, that is of the same order of magnitude as that for the Trombe wall.

But the "Hollow Wall" has a lot of advantages, very interesting and useful for dwelling. With this system it is possible to control the passage of heat, to regulate the heat storage and also the time of the heat transfer. Moreover the "Hollow Wall" permits the air ventilation in some wheather conditions, besides to allow the light passage and to be an aesthetically attractive wall.

All these advantages of the "Hollow Wall" over the Trombe wall render it extremaly versatile and capable of a great variety of applications in dwelling.

4. Research Execution Plan

The research will be executed according to the following program :

First Step : Definition of the characteristics

Hollow blocks in various materials and shapes with different physical characteristics can be taken into consideration.

On the basis of the estimated thermal conductivity (K) of the hollow block material, of its thermal inertia, of the solid/void ratio and of the geometrical shape of the holes, the theoretical performances of each single component and of the entire wall will be determined.

It will be thus be possible to check the theoretical correspondance to the foreseen requirements as an alternative to the Trombe wall.

Second Step : Construction

Design and construction of the "Hollow Wall" prototype

Third Step : Check of the system performances

After the work is completed a monitoring program of several months will be carried out in order to check project data against experimental data.

The "Hollow Wall" characteristics and performances will be compared with the well known data of the traditional Trombe wall, so as to highlight similarities and differences.

For this purpose, measuring instruments will be installed on the "Hollow Wall" and inside the adjacent room in order to find out what the thermal performances are.

Research in and development of a passive solar energy facade in the "council housing" sector

Authors : H.Th. Talle / J. Lammers

Contract number : ESA-PS-148-NL

Duration : 24 months, from January 1982 to December 1983.

Total budget : Dfl. 663.180,--
CEC contribution: 38 %

Head of project : J.A. van Lit

Contractor : ERA Bouw b.v.

Address : P.O. Box 62, 2700 AB Zoetermeer
The Netherlands

Subcontractors : Bouwcentrum Techno.

Architects'Community van den Broek en Bakema.

Summary.

This paper summarises the work carried out the first months of contract no. ESA-PS-148-NL on the subject of "the research in and the development of a passive solar energy facade in the council housing sector".

The uniformity of the design and the use of components ordinarily available from the trade will make this facade suitable for mass-production, whereby the costs can be relatively low.
At this very moment we almost finished designing the test facility.

Conclusions sofar are that we very much simplified the first designs because of the usability for future inhabitants of this solar dwellings and the cost-effectiveness analyses we made.

1.1 Introduction.

The aim of this project is the research into and development of a passive solar energy facade consisting of self supporting elements for dwellings, which are suitable for supplying solar energy inside the dwelling.
Contrary to the current technical ways in which heat is obtained by relatively small wall areas of facades facing south, the proposed system will make use of practically the whole facade walls and windows to collect solar energy.
To carry out measurement in practical use and circumstances, and about the effectiveness of the elements a project of 12 identical single-family houses will be built in the nearby future (starting March 1983), after a thourough period of theoretical research and the testing of a prototype of the passive solar energy facade itself in practice.
One block, the 6 solar (energy) houses, is provided with special energy saving measures and the passive solar energy facade.
The other block, the 6 reference houses, is identical to the first block without, however, the passive solar energy facade.
All dwellings are situated on the south with unobstructed light acces.
An auxiliary air heating system with central air supply is chosen because of the air tightness of the dwelling and the high thermal insulation of the walls.

1.2 Description of the passive solar energy facade.

The south facade of the solar dwellings consists outof two different solar energy collectors.
The passive wall collectors (see fig. 1) and the window collectors (see fig. 2).

In our first design:

- The passive wall collectors consist of a 8 cm thick concrete slab with at the one side triple glass and a roller blind on the outside, and on the inner side roller blinds to control the heat radiation into the dwelling.
  The window side of the concrete slab is painted dark and tubes are cast in to guide the air-flow.
  In this design the collector acts as an independent direct space heating element.

- The window collectors consist of a triple glass-window, aluminium venetian blinds on the backside and on the interior side a moveable heat insulating pane.
  During periods of direct or diffuse sun radiation, the air of the auxiliary heating system, and the ventilation air of the dwelling are led through the window collector and so being pre-heated and distributed through the dwelling (see fig. 3).
  The air-stream through the window collector is shut off when there is not enough sun radiation.

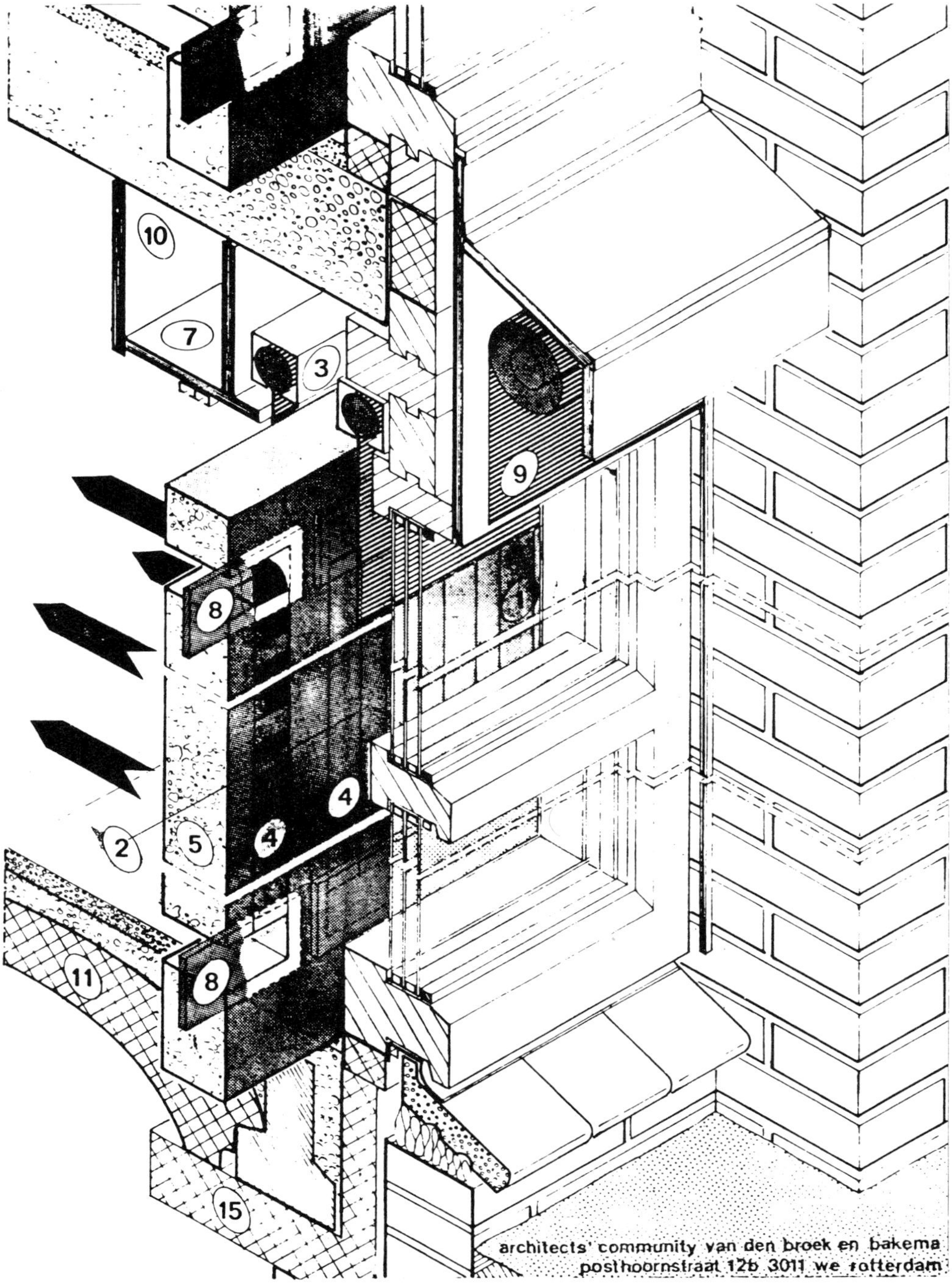

isometric cross-section passive facade collector

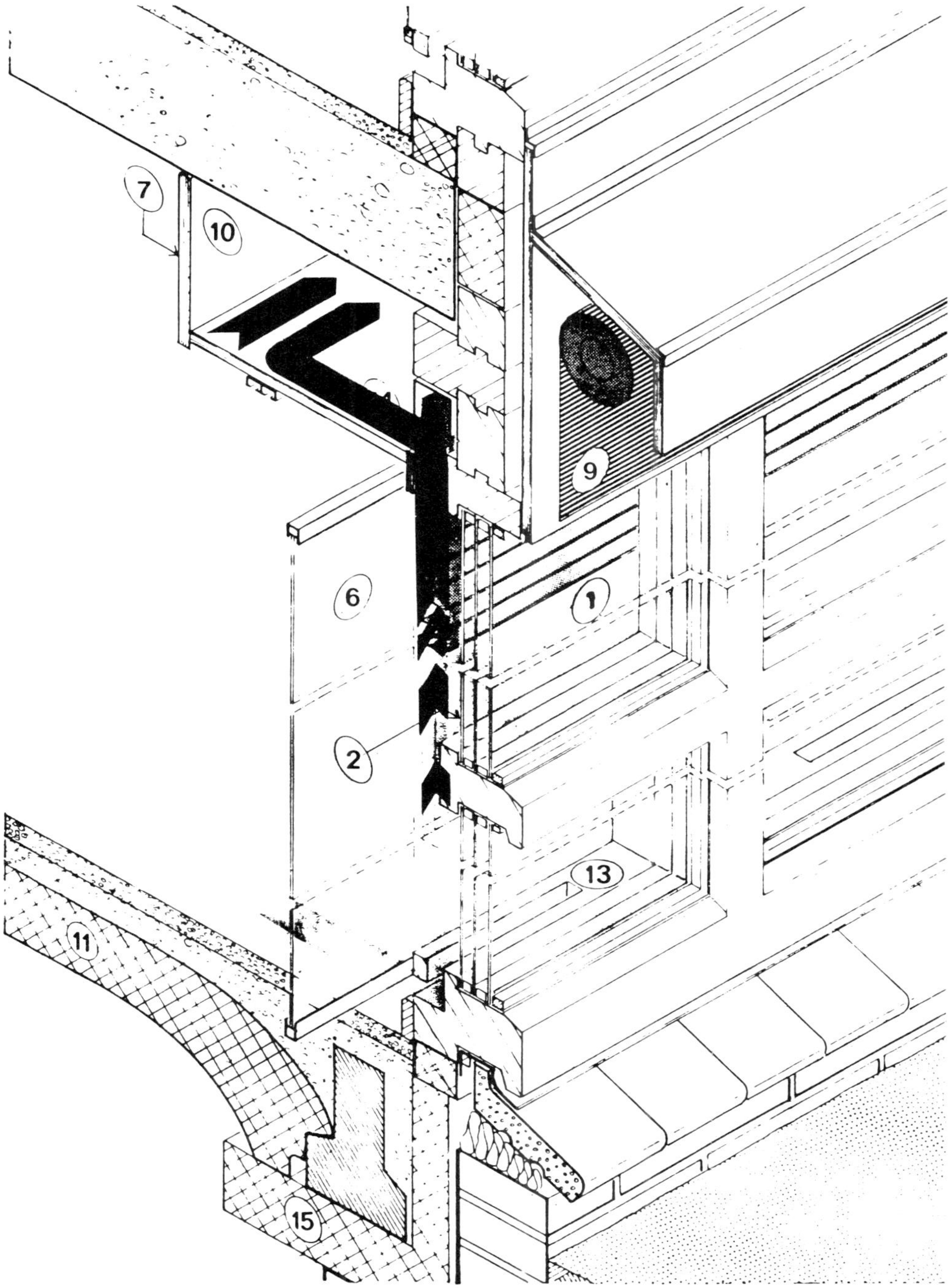

window collector isometric cross-section
architects' community van den broek en bakema posthoornstraat 12b 3011 we rotterdam

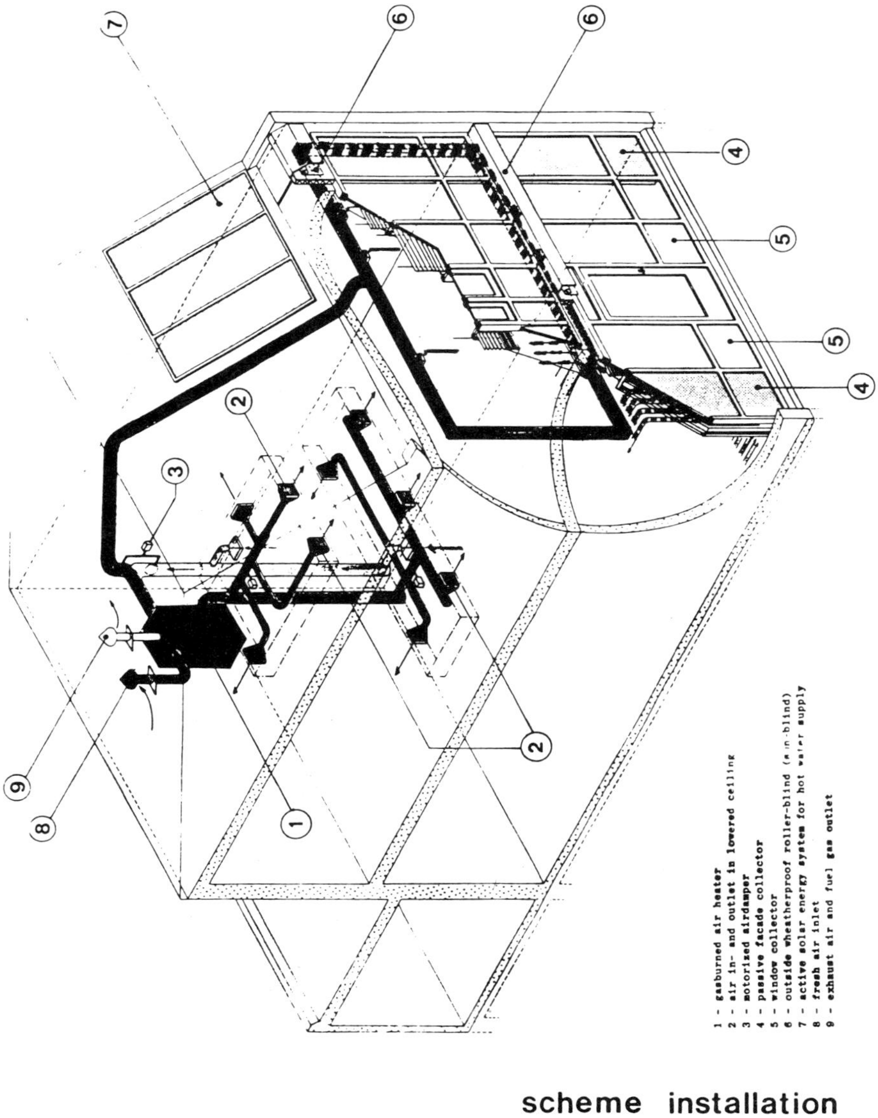

1 - gasburned air heater
2 - air in- and outlet in lowered ceiling
3 - motorized airdamper
4 - passive facade collector
5 - window collector
6 - outside wheatherproof roller-blind (sun-blind)
7 - active solar energy system for hot water supply
8 - fresh air inlet
9 - exhaust air and fuel gas outlet

scheme installation

1.3 Conclusions.

We calculated a gas-saving, in comparison with a standard built simular dwelling according the regulations in the Netherlands nowadays, of approximately 80%.

Cost-effectiveness analyses and the usability for future inhabitants, however, led to a simplification of the design and construction of the passive solar facade.

2.1 Latest design.

At this moment no drawings are available of the latest composition of the passive solar energy facade, but we will just describe the major changes in the design.

In contrast with our first design we now constructed the passive solar facade entirely as an indirect pre-heating air system, which is connected to an auxiliary air-heater.

Space heating and lay-out of the installation

2.2 Lay-out.

A hot air heating system with central air supply has been chosen because of the air tightness of the dwelling and the high thermal insulation of the walls.
The heating of the air, necessary for heating the dwelling, can be achieved in three ways.
Two of these are solar assisted systems, the third is a gasfired air heater.

a) Passive wall collector.

The wall collector consist of a concrete slab with on one side an airgap and a glass window, and on the other side insulation.
The windowside of the slab is covered with a spectral selective absorber coating. The air tubes are cast in the concrete slab.

b) Air return by triple glass window.

The window consists of a triple glass window.
Aluminium venetian blinds (sunside painted dark, interior side high polished) will be placed between the second and third glass window.

2.3 Space heating.

During periods with direct or diffuse radiation, the return air and ventilation air of the hot-air-heating-system will be led through the window collector along the venetian blinds and so being pre-heated and distributed through the dwelling.
The gasfired heater will be used as an auxiliary heater.
The air flow through the wall components is shut off when there is no sun radiation and the temperature of the concrete slab is not sufficiently high.

2.4 Summary.

Excellent airtightness of the dwelling an high thermal insulation will give a high energy saving.
Additional savings are obtained by using the (south) facade as a double acting solar collector.
In one of the collectors solar energy is directly gained, maintaining daylight transmission, in the other solar energy is partially gained directly and partially stored.
Compared with a normal dwelling facade, the heat gain by solar energy is improved with about hundred percent while the thermal insulation of the facade is improved with about fifty percent.
The collectors and the improved thermal insulation of the walls, will reduce the energy demand for heating the dwelling with at least sixty percent.

3.1 Actions at this moment and the nearby future:

- finishing the definite models of calculation about the performances of the passive solar facade,
- the testing of the prototype of the passive wall collector in practice circumstances and interpret the results of these tests,
- the final design of the passive solar energy facade, and
- the design of the auxiliary air heating system in connection with the dwelling and this facade.

HIGH PERFORMANCE PASSIVE SOLAR HEATING SYSTEM WITH HEAT PIPE ENERGY TRANSFER AND LATENT HEAT STORAGE MODULES

Author : H.A.L. van Dijk

Contract number : ESA-PS-140-NL

Duration : 18 months, 1 January 1982 - 1 July 1983

Total budget : Dfl. 506.200,-- CEC-contribution: Dfl. 241.200,--

Head of project : H.A.L. van Dijk

Contractor : Institute of Applied Physics TNO-TH

Address : P.O. Box 155, 2600 AD DELFT, The Netherlands

Subcontractors : 1. Technical University of Eindhoven (THE)
2. Technical University of Delft (THD)
3. Koninklijke Maatschappij "De Schelde", Vlissingen.

Summary

The aim of the project is to develop a passive solar heating system with a higher eficiency than conventional concrete thermal storage walls regarding accumulation and transfer of solar heat into dwellings and with restricted extra costs for manufacturing the system.
This is to be achieved by the introduction of two special components:
a). a heat pipe as a thermal diode tube for the efficient transfer of collected solar heat from the absorber plate to behind a insulation layer,
b). a latent heat storage section with high storage capacity at moderate operating temperatures. To restrict manufacturing costs and to stimulate the application of the system under a wide range of conditions the system will be designed as small modular elements which can be manufactured uniformly and can be easily arranged in proportions and dimensions as required by the specific situation.
The project is divided into three phases: in the first phase the research is aimed at the global design of the system, in the second phase the characteristics of the various components are examined in detail. The project will be concluded with a third phase, the design of a prototype system.
This paper describes the initial phase of this research.

1.1 Introduction

Within the passive solar approach, the direct gain and convective loop systems are the most simple and therefore cheapest solutions. These kind of systems, however, are characterised by high indoor temperature swings and high solar heat supply on the hours of least demand.

A thermal storage wall has a advantageous accumulation of solar heat which results in low temperature swings and a time delay between the absorbtion of solar energy and the heat supply to the building.

The main disadvantage of a conventional thermal storage wall, however, is the higher temperature of the solar collecting surface which in combination with the high thermal capacity of the wall can lead to considerable heatloss to the outside. Multi-glazing can limit this heatloss to a certain extend, but is costly. Movable insulation is also expensive, has a slow response and needs careful attention for a reliable and effective operation. Furthermore, the extra mass introduced into the building may sometimes lead to extra costs particularly in multi-story buildings and valuable space within the building is occupied by the system.

1.2 Aim of the project

This project aims to develop a passive solar heating system which has the advantages of the conventional thermal storage wall, without sharing its disadvantages as mentioned above. This can be achieved by the introduction of two special components:

a. A thermal diode tube, which acts with small temperature difference already as a high performance thermal conductor. It transfers heat from the collector to behind a insulation layer, but does not transfer heat in the reverse direction. With a low capacity collector the system responds rapidly to changes in outdoor conditions, while needing no movable parts or manually or automatically operated control equipment.

b. The latent heat storage. With thermal storage in elements containing phase-change material a reduction in storage volume compared with concrete down to 20% can be achieved at a moderate storage temperature level. The thickness and weight of the complete system can then be restricted.

To restrict the manufacturing costs and to stimulate application of the system within a wide range of conditions the system will be designed as separate functional components, which can be manufactured uniformly as identical small modular elements. The proportions and dimensions of the components can be chosen in accordance with the specific situation. Only well-known building techniques are under consideration for the installation of the system from its constitutive elements.

1.3 System description

The design of the passive solar element is based on the following mechanisms (see figure 1):

- Solar radiation is transmitted by one or more glass panes acting as a cover against too high heat losses (1).
- The radiation is absorbed by an absorber plate (2).
- Heat pipes as thermal diode tubes transfer the absorbed heat through a insulation layer (3).
- A transfer plate (4) distributes the heat from the diode tube to the latent heat storage elements (5) and with natural convection directly to the room air (6) which flows under the buoyancy force within the cavity (7).
- A common insulation sheet acts as a separation between the storage

elements and the room (8).

The thermal characteristics of a heat pipe in inclined position can be shortly described as follows: the heat pipe is basically a closed hollow tube partially filled with a working fluidum, in equilibrium with its vapor. When the lower end is heated the fluid absorbs this heat by evaporating. At the -colder- upper end the vapor will condense and release the latent heat, the liquid will return to the lower end by gravity force. With reverse temperature difference the upper end will dry out and no heat transfer takes place in this element (thermal diode effect). The working fluid and the heat pipe tubing are chosen on the basis of the operating temperature and their mutual compatibility. For the relevant temperature range e.g. water and copper is a promising combination.

The latent heat storage elements are fitted to the transfer plate in the air layer. The elements are filled up by (e.g.) hydrated sodium phosphate with additives (to prevent segregation) and to start crystalization).

The heat transfer to the storage elements combined with direct transfer to the circulating air has the advantage of a) admitting cheap construction and b) providing direct gain when required, which in moderate climates forms a great percentage of all solar gains. The direct gain and the heat from the storage is transferred by natural convection with manually operated dampers according to the schemes for different situations, given in figure 2.

1.4 Workplan

The research will consist of the following phases:

phase 1: Global design
- a) defining the boundary conditions and criteria with respect to temperature levels, solar radiative energy and heat transfer.
- b) global design of the system using available components and known techniques.
- c) a sensitivity analysis with a simplified calculation model.

phase 2: Optimization and testing of the innovative components
- a) optimization of material choice, dimensions and constructive details for 1) the sandwich modules, 2) the latent heat storage elements and 3) the heat transfer mechanisms in the ventilated air cavity.
- b) construction of prototype heat pipes and storage.
- c) performance testing of the prototype components.

phase 3: Design of a prototype high performance system
- a) design of a prototype.
- b) prediction of the heat gains with a unsteady state computer model.

After this concluding part of the project the results will be described and evaluated in a report. Then, based on the results, a phase 4 will be proposed:

phase 4: Construction and testing of the prototype high performance system
- a) construction of the prototype system.
- b) performance testing of the prototype system on laboratory scale.
- c) performance testing of the prototype system under real climate conditions.

In this paper only first results from phase 1 are presented.

1.5 Global design of the system; first results

Figure 3 shows a schematic view on one of the global designs under consideration at this moment. The system consists of the following separate components:

1) One or more transparant panes.
2) Small airgap.
3. Sandwich building blocks, consisting of a) collectorplate, b) a insulation layer perforated by a S-shaped heat pipe and c) a transfer plate.
4. Rectangular storage tubes which can be fitted to the sandwich blocks in a variable number per unit of area.
5. Insulation sheet.

Preliminary calculations have been performed on such a design, which was therefore simulated as a resistance network under assumed average conditions. These calculations indicate a range in performance as shown in table I.

In figure 4 an example from preliminary calculations with a steady state hour by hour model is shown. The figure shows the net heat gain by the system under the preliminarily assumed conditions for an average Dutch heating season. The heat demand is taken hourly from a well-insulated living room, with consequently quite low annual heat consumption. For comparison a similar (but unsteady state) calculation has been performed with a active system under the same climatic conditions.

It should be emphasized, that the presented net heat gains should be considered against a (well-insulated) normal opaque wall with a annual heatloss of about 50 kWh/m^2.

1.6 Discussion of results and outlook on future work

Although a number of assumptions are still preliminary, these first calculations indicate that the chosen high performance passive design can only provide satisfactory heat gain when it is optimized on a number of details.

In this optimization proces a balance between heat gain and manufacturing costs has to be found, e.g. in the choice of number of heat pipes per m^2, in the detailing of the ventilated air cavity and in the selection of collector surface and glass covering.

Leaving more accurate calculations on the heat gains for a later stage, the determination of the manufacturing costs is the next step in this research.

When considering the costs against the heat gains one should note that the passive system replaces a normal wall, the building costs of which are saved.

When considering the costs in comparison with a conventional thermal storage wall the possible extra costs to gain extra heat may be compared with the extra costs involved in deminishing the high heat losses from the latter.

When considering the costs in comparison with an active solar system one should realize that a) the chosen passive design is composed of separate elements which can be applied under a wide range of conditions and which can easily be installed with well-known building techniques, b) the passive design has no movable parts except for a couple of manually operated air valves, thus maintenance costs are expected to be very low and c) in a comparable active design the costs to transfer the heat from the storage vessel to the indoor air have to be included.

Other items for further investigation are the heat transfer by the thermal diode tubes, the accumulation in the latent heat storage elements and the heat transfer in the vented air cavity, as mentioned above (section 4). For the latter subject a program of experiments is about to start.

The results from these studies will be used in the third part of the project, the design of a prototype high performance system. In this final phase the heat gains will be calculated more accurately for a number of situations, with a unsteady state calculation model.

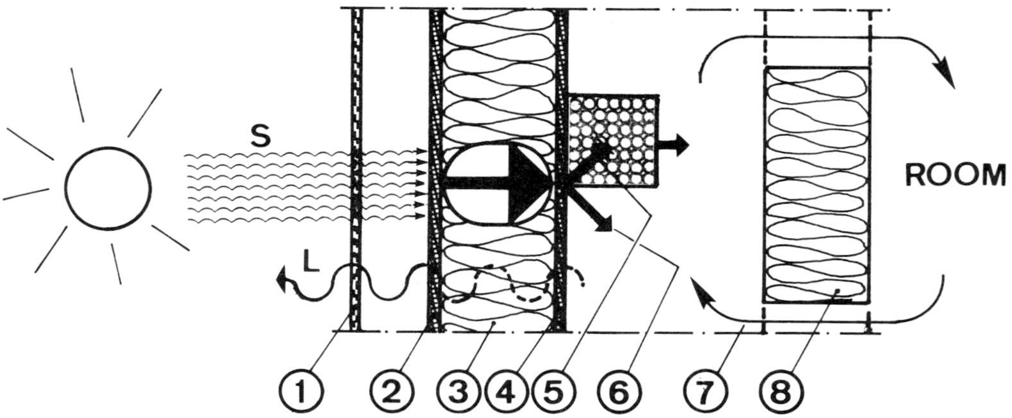

(1). transparant pane (-s)

(2). absorber plate

(3). thermal diode tubes in insulation layer

(4). transfer plate

(5). heat supply to storage elements

(6). heat supply directly to the air

(7). natural ventilation of room air through the cavity

(8). thermal insulation sheet.

S: solar irradiation

L: heat loss.

Figure 1: Schematic drawing of the basic mechanisms of the high performance passive solar heating system.

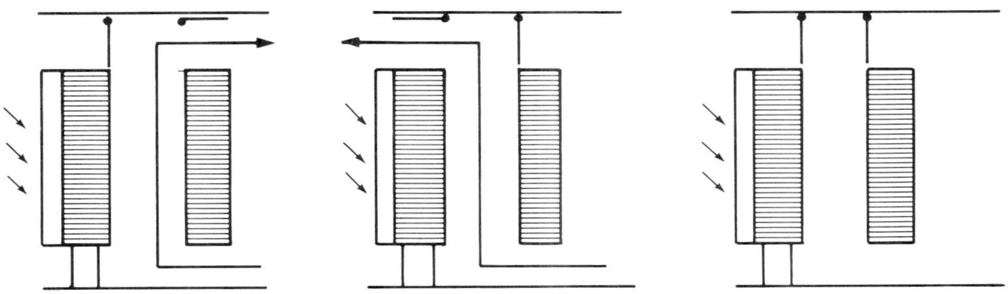

a. heating, wintercondition.

b. ventilating, summercondition; necessary when overheating of the system must be prevented, also possible when cooling of the building is desired.

c. storage.

Figure 2: Schematic view of the possible situations with regard to the convective heat transfer from the storage space.

Table I: Some indication on the range in performance of the h.p. passive design, according to preliminary calculations.

Definitions: Heat gain $Q = F_r (AT \times S - U \times (T_i - T_e))$ W/m^2 collector

F_r	= Heat removal factor	(-)
U	= Topp loss coefficient	(W/m^2K)
T_i	= Indoor temperature	(C)
T_e	= Outdoor temperature	(C)
AT	= Effect. abs. transm. factor	(-)
S	= Solar irradiation	(W/m^2).

	single glass black absorber plate (U = 5.6)		single glass spectral selective layer (U = 4.0)		single glass spectral selective layer (U=3.1)	
resistance in air cavity (natural ventilation) m^2K/W	0.09 1)	0.21 2)	0.09 1)	0.21 2)	0.09 1)	0.21 2)
resistance of diode tubes (positive direction) m^2K/W						
0.10	0.49	0.37	0.57	0.45	0.63	0.51
0.20	0.39	0.31	0.47	0.38	0.53	0.44
0.30	0.33	0.27	0.40	0.34	0.46	0.39

1): cavity provided with extra heat exchange surface
2): plain rectangular cavity.

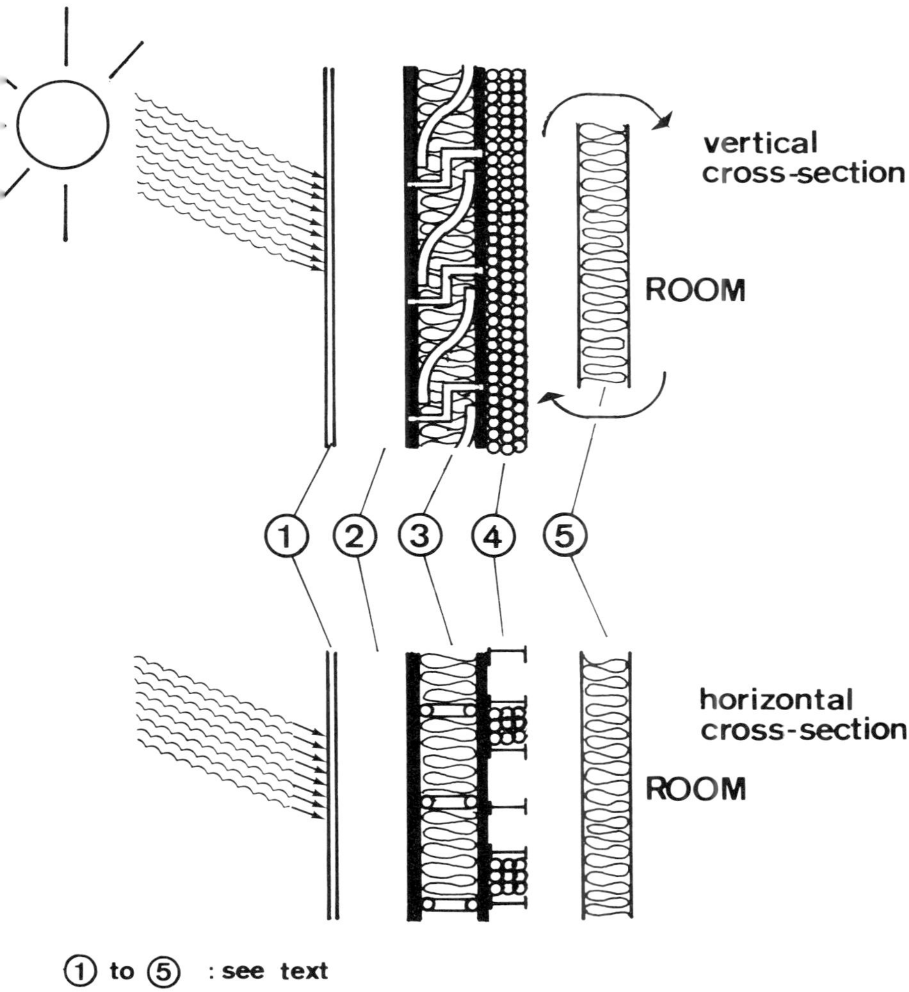

Figure 3: Schematic view on one of the global designs under consideration.

Figure 4: Preliminary calculation of the net heat gain with the h.p. passive design and a first comparison with an active system.

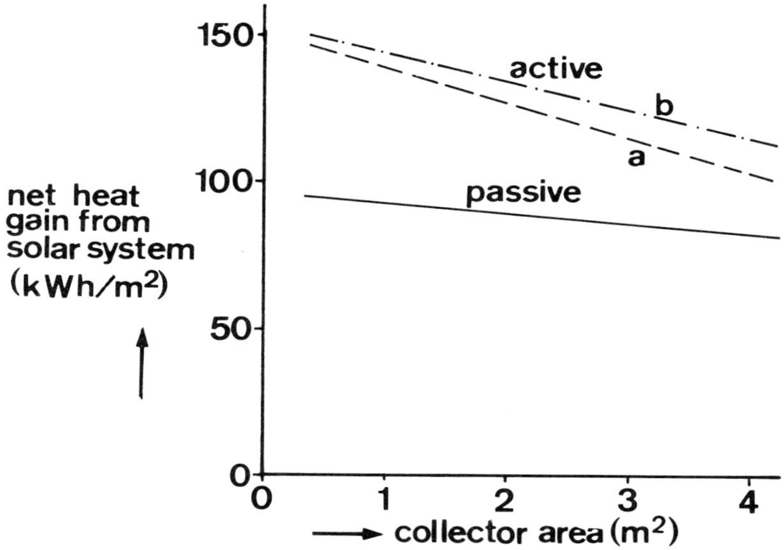

Assumptions:

Climate and period: an average Dutch heating season
Orientation : south, vertical
Annual heat demand living room Q_d = 3698 kWh
Total solar irradiation Q_s = 412 kWh/m^2

		passive system	active system	
Heat removal factor	F_r:	0.65	0.95	(-)
Topp loss coefficient	U :	4.00	4.00	W/m^{2}K
Effect. abs. transm. factor	AT:	0.80	0.80	(-)
Leakage factor storage system	LF:	3.00 %/h		
U-value storage system			0.5	W/m^{2}K
Storage heat capacity	Cs:	8	a: 7	kWh
			b: 17	kWh
Collector area		0.5-4*	0.5-4*	m^2

* 3 to 4 m^2 is a realistic choice for the chosen situation.

TRANSPARENT INSULATION FOR THERMAL STORAGE WALLS

Author	: L. Olsen
Contract number	: ESA-PS-146-DK(G)
Duration	: 18 months 1 January 1982-30 June 1983
Total budget	: Dkr. 1.250.000,- CEC contribution: 50%
Head of project	: Vagn Korsgaard
Contractor	: Thermal Insulation Laboratory
Address	: Technical University of Denmark Building 118 DK-2800 Lyngby, Denmark

Summary

The aim is to develop a permanent, transparent insulation primarily for thermal storage walls in order to improve the ratio of solar transmission and heat loss coefficient. In this project different types of transparent insulation for thermal storage walls are investigated.

- Triangel-shaped cylinders of insulation placed in rows at the exterior side of the plane storage wall. The cylinders are covered with a reflective surface in order to reflect the irradiation to the absorber area between the cylinders.
- Slatted and V-corrugated honeycomb construction which slopes to give optimal solar transmission.
- Other devices giving better performance, for instance selective absorbers and transparent film.

The heat transfer and solar transmissions in these structures are analysed. The convective heat losses are estimated dependent on the plate spacing, aspect ratio, slope and temperature difference. A raytracing model for calculation of the solar transmission has been developed. With this model the dependence of the acceptance angle and angle of incidence is estimated.

1. Introduction

The building envelope is a part of the building which in most cases is only used for protection of the inhabitants against the elements. But it is also possible to use it for an energy producing system. In the heating season, a considerable amount of solar radiation is striking a vertical southfacing wall. This potential can be exploited by using the wall as a combined collector, storage and space heater. A thermal storage wall combines these characteristics.

Solar walls are an integrated part of the building structure. The central part is a wall made of a material which is able to accumulate heat. Towards the outer side of the wall a collector cover is mounted, which at the same time can transmit solar radiation and insulate against heat loss. Both with and without incident solar radiation it is necessary to reduce the heat transfer through the wall from the room to the outside air. This can be done by different kinds of transparent insulation.

The absorber is placed at the outer side of the storage and can be of black paint or a selective foil.

When there is solar irradiation at the wall, the radiation is transmitted through the collector cover and is absorbed on the outer side of the wall. The absorber heat is conducted into the wall and accumulated. The accumulated heat can be stored for later use, typically with a time lag of 6 hours to 2 days.

Heat transfer from the wall will take place by heat conduction through the wall combined with natural convection and radiation from the inner side of the wall to the room.

2. Objective

The aim of this project is to develop transparent insulation systems which improves the ratio of collected solar energy and heat losses. It is especially important to improve the heat resistance in northern cloudy climates or at other wall orientations than south.

In this project different solutions are proposed:

3. Triangle shaped cylinders

A cross section of this type is shown at fig. 1. The reflective surfaces will concentrate the incident solar radiation. The advantage of this solution is that the absorber surface is limited, and this will provide a reduction of both the convection and the radiation loss.

The convective losses have been investigated [1]. Fig. 2 shows the convective heat transfer as a function of plate spacing (spacing between absorber and cover) and a slope. The reduction of the convective heat transfer for plate spacings more than 70 mm is significant. The reduction of the convective heat transfer for this solution compared with the flat plate collector is up to 25%.

The radiation heat exchange is also reduced because of the reduction of absorber area. The problem in this solution is that the temperature at the absorber is larger than in the flat plate case. This is due to the concentration of the rays.

If the storage is of water or a good conductor, this problem will be of minor importance.

4. Honeycomb

The heat transfer in different types of honeycomb has been analyzed. Three types can be identified:

- Rectangular celled honeycomb.
- Slatted honeycomb.
- V-corrugated honeycomb.

The rectangular celled honeycomb has the advantage of low convective heat transfer, but it cuts off more of the solar radiation than the other types. This is due to the daily movement of the sun.

The slatted honeycomb is made of parallel walls. The walls can be sloping and having a horizontal line of intersection with the vertical storage wall.

Such a honeycomb type can provide an excellent solar transmission because there are no vertical honeycomb walls.

The convective losses in honeycombs are dependent on the aspect ratio, shape, slope, temperature difference and plate spacing. Fig. 3 and 4 show the convective heat transfer in relation to the plate spacing [3]. The two sets of curves are based on a temperature difference of respectively 25 K and 50 K and an average temperature of 290 K. It can be seen that each of the curves has a minimum heat transfer. This minimum heat transfer is very dependent on the aspect ratio and the slope. The larger the aspect ratio is, the more it is possible to suppress the convective flow. The dependence on the slope is also indicated. For large plate spacings there is a large increase in the heat flow with an absorber slope of 75^0 instead of 90^0 (vertical).

The convective heat loss of a flat plate collector is shown. It can be seen that the convective losses can be reduced to one forth for slatted honeycomb compared with the flat plate case. From the curves it can be seen that an aspect ratio of A=10, and a plate spacing between 50 and 100mm seems to be an appropriate optimum.

The V-corrugated honeycomb is another possibility (fig. 5). The convective flow of this type is not investigated before, but the results from the slatted honeycomb can be used with caution.

The solar transmission through the V-corrugated honeycomb has been investigated by a computer simulation model using a Monte Carlo raytracking method. With this model it is possible to calculate the solar transmission in dependence on the angle of incidence. Fig. 6 shows the solar transmission of a V-corrugated honeycomb for different angles of incidence.

Also the infrared radiation is important. A model simulating this heat transfer is being developed.

The advantage of V-corrugated honeycomb is that the air is divided into two cells, that this type maintains a high solar transmission and that it is easy to support structurally.

5. Future work

A prototype will be built in the nearest future. These prototypes will be tested as it was a cover in a solar collector. The collector efficiency with and without the transparent insulation in front of a collector will be measured. Also the heat loss coefficients will be estimated.

The most promising types will be built into a thermal storage wall in a test house. A honeycomb type insulation will be compared with a reference cover of the flat plate type.

6. References

1. B.A. Meyer, J.W. Mitchell and M.M. El-Wakil
   Convective Heat Transfer in Vee-Trough Linear Concentrators.
   Solar Energy, Vol. 28. No. 1, pp. 33-40, 1982.
2. K.I. Guthrie and W.W.S. Charters
   An Evaluation of a Transverse Slatted Flat Plate Collector
   Solar Energy, Vol. 28. No. 2. pp. 89-97, 1982.
3. K.G.T. Hollands, K.N. Marshall and R.K. Wedel
   An Approximate Equation for Predicting the Solar Transmittance of Transparent Honeycombs
   Solar Energy, Vol. 21. pp. 231-236, 1978.

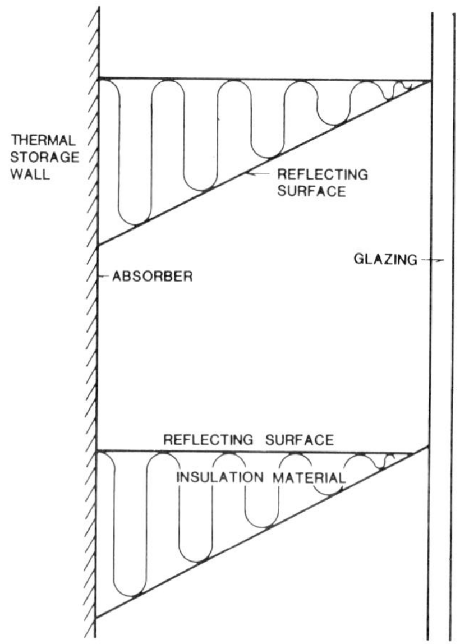

THERMAL
STORAGE
WALL
REFLECTING
SURFACE
ABSORBER
GLAZING
REFLECTING SURFACE
INSULATION MATERIAL

Fig. 1

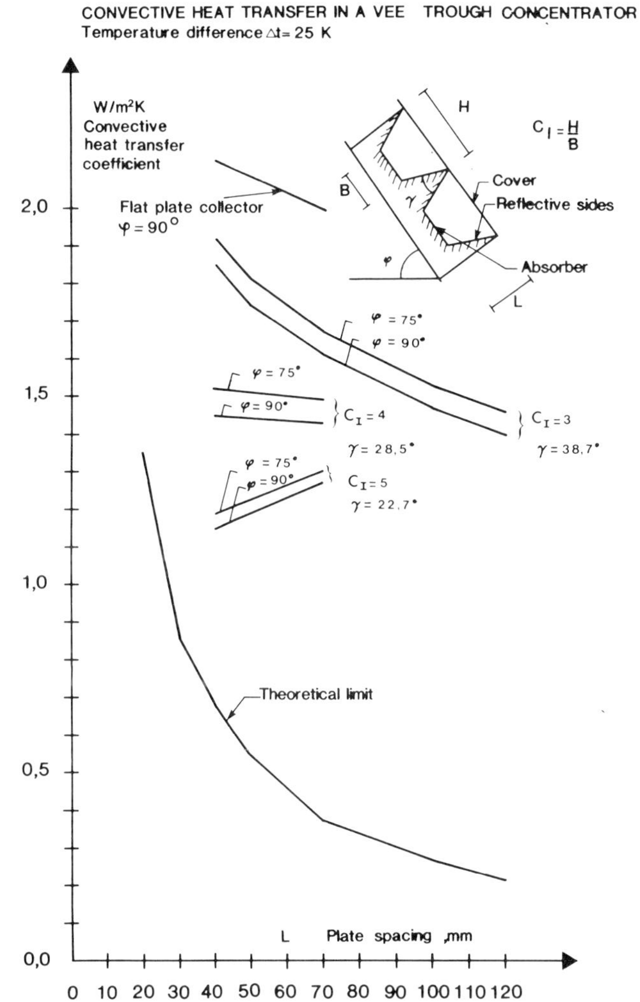

CONVECTIVE HEAT TRANSFER IN A VEE TROUGH CONCENTRATOR
Temperature difference $\Delta t = 25$ K
W/m²K
Convective
heat transfer
coefficient
Flat plate collector
$\varphi = 90^\circ$
H
B
$C_I = \frac{H}{B}$
Cover
Reflective sides
Absorber
L
$\varphi = 75°$
$\varphi = 90°$
$C_I = 3$
$\gamma = 38,7°$
$C_I = 4$
$\gamma = 28,5°$
$C_I = 5$
$\gamma = 22,7°$
Theoretical limit
2,0
1,5
1,0
0,5
0,0
0 10 20 30 40 50 60 70 80 90 100 110 120
L Plate spacing ,mm

Fig. 2

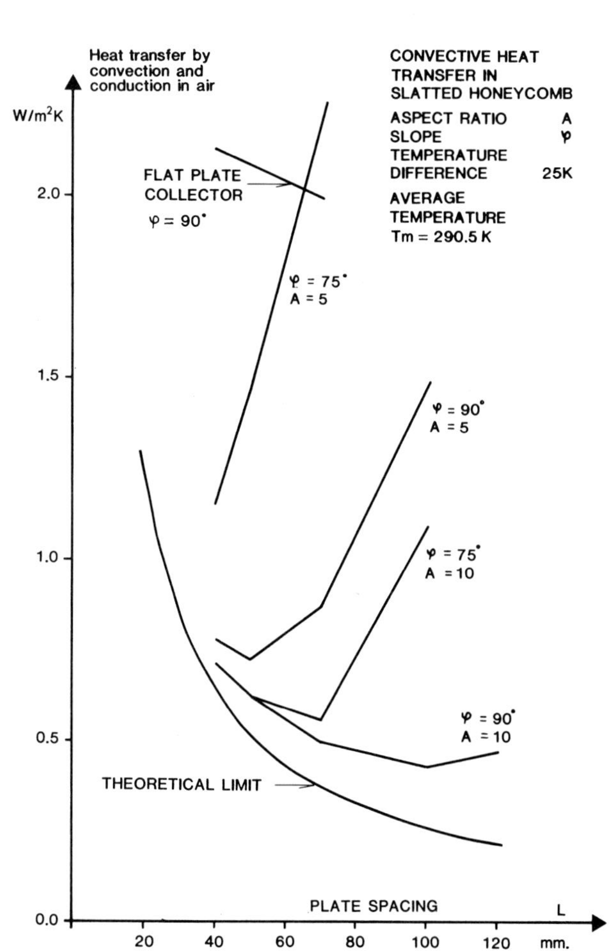
Heat transfer by convection and conduction in air
W/m²K
CONVECTIVE HEAT TRANSFER IN SLATTED HONEYCOMB
ASPECT RATIO A
SLOPE φ
TEMPERATURE DIFFERENCE 25K
AVERAGE TEMPERATURE Tm = 290.5 K
FLAT PLATE COLLECTOR
φ = 90°
φ = 75° A = 5
φ = 90° A = 5
φ = 75° A = 10
φ = 90° A = 10
THEORETICAL LIMIT
PLATE SPACING
L
mm.
2.0
1.5
1.0
0.5
0.0
20
40
60
80
100
120

Fig. 3

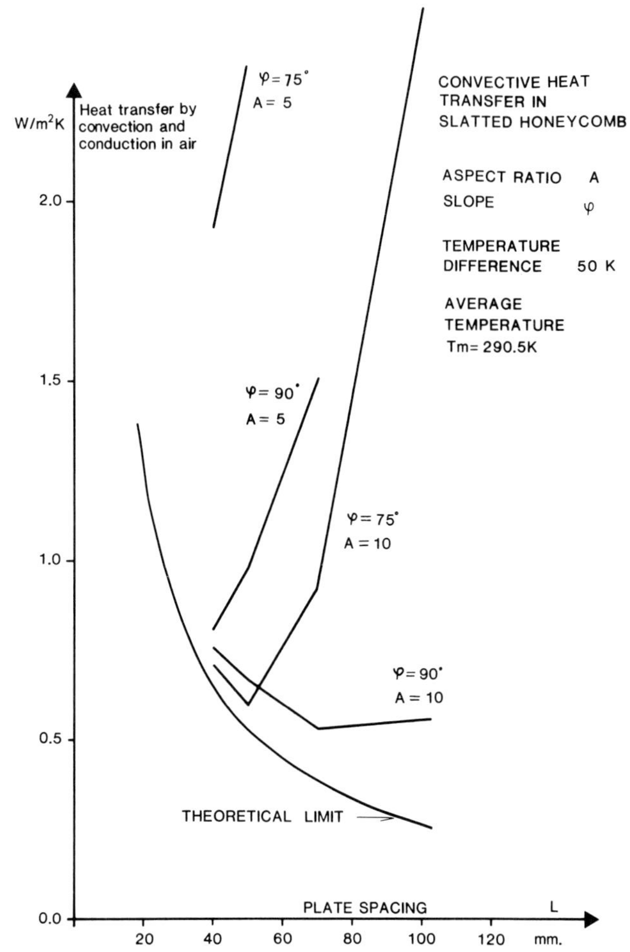
W/m²K
Heat transfer by convection and conduction in air
CONVECTIVE HEAT TRANSFER IN SLATTED HONEYCOMB
ASPECT RATIO A
SLOPE φ
TEMPERATURE DIFFERENCE 50 K
AVERAGE TEMPERATURE Tm= 290.5K
φ = 75° A = 5
φ = 90° A = 5
φ = 75° A = 10
φ = 90° A = 10
THEORETICAL LIMIT
PLATE SPACING
L
mm.
2.0
1.5
1.0
0.5
0.0
20
40
60
80
100
120

Fig. 4

CROSS-SECTION OF A V-CORRUGATED HONEYCOMB.
An example of raytracks are shown.

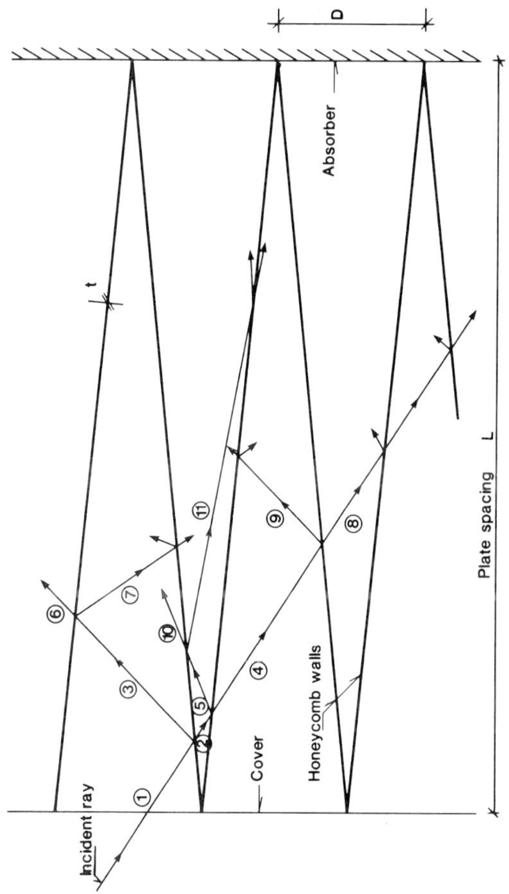

Fig. 5

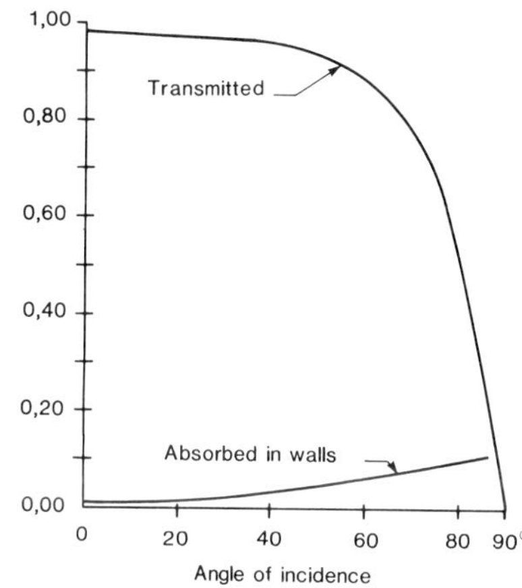

TRANSMISSION OF V CORRUGATED HONEYCOMB

Dimensions L=100 mm,D=20 mm,t=0.1mm

Optical properties as glass

Fig. 6

A VENTILATED TROMBE WALL UNIT

Author : F. BOURDIN
Contract number : ESA PS - 158 F
Duration : 21 months 01.10.81 - 30.06.83
Total budget : 1 147 700 F.F. - CEC Contribution : 450 000 F.F.
Head of project : F. BOURDIN
Contractor : SAINT-GOBAIN VITRAGE FRANCE
Adress : Centre de Développement Industriel
15, rue Rouget de Lisle
92400 COURBEVOIE

Summary

The aim of this research program is to complete the results which have been obtained throughout 3 years of study of the Trombe wall ventilated by fresh air. That completion should be achieved with the experimentation of a transparent Trombe wall and with the realization of a full range of industrialized modular units.

A water Trombe wall ventilated by fresh air has a mean captation efficiency of 60 %, better than the standard concrete Trombe wall (35 %). The association with a heat pump on extracted air provides fuel savings up to 70 %.

To insure the commercialization of this product, it will be necessary to create a full range of modular units which will meet the architects'demands i.e. to associate opaque modular units to transparent ones with a double glazing filled with aqueous gel used as thermal storage.

Those wall components should include the solar storage, the double glazing, the insulating curtain and the air circuit in a complete system.

They will have to comply with reduced building costs and good liability in use.

In the following paper, we will discuss our present knowledge of the system performances and the possible ways of setting up the manufacturing process.

1. Introduction

Le mur étudié par le Professeur Trombe était constitué de 45cm de béton, le chauffage des locaux était assuré par le rayonnement du mur et par une circulation en thermosiphon d'air intérieur. Le système avait plusieurs inconvénients. Durant l'ensoleillement la température superficielle du mur peut atteindre une valeur très élevée, jusqu'à 60°C, au détriment du rendement de captage. Durant la nuit même avec un double vitrage les déperditions sont importantes et le rendement moyen pendant la période de chauffage est d'environ 35 %. Durant les demi-saisons et l'été la température du mur capteur rend inconfortable les locaux adjacents.

Pour pallier ces inconvénients, nous avons amélioré le système en prévoyant :

- de préchauffer sur la façade Sud tout l'air de renouvellement de l'habitation.
- de remplacer le béton par un réservoir d'épaisseur 7cm rempli d'eau.
- de prévoir une cloison isolante derrière le stockeur avec possibilité de réaliser une circulation d'air entre les deux.

Toutes ces mesures concourent à abaisser la température de la face extérieure du capteur-stockeur, donc à améliorer le rendement.

Pour assurer le confort d'été et diminuer les déperditions de nuit, nous avons prévu un store à faible émissivité : toile aluminée deux faces avec protection polyester. Le store descend automatiquement au coucher du soleil ou quand la température de l'eau devient trop forte.

Malgré les bons résultats obtenus sur le plan technique : le rendement moyen de captage pendant la période de chauffage est de 60 %, le système associé avec une pompe à chaleur sur l'air extrait assure des économies de 70 % par rapport à une solution classique, les tentatives de lancement dans le commerce ont buté sur deux obstacles :

- Le premier est l'aspect architectural, une façade Sud doit comporter des parties éclairantes : nous avons donc étudié un mur Trombe transparent.
- Le second est économique et pratique : il faut réaliser une construction monobloc en usine, moins chère et plus fiable qu'un assemblage sur le chantier.

2. Description des produits

2.1 Le mur Trombe ventilé opaque

Le mur capteur-stockeur (figure 1.) est constitué de réservoirs modulaires d'épaisseur 7cm, de largeur 30cm et de hauteur 220cm. Ils sont réalisés en acier inox. La surface captrice est revêtue d'une couche absorbante peu émissive réalisée par dépôt électrolytique. Les réservoirs sont remplis d'eau additionnée d'antigel.

L'isolation thermique de jour est assurée par un double vitrage étanche 4 (12) 4 monté dans un ouvrant permettant le nettoyage. La condamnation s'effectue de l'intérieur.

Le store assurant un complément d'isolation nocturne et le confort thermique d'été est réalisé en "Reflex-Rol" : tissu de verre aluminé et protégé. La réflexion entre 0,4 et 8 μ est de 80 %, la transmission énergétique est nulle. Le store peut être équipé d'une commande électrique asservie à un capteur de luminosité et à une sonde de température d'eau ou d'air.

L'air de renouvellement est admis en partie basse, s'échauffe sur la face absorbante du capteur et est admis dans l'habitation en partie haute. Le système fonctionne grâce à la dépression créée par la ventilation mécanique contrôlée. Les sections de passage d'air sont largement dimensionnées pour limiter la perte de charge.

L'ensemble est monté dans une menuiserie monobloc. La fabrication initiale est prévue en bois pouvant évoluer vers le PVC. On rapporte ensuite une contre-cloison isolante destinée à assurer un meilleur confort thermique. Eventuellement, une circulation en thermosiphon, avec bouches obturables, peut être prévue entre la cloison et la face arrière du capteur. Ceci nécessite, cependant, une étanchéité à l'air très soignée à la périphérie du capteur-stockeur.

2.2 Le mur Trombe ventilé transparent

La seule différence est le remplacement du capteur-stockeur opaque par un double vitrage 4 (32) 4, dont l'espace entre les deux vitrages est rempli d'un gel constitué de 80 % d'eau et de 20 % d'une résine acrylique (figure 1.).

Le but est de réaliser un mur dont la transmission lumineuse est de 50 %, la transmission énergétique de 30 % et le facteur solaire de 40 %. A cet effet, le verre intérieur du vitrage à gel est un vitrage coloré bronze (Parsol) qui augmente l'absorption du gel à une valeur voisine de 40 %. L'énergie absorbée pendant le jour est restituéeen grande partie la nuit par rayonnement vers l'intérieur et par échauffement de l'air de renouvellement.

L'avantage architectural est certain et, par rapport à une surface vitrée normale, le confort thermique est bien meilleur grâce à l'étalement de l'apport solaire sur 24 heures.

Le rideau isolant est légèrement différent du précédent ; il est réalisé en "Reflex-Rol" dont la transmission lumineuse est de 10 %, la réflexion étant de 70 % entre 0,4 à 8 μ. En été, la transmission lumineuse est suffisante et le confort thermique assuré.

3. Les essais

Ils sont réalisés sur une maison préfabriquée à ossature métallique de 50m2 et 125m3. La façade Sud est en pignon et la surface de captage utile est de 17m2. Les prototypes de mur Trombe ventilés ont été réalisés en acier et assemblés sur le chantier.

Une partie des panneaux se trouve située derrière une véranda.

Pour chaque module, on mesure les températures d'entrée et de sortie d'air ainsi que différentes températures dans les stockeurs.

Le site est équipé d'une centrale de mesures et l'on relève toutes les heures les différentes valeurs d'ensoleillement, les températures, les consommations électriques. Le dépouillement est effectué sur calculateur. Les mesures sur la maison de 50m2 n'ont débuté qu'en Mars 82. Cependant, depuis Octobre 1981, nous mesurons les performances d'un mur transparent de 1,80m2 monté sur une maquette de 8m3, mais non équipé d'un store isolant.

4. Les résultats

Pour les capteurs opaques, nous avons plusieurs années de mesures. Pour le capteur transparent, on peut donner des approximations à partir des résultats enregistrés sur maquette réduite sans store isolant.

Les résultats donnés dans les tableaux suivants sont des moyennes sur trois ans pour le climat de notre station d'essai située à 100 km au Nord de Paris.

4.1 Rendement de captage du mur Trombe ventilé opaque

Oct.	Nov.	Déc.	Janv.	Fév.	Mars	Avril	Mai	Moyenne
60 %	55 %	46 %	43 %	54 %	60 %	62 %	62 %	55 %

Ce rendement est le rapport de l'énergie solaire ayant effectivement contribué au chauffage et de l'énergie incidente sur un plan vertical Sud. Il intègre donc la récupération de la fuite thermique par isolation dynamique du circuit d'air ainsi que les pertes par surchauffe.

Un modèle mathématique, calé sur nos résultats expérimentaux, nous a permi de vérifier que l'épaisseur de 7cm du stockeur était bonne. Pour un logement situé en région parisienne équipé de 9 à 18m2 de capteurs, on peut compter sur un apport solaire de 220 kWh/m2 entre le 1er Octobre et le 30 Avril.

4.2 Rendement de captage du mur Trombe ventilé transparent

Comme indiqué au paragraphe 2.2, la transmission directe vers le local est de 40 % (facteur solaire), les variations de rendement du système ne peuvent être dues qu'à la répartition des 30 % d'énergie absorbée par le gel et aux pertes par surchauffe du local. On pouvait donc espérer une amélioration du rendement par rapport au mur opaque, c'est ce que l'on constate dans le tableau suivant estimé en fonction des résultats sur maquette.

Oct.	Nov.	Déc.	Janv.	Fév.	Mars	Avril	Mai
65 %	63 %	59 %	58 %	63 %	65 %	65 %	65 %

4.3 Relevé des températures

Dans les capteurs opaques, nous avons relevé en Mars une température de 50 °C en partie haute, store ouvert. L'air entrait à 45 °C ce qui est inacceptable. Grâce au store, nous limitons la température d'air à 35° en hiver et 30° en mi-saison. La température minimum d'air relevée a été de 7°C pour -3°C à l'extérieur, le gain dû à l'isolation dynamique est de 6°C.

Dans les capteurs transparents, nous avons relevé en Avril une température maximum de 40°C en haut du vitrage ; l'air entrait à 26°C. Sur maquette la température minimum dans le gel a été de 2,5°C, sans store, pour - 10°C à l'extérieur ; le gel résiste jusqu'à - 15°C.

Nous donnons, sur la figure 2, un relevé des températures pour deux jours consécutifs.

5. Difficultés à résoudre

Le principal problème est l'esthétique et le fonctionnement du store. Les toiles "Reflex-Rol" ont apporté un progrès par rapport aux "Mylar" aluminés qui existaient auparavant. Cependant, il subsiste des déformations de la toile visibles sur la figure 3.

Des progrès peuvent être réalisés et de nombreux industriels se penchent sur le problème. Par contre, la transparence et la qualité optique du gel sont très bonnes (Figure 4 à l'extrême gauche).

L'autre problème est la réalisation d'un produit industriel à prix suffisamment bas pour que le temps d'amortissement soit raisonnable.

6. Industrialisation et commercialisation des produits

C'est l'étude qu'il nous reste à exécuter. Elle est en cours chez un menuisier industriel, un storiste, une Société de chauffage. Nous avons également consulté les bureaux d'études de grandes Sociétés Françaises de construction d'habitation.

On peut envisager une diversification du produit avec des dimensions plus faibles en hauteur pour réaliser des châssis fixes accolés à une fenêtre ou des fenêtres de toit.

7. Conclusion

La détermination des performances des systèmes est largement avancée. L'intérêt économique et architectural est certain. Les problèmes technologiques devraient être résolus de façon satisfaisante. Le principal frein au développement sera sans doute l'adoption du produit par les constructeurs et les clients. L'atout majeur devrait être le confort, bien meilleur qu'avec les autres systèmes de chauffage solaire passif.

Fig. 1

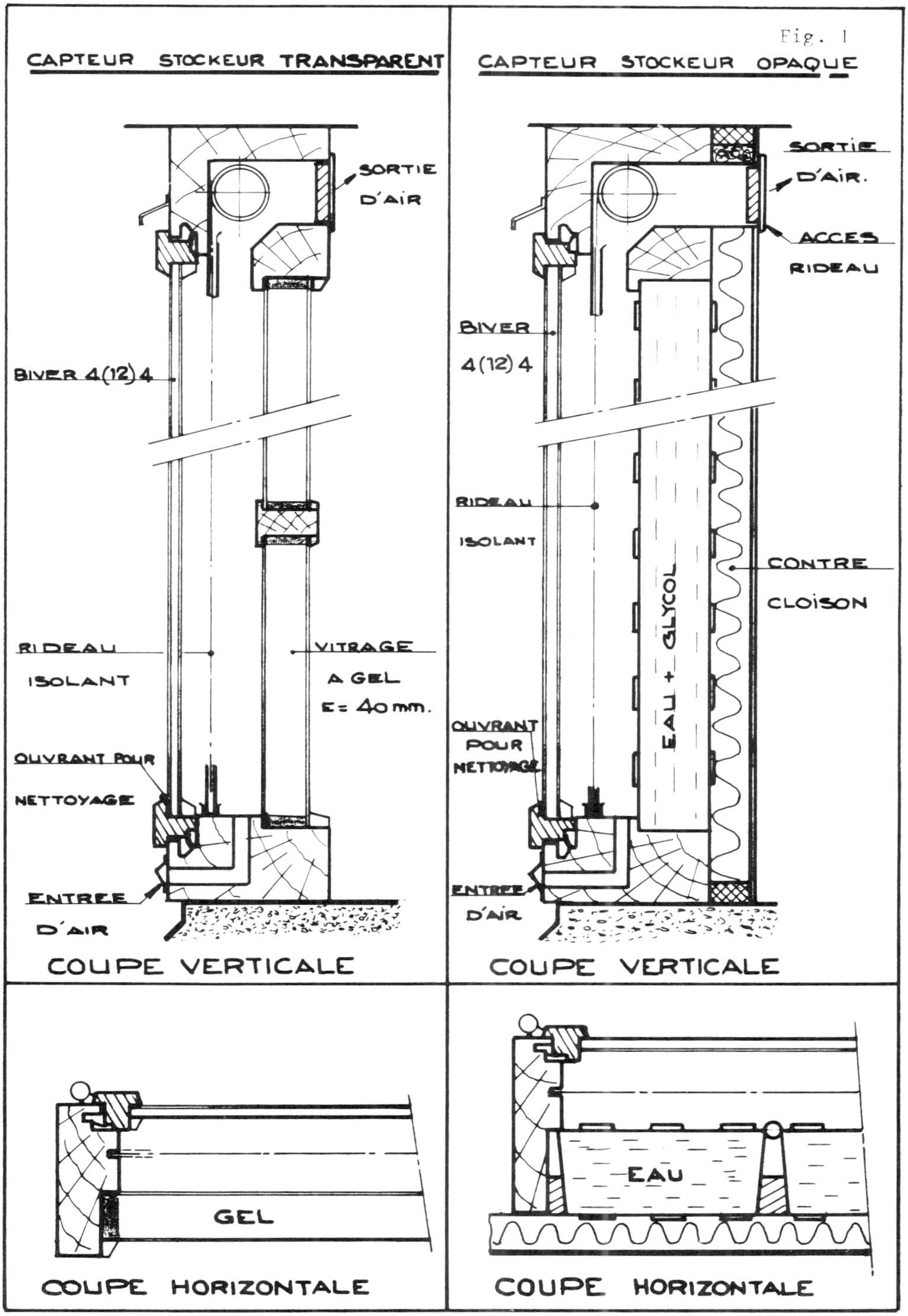

CAPTEUR STOCKEUR TRANSPARENT
SORTIE D'AIR
BIVER 4(12)4
RIDEAU ISOLANT
VITRAGE A GEL E= 40 mm.
OUVRANT POUR NETTOYAGE
ENTREE D'AIR
COUPE VERTICALE
CAPTEUR STOCKEUR OPAQUE
SORTIE D'AIR.
ACCES RIDEAU
BIVER 4(12)4
RIDEAU ISOLANT
CONTRE CLOISON
EAU + GLYCOL
OUVRANT POUR NETTOYAGE
ENTREE D'AIR
COUPE VERTICALE
GEL
COUPE HORIZONTALE
EAU
COUPE HORIZONTALE

Fig. 2

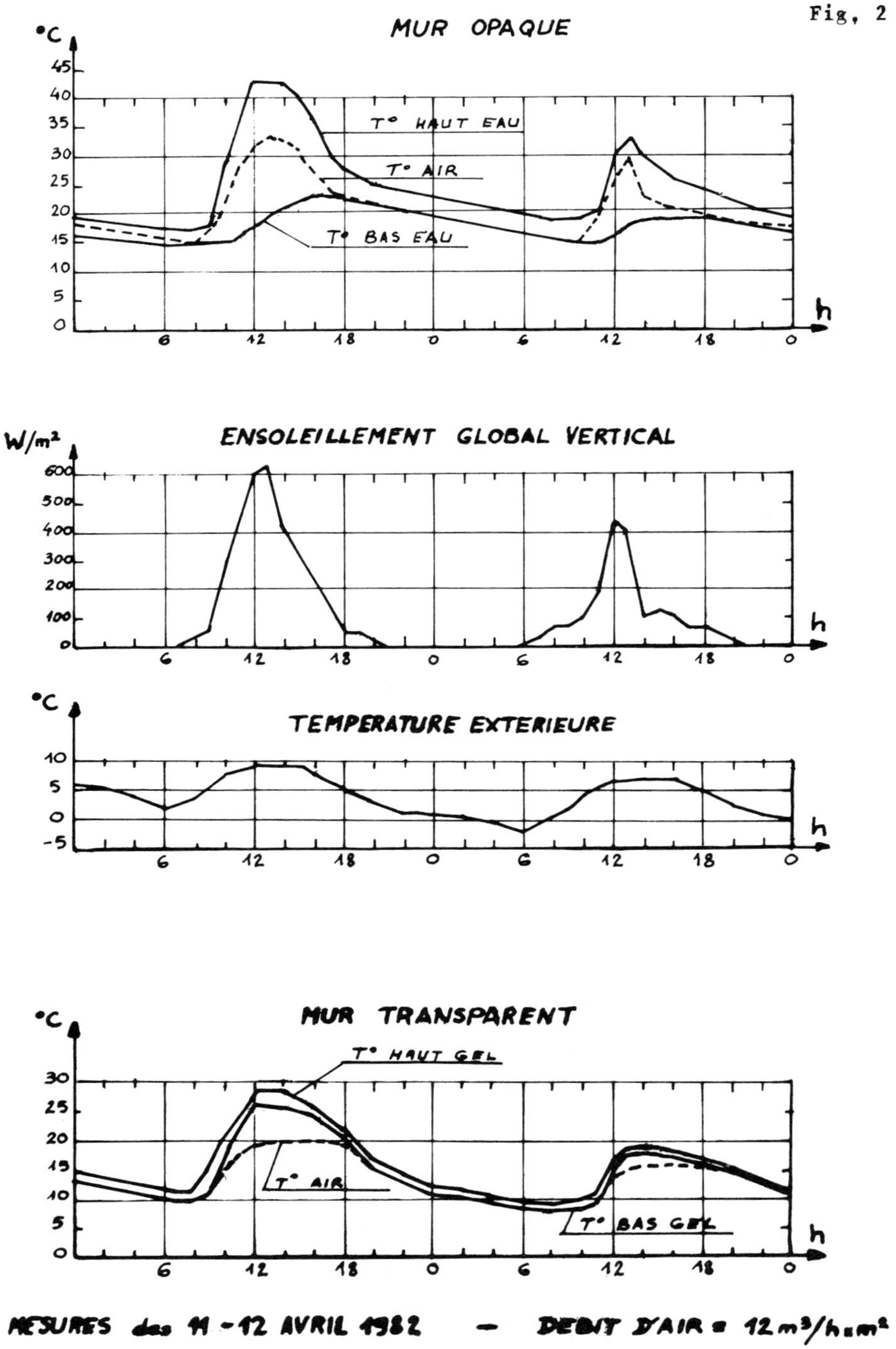
MUR OPAQUE
°C
T° HAUT EAU
T° AIR
T° BAS EAU
h
ENSOLEILLEMENT GLOBAL VERTICAL
W/m²
h
TEMPERATURE EXTERIEURE
°C
h
MUR TRANSPARENT
°C
T° HAUT GEL
T° AIR
T° BAS GEL
h
MESURES des 11 - 12 AVRIL 1982 — DEBIT D'AIR = 12 m³/h.m²

Figure 3.

Figure 4.

STUDY TO DESIGN AND MANUFACTURE AN INSULATING TRANSLUCENT GLASS PANEL MADE WITH EVACUATED PYREX TUBES

Authors	:	CORNING FRANCE and M.M. LIEBARD and ALEXANDROFF
Contract number	:	ESA-PS-139 F (S.D.)
Duration	:	18 months 01.01.82 - 30.06.83
Total budget	:	FF 540.750 CEC contribution FF 270.375
Heads of Project	:	A. COURTIER A. LIEBARD
Contractors	:	Corning France, 44 Avenue de Valvins B.P. 61 77211 Avon Cédex France
		Mr. LIEBARD Mr. ALEXANDROFF } 7 rue d'Argenteuil, 75001 Paris

Summary

The net gain between absorption during the day and losses during the night of a glass panel is a function of the transmission of the light and the K coefficient of heat loss.
In order to reduce this coefficient K, several products are now sold on the market :

Double glazing	K = 3,2 W / M2°C
Triple glazing	K = 2,4 W / M2°C
Ref single glass	K = 5 W / M2°C

The purpose of this project is to develop a translucent panel which can both transmit solar radiation and be a permanent insulation, using tubes of Pyrex in which a relative vacuum has been created. The objective is to confirm that such a panel will have K coefficient between 0,6 and 1,3 W / M2°C.

I. INTRODUCTION

The result of the direct gain supply/loss of a glazing wall is affected by the transmission of the light and by the K coefficient which determines the losses.

In order to reduce this K coefficient, different research has been made on glazing walls with temporary insulation as :

- System of window which is completely air-tight, using outside or inside shutters.
- Double glazing in which we use Venetian blinds with an absorbant black side and a reflective side in order to absorb and reflect the solar radiation.
- Double glazing with a gap of 15 CM of air, in which we inject polystyrene balls during the night in order to improve the night insulation (American system).

All these types of system have good transmission during the day and have K/day coefficient between 2,8 and 3,5 W/M2 C°. Other manual or automatic systems have achieved 1,7 W/CM2 C°, which is the minimum K coefficient possible.

II. GENERAL ASPECT OF THE WORK UNDER CONTRACT

The general objective of the present project is to design and industrialise insulating translucent glass panels, which will use solar energy in its passive forms (light and heat). These panels will be of durable construction, modular, of a great architectural flexibility and aesthetical.
The first objective is to obtain a glazing wall translucent to the light and radiations of the sun, and offering a permanent insulation, using a relative vacuum inside Pyrex tubes.
The K performance of these panels following the way of construction will be between 0,6 and 1,3 W/M2 C°, performances which are much better than any other products presently sold on the market.
Another purpose of the panel is to obtain a thermal insulation as close as possible to opaque panel, but permitting the transmission of the light and the radiation of the sun in order to light and to warm the rooms in a passive way.

III. ASPECT OF THE WORK UNDER CONTRACT

Phase 1

Industrial investigations realised by method - engineers with a value analysis and selection of different components, which compose the wall, that is to say :

- design and nature of the joints extruded in vacuum with study of the extruder screw.
- design and nature of the frames, fixed and mobile.
- architectural integration according to the different position in the buildings.

Phase 2

Fabrication of two basic protopytes, using components of a different nature, especially concerning the joints.

Phase 3

- First test series with successive tests to improve the joints.
- Comparative thermal measurements and preliminary thermal qualification.
- Determination of the tests :
  a) Retention of two different climatic atmospheres on both sides of the wall, and measurement of the thermal flow in transit.
  b) Measurement of the transparency coefficient of the solar radiation by photo-electrical cell.

Phase 4

- Study of the final applications and choices of the nature and the dimensions of the components.
- Study of the variants with an additional glass and with interior mobile shutters.

Phase 5

Fabrication in prelimary series of two tubular elements, one with additional glass and the other with interior, mobile integrated shutters.

Phase 6

Final tests of thermal qualification in order to obtain the official agreement.

Phase 7

- Investigation of the industrial process.
- Final choice of the production program issued from two basic hypotheses of quantative development.
- Investigation of the fabrication of the glass components and of their application as modular panel elements.

INSULATED TRANSLUSCENT PANELS

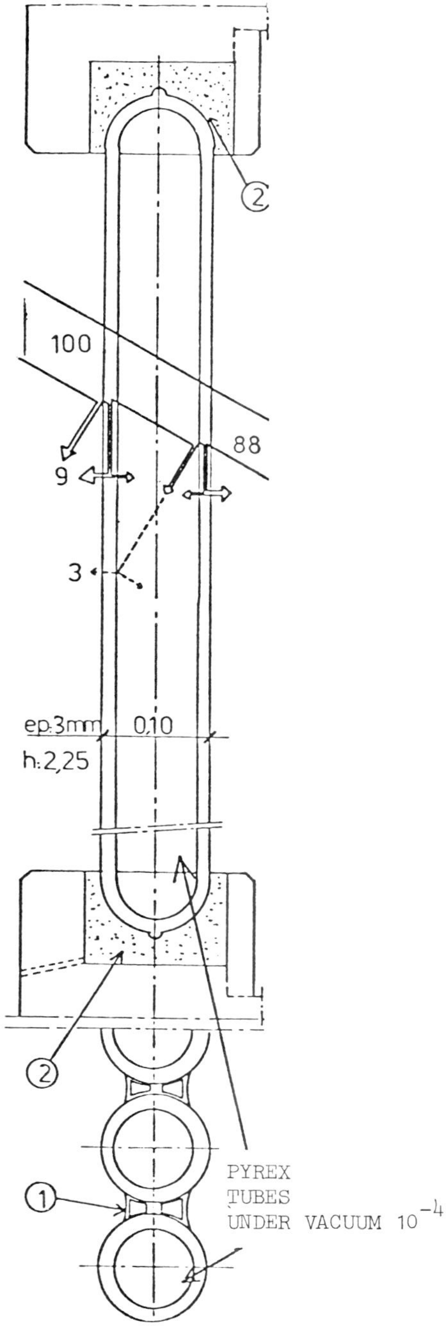

RESEARCH AND DEVELOPMENT OF A PASSIVE SOLAR FACADE INDUSTRIAL COMPONENT

Authors : J.P. BAILLON, S. SIDOROFF (SOREIB)

Contract number: ESA-PS-151-F

Duration : 14 months (expected)

Total budget : FF 420 710 CEC contribution 50 %

Head of project: Coordination M. BELLANGER , Design J.P. BAILLON
Engineering S. SIDOROFF , Manufacturing L. SALVAIRE

Contractor : ROSSIGNOL S.A.

ADDRESS : Route de St-Céneré
B.P. 3
53150 MONTSURS (FRANCE)

Summary

The aim of the project is to design a passive solar component as a part of a building façade, with (among others) the following functions :

* closure of the building, from floor to ceiling, over the width of a window,

* Enlighting" and solar gain collection through glazing materials,

* Solar control with shading devices,

* Good insulation for the night,

* Architectural and constructional qualities will raise the component to the level of classic architecture.

The main task is twofolded :

- From a technical viewpoint, to offer a stand-alone industrial product, whose functions were up to now provided by several different devices.
Moreover the quality and reliability will be better with an equivalent price. This is now possible thanks to the manpower cost reduction allowed by the industrialization of several operations which were previously manual.
Improved reliability can be obtained by the transfer into the workshop of assembly operations previously taking place on site in bad conditions (rain, temperatures differencies, dirt), and involving more efficicient tools.

- From an architectural viewpoint, to fecondate the design and the architectural aspect of an essential element of the building envelope by industrial manufacturing techniques.

The specification and requirements list will be set up by the 1st of July 1982.
The conception phase is planned to last up to November 1982.
The prototype will be manufactured in Januar 1983, and the tests will be conducted during the year 1983.
A serial production is supposed to start in the middle of the year 83.

1. PRESENTATION DE LA DEMARCHE DE CONCEPTION

1.1 Préambule

Dans le domaine dit "des technologies légères du bâtiment", le groupe GS6 des fenêtres, fermetures et éléments de remplissage, prend un regain d'intérêt avec la nouvelle réglementation sur les économies d'énergie. Et à la différence des façades légères, tombées en complet discrédit depuis quelque temps, l'activité de ce groupe ne cesse de s'affirmer chaque jour davantage.

1.2 L'industrialisation du bâtiment

Il faut dire qu'une motivation de fond sous-tend cette évolution : touchant à l'économie générale du secteur du bâtiment, handicapé par ses méthodes archaïques de production, elle postule qu'il faut reporter le maximum de tâches en usine ou en atelier afin de simplifier l'activité du chantier.

Grâce à quoi on obtient une rationalisation du travail et une précision accrue par l'emploi de machines à demeure, ainsi que des conditions de sécurité et de confort nettement améliorées.

1.3 Conséquence du report en usine

1.3.1 Résistance du marché

Ce report à l'usine ne va pas sans entraîner quelques problèmes de marché, tant au niveau de l'extension et de la distribution du produit nouveau* qu'à celui de son impact sur les opérateurs traditionnels du bâtiment. De nombreuses entreprises "ressentent en effet l'apparition de chaque nouveau composant plurifonctionnel comme une menace dans leur existence même", (1) car la part de valeur ajoutée introduite dans le produit est transférée du chantier à l'usine, ce qui les prive d'une source de plus-value.

1.3.2 Réduction du marché potentiel

A cette première difficulté liée à la structure du marché, s'en ajoute une seconde, attachée à la nature même du produit. Car le report en usine de sa fabrication s'accompagne très logiquement d'un enrichissement, donc "d'une augmentation de sa spécificité". Ce qui a pour conséquence de réduire son marché potentiel.

1.4 La notion de composant

C'est dès l'origine que la démarche de conception d'un tel produit doit prendre en compte ces facteurs restrictifs. Mais que ce produit soit conçu comme composant, voilà qui aide à tourner les difficultés évoquées. Qu'est-ce qu'un composant en effet ?

Selon la conception qui prévaut en France, à la Direction de la Construction, ce n'est pas un objet. C'est avant tout un concept économique. Un composant naît, nous l'avons vu, lorsqu'il est possible de dégager une plus-value grâce au regroupement d'un ensemble de tâches (ou d'opérations) auparavant distinctes qui, chacune, réclamait un prix de revient élevé souvent fonction du temps passé.

---

* Ce problème sera abordé ultérieurement.

1.5 Conséquences du regroupement des fonctions

Ce regroupement met en jeu des corps professionnels très différents, ce qui ne présume en rien de la combinatoire à adopter lors de la conception du produit.

Par contre, le choix des fonctions regroupées va modifier le chemin critique du chantier dans la mesure où il médiatise l'intervention des divers corps d'état. Des opérations qui sur le site devaient se succéder pourront être effectuées simultanément en usine, ce qui va raccourcir la durée du chantier, et ce à proportion du nombre d'intervenants chargés jusqu'ici de la mise en place des fonctions regroupées par le composant.

L'instant de la pose du bloc sera déterminé par le choix des fonctions qu'il regroupe et va déterminer à son tour son mode de conditionnement et de stockage. Inversement, l'examen du chemin critique aiderait à choisir les fonction regroupées.

1.6 La démarche de conception : localiser les noeuds de complexité

Mais de même que dans un téléviseur ou une automobile, le regroupement des tâches et des fonctions ne s'effectuera pas dans le bloc d'une manière homogène quant à sa complexité. Les compétences qu'il réclame sont nombreuses, de même que leurs points d'application.

Des zones vont apparaître, des temps forts où se concentrera l'investissement intellectuel et technologique. Et la mise en forme de ces temps forts ainsi que leur localisation précise constituent l'une des sources architecturales les plus pures et l'un des guides essentiels de notre démarche de conception. Le produit sera d'autant plus probant que des temps forts, peu nombreux, regrouperont chacun le maximum d'opérations, issues du plus grand nombre d'intervenants.

1.7 Remèdes aux conséquences du report en usine

Dès lors la seconde difficulté évoquée plus haut qui, agissant en facteur restrictif du marché potentiel, pousse à la banalisation, trouve ici sa résolution en ce qu'il suffit de faire correspondre à la discrimination temps forts/temps faibles, une ségrégation des sites concernés : les plus complexes devenant les pièces principales du produit et les plus simples -celles qui détiennent la part "d'enrichissement" la plus faible- s'individualisant en pièces secondaires.

Ainsi chaque pièce principale est conçue comme une sorte de "prise multiple" dont on exploite ou non, selon la circonstance, toutes les potentialités d'usage. De sorte que,bien que complexe, le produit ne sera pas spécifique en ce qu'il sera utilisable de diverses manières. A condition que son prix reste raisonnable au regard des fonctions réduites qu'il peut être amené à assurer suivant les désirs de l'entrepreneur, sa plurifonctionnalité et sa richesse ne devraient pas être des obstacles à sa diffusion, chacun exploitant le degré de richesse fonctionnelle du produit qui lui convient.

Quant à la levée de la première difficulté, elle s'effectue d'elle-même si l'on s'efforce de rendre universelles les pièces principales, tandis que les pièces secondaires prennent le rôle d'accessoires chargés d'assurer la diversité du produit.

1.8 Interface et compatibilité

On notera que la ségrégation des sites "actifs" -véritables noeuds de complexité nés de l'opposition temps forts/temps faibles- introduit la problématique de l'interface qui apparaît toujours lorsqu'il s'agit de relier deux systèmes autonomes.

Dans le courant de cette étude, deux niveaux d'interface, seront considérés selon qu'il s'agisse d'associer des pièces compatibles ou non compatibles. Les pièces dites compatibles seront les pièces conçues dans le cadre du composant que nous proposons. Elles seront associées par clipsage ou boulonnage. Les non compatibles comportent les accessoires issus du marché ainsi que les liaisons du produit avec le gros-oeuvre. L'interface est alors assurée par des pièces intermédiaires venant s'emboiter dans des réservations ménagées dans les pièces universelles, ce qui permet d'étudier le produit à peu près indépendamment du problème des liaisons, pourvu qu'il se place dans le cas de figure le plus général possible, c'est à dire entre planchers de l'édifice.

1.9 Enfin, s'agissant ici d'un organe particulièrement exposé au sein du bâtiment, il semble intéressant d'attribuer aux pièces accessoires une fonction d'enveloppe et de protection afin d'écarter les pièces les plus complexes (au relief tourmenté), donc les plus fragiles, des agressions occasionnées par les agents naturels.

1.10 Considérations architecturales, constructives et thermiques

1.10.1 Architecturalement, nous avons été amenés à considérer que le choix du matériau était indissociable de la problématique du détail, c'est-à-dire des divers reliefs apparaissant sur la surface du bloc.

Or la logique du composant qui nous a conduit à proposer :

- La polyfonctionnalité du produit,
- La concentration de nombreuses fonctions en certains sites privilégiés,

intervient au premier chef sur ce choix.

En effet, si l'on songe à employer des matériaux qui se travaillent par enlèvement de matière, comme le bois, deux facteurs limitatifs interviennent rapidement, qui sont l'impossibilité physique de dépasser une certaine densité d'interventions sans nuire à la stabilité du matériau , et l'importance du travail qui doit être effectué sur chaque pièce.

C'est pourquoi nous nous sommes tournés vers des matériaux se travaillant par moulage ou par extrusion, où la complexité est reportée sur une matrice et n'appartient plus à l'objet produit. On peut bien sûr imaginer une solution de compromis, qui sertirait des pièces moulées (acier, plastique, nylon) dans un bâti de bois.

Mais outre les coûts importants (actuellement du moins) des matériaux traditionnels, des considérations techniques sont intervenues pour fixer les choix, parmi lesquelles les considérations thermiques se sont révélées les plus contraignantes.

1.10.2 Justification du choix du PVC

Ce dernier, malgré de moins bonnes qualités isolantes, a été préféré au polyuréthane parce que son comportement et sa tenue dans le temps, sont mieux connus. De plus, il autorise une recherche plus souple de la forme optimale des profilés en raison du faible coût d'une filière comparé au prix des moules utilisés pour le polyuréthane.

Par ailleurs, le PVC a une meilleure tenue au feu (M_2 contre M_4).

Cependant son faible module d'élasticité (∼ 3000 N/mm^2), rend nécessaire l'adjonction de renforts métaliques dès que la largeur des fenêtres dépasse une certaine valeur (problèmes de gauchissement).

Et sa faible conductivité thermique (∼ 0,2 W/m/°K) provoque de forts gradients de température entre les parois interne et externe (pouvant atteindre 40K en plein soleil) qui créent des tensions importantes dans le profilé et installe des déformations nuisibles à l'étanchéité de la fenêtre. On peut pallier cet inconvénient par la mise en oeuvre "d'un bouclier thermique".

Enfin son fort coefficient de dilatation thermique (10^{-4} K^{-1}) pose des problèmes de rattrapage des jeux dûs aux allongements différentiels du PVC, du verre et des autres matériaux mis en oeuvre.

1.10.3 Mais si le PVC est facile à entretenir et s'altère peu, sa texture et ses qualités de surface sont jusqu'alors peu satisfaisantes et confèrent au produit un aspect qui s'harmonise difficilement avec les matériaux traditionnels du bâtiment. Il s'agit donc de tenir la gageure, du point de vue de l'aspect final du bloc LT, que représente l'emploi du PVC comme matériau de base, ce qui suppose :

- Une étude de la texture du PVC et de sa "charge" éventuelle en adjuvants divers, ceci en coordination avec le fabricant ;
- Une étude des profilés, non seulement aux niveaux thermique et constructif , mais aussi quant à leur insertion harmonieuse dans l'ensemble du bloc (échelle des détails, dessins des liaisons).

On connaît le bénéfice que des matériaux comme la pierre, la brique ou le bois tirent d'un vieillissement maîtrisé (par l'entretien). Il reste à définir les traitements de surface et le jeu d'éventuelles juxtapositions de matières aux textures riches qui conféreront au PVC les qualités visuelles et tactiles souhaitées.

1.11 <u>Terminologie</u>

L'expression "BLOC LUMITHERM" (en abrégé, bloc LT) a été choisie de préférence à bloc-baie ou composant-fenêtre pour éviter le mot fenêtre et sa connotation restrictive qui n'évoque pas la polyfonctionnalité du composant, notamment ses fonctions thermiques.

2. LES FONCTIONS CHOISIES ET LEUR REPARTITION DANS LE BLOC LUMITHERM

2.1 Tel un bloc-baie ou un bloc-fenêtre traditionnel, le bloc LT est chargé d'assurer les fonctions suivantes :

<u>Constructives</u> :
- Clôture du bâtiment,
- Liaisons avec le gros-oeuvre et les autres constituants de façade,
- Stabilité sous les sollicitations propres (ouverture) et du vent,
- Etanchéité à l'eau et à l'air.

Architecturales : . Vue (d'où la hauteur d'implantation et le problème des masques),

. Eclairement (dimensionnement),

. Communication avec l'extérieur (ouverture),

. Protection contre les intrusions en rez-de-chaussée.

Mais aussi . Contrôle de la lumière (stores et volets).

Mais sa participation au programme de recherches sur les composants solaires passifs se justifie par l'adjonction d'exigences thermiques qui faciliteront d'autant le respect de la récente législation (coefficient B) :

. Isolation phonique

. Controle thermique d'été

. Isolation nocturne

. Captation des apports solaires

Ces deux dernières exigences devant assurer au bloc LT un bilan thermique positif sur 24 h. Ceci implique en particulier une isolation nocturne particulièrement soignée. D'autant que des gains solaires importants (surtout en hiver et demi-saison) ne peuvent être obtenus que par un léger surdimensionnement des vitrages. C'est pourquoi, de même que l'occultation, la protection contre l'ensoleillement acquiert une fonction thermique importante en plus de sa fonction architecturale d'usage.

Enfin, le bloc LT, en tant que composant polyfonctionnel est chargé en outre des fonctions optionnelles suivantes :

. Dispositif permettant de l'intérieur le nettoyage complet des vitrages,

. Rangement,

. Eclairage artificiel et prise de courant,

. Cimaise pour rideau et voilage,

. Porte-fenêtre avec ou sans balcon,

. Contrôle de la ventilation (prise d'air ou évacuation,

. Appoint de chauffage (éventuellement couplé à un absorbeur)avec thermostat electronique,

. Dispositif de coupure des convecteurs à l'ouverture de la fenêtre.

2.2.1 Dimension constante, la largeur du bloc est fixée à 1,40 m.

Les statistiques montrent que cette largeur est l'une des plus répandues et sans doute l'une des plus demandées aujourd'hui encore.

La modulation des surfaces vitrées se fera donc en jouant seulement dans le sens de la hauteur. Elle peut s'effectuer de deux manières :

* Soit selon 5 paliers : - Porte-fenêtre avec imposte vitrée,

- " " " " opaque

- Fenêtre avec imposte vitrée
- Fenêtre avec imposte opaque
- Imposte seule (haut-jour).

* Soit de façon progressive en jouant sur la dimension même du châssis central.

2.2.2 Rôle de l'épaisseur

Le fort accroissement de contraintes résultant de la multiplication des fonctions rend très délicate leur juxtaposition dans un même plan (celui de la façade). Un tel agencement risque en outre de nuire à la tenue dans le temps des performances du bloc. La recherche effectuée par ailleurs et décrite dans les § 1.2 à 1.5, a conduit à désigner les joues latérales comme étant les éléments sur lesquels peuvent se concentrer les efforts de conception. Enfin, la nécessité d'obtenir un bloc auto-porteur et auto-stable, impose une certaine épaisseur.

Ces trois considérations amèent à traiter le bloc LT non comme une surface que se répartiraient tant bien que mal les diverses fonctions, mais comme un volume chargé d'assurer la relation entre le bâtiment et l'extérieur, et donc à travailler dans l'épaisseur.

2.2.3 La relation intérieure/extérieure ainsi médiatisée par l'épaisseur exige qu'une attention particulière soit portée aux aspects de surface et à la couleur des éléments constitutifs du tableau, qui doivent en outre permettre une large diffusion de la lumière incidente.

Outre sa fonction d'appui, une tablette horizontale (à 95 cm du sol) assure cette diffusion vers le plafond.

2.2.4 Il faut se rappeler que l'un des objectifs majeurs du projet est de féconder l'architecture des percements (baies, portes et fenêtres) par la logique des composants et grâce au savoir-faire industriel. Cette démarche conceptuelle nous ayant conduit à travailler dans les trois dimensions de l'espace, ouvre du coup l'épaisseur du bloc à des préoccupations architecturales de fond qu'une simple surface n'aurait pu solutionner de façon satisfaisante. Ces préoccupations touchent essentiellement aux notions d'existence et de mise en situation des fenêtres d'une part, de partition des façades d'autre part. Et s'il n'entre pas dans notre propos de composer des façades-types, il est toutefois nécessaire d'en bien comprendre les lois de composition au travers des choix effectués.

2.2.4.1. Notion de fenêtre : une fenêtre n'est pas un trou dans un plan abstrait ; c'est un lieu d'émergence d'un espace intérieur dans l'environnement extérieur. Dans l'architecture universelle, de même qu'en physique, l'émergence d'un milieu dans un autre donne naissance à une perturbation. Elle va se traduire par une zone de transition entre le plan du mur et l'ouverture proprement dite. Cette approche par la perturbation présente un intérêt pratique immédiat : celui de laisser le problème de la jonction bloc LT/logement à la charge de la zone de transition (joint). Mais elle interdit tout traitement en rideau de la façade : celle-ci doit paraître continue autour des percements.

2.2.4.2. Partition des façades :

. Verticalement, trois zones sont à considérer qui sont déterminées par la relation de n'importe quel édifice à l'environnement. Ce sont le soubassement, les étages courants et le couronnement (ou le toit). Afin de respecter cette organisation naturelle de la façade, les dispositions et l'enfoncement de la fenêtre dans la façade seront différents en rez-de-chaussée (relation au sol), en étage courant et en attique (relation au ciel). Les dispositifs d'occultation et de protection seront étudiés en conséquence.

. Horizontalement, le bloc doit pouvoir être seul, jumelé ou disposé suivant un rythme . En ce cas, l'uniformité née de la répétition sera combattue par l'emploi de variations thématiques portées par un ensemble de détails de très petite échelle. Choisis en fonction du style du bâtiment, ces détails sont portés par un petit nombre de sites, toujours les mêmes.

3. CAHIER DES CHARGES : CONTRAINTES ENVELOPPE

3.1 Etanchéité

- Etanchéité à l'air renforcée (A3)
- " à l'eau améliorée (E2)
- Résistance au vent : Classe V2

Selon la norme NF P 20-302

3.2 Isolation Acoustique

Conforme au label Acotherm (Affaiblissement acoustique de 30 à 35 dB Route).

3.3 Sécurité

3.3.1 Protection de l'occupant contre les chutes :

L'allège doit répondre aux normes NF P01-012 et NF P01-013.

3.3.2 Incendie :

Le composant devra satisfaire aux exigences de la réglementation en vigueur (cf. 10, p. 27 à 32). En particulier le classement au feu du panneau d'allège et le "C+D" devront permettre l'utilisation de bloc dans des ERP (Etablissements Recevant du Public) au moins de 4e catégorie.

3.4 Isolation Thermique

3.4.1 Fenêtre

Selon les prescriptions de l'appel d'offres H2E85, rappelées Fig. II.

3.4.2 Allège

Le coefficient de transmission thermique utile moyen devra, conformément à la législation, être inférieur à 1,15 W/m2/K.

4. CONTRAINTES STRUCTURELLES

4.1 Dimensions

Après étude des gammes de constructeurs de fenêtre et des normes existantes, on a retenu une largeur unique de 1,40 m hors tout , en se réservant la possibilité de varier la répartition sur la hauteur de l'allège et des ouvrants.

L'épaisseur devra être suffisante pour assurer une largeur de joues latérales intérieures de 30 cm, afin de pouvoir accueillir d'éventuels volets rigides en 3 parties.

La hauteur totale sera de 2,50 m, la distance linteau-plafond étant éventuellement remplie par une imposte opaque.

4.2 Choix du Vitrage

Les contraintes d'isolation thermique imposent, d'après le Tableau I, au moins un double vitrage. Grâce aux qualités isolantes du PVC, une lame d'air de 6 mm sufit pour obtenir un K_j global de fenêtre inférieur à 3 W/m2/K. Le Knuit sera ajusté par le pouvoir isolant de l'occultation nocturne. Le choix de l'épaisseur de la lame d'air et l'adoption éventuelle d'un vitrage non-émissif se feront en fonction des performances thermiques de la menuiserie et des contraintes de coût.

4.3 Mise en Oeuvre du Vitrage

4.3.1 Feuillure

Des contraintes réglementaires (2) et techniques (3) imposent une hauteur de feuillure de 20 mm, le drainage de celle-ci, et interdit la feuillure "portefeuille".

4.3.2 Calage

La réglementation (2), résumée par (3), précise l'utilisation des cales. Les professionnels (4) préconisent l'emploi de cales d'appui évidées pour faciliter la circulation de l'air en feuillure et éviter l'accumuation d'eau au niveau des cales.

4.3.3 Etanchéité Vitrage-Châssis

Le bourrage complet de la feuillure et l'emploi de mastics oléoplastiques sont interdits. Le Cahier du Moniteur consacré aux fenêtres reprend les différentes garnitures d'étanchéité utilisables d'après le DTU 39.4 (2), qui fournit par ailleurs sous forme de spécifications provisoires, les cahiers des charges relatifs à ces produits.

4.4 Menuiseries

Le choix du matériau (le PVC) et ses caractéristiques ont été exposés plus haut.

L'extrusion des profilés sera sous-traitée à un fabricant de matières plastiques, les filières étant dessinées par Rossignol qui prendra en charge par ailleurs l'usinage et le montage des châssis.

Le fabricant est responsable de la qualité des profilés et du respect des caractéristiques annoncées.

Divers fabricants de machines ont été contactés afin de réunir une documentation sur les machines utilisées pour l'usinage et l'assemblage des profilés.

4.5 Fermeture et Occultations

L'isolation thermique recherchée ($\frac{2K_i + K_n}{3} \leq 2,35\ Wm^{-2}\ K^{-1}$) conduit à adopter une isolation nocturne par l'intérieur, ce qui amène à dissocier cette fonction de celle de protection contre les intrusions.

Lorsque cette dernière fonction est requise, une claustra placée en allège et coulissant verticalement dans des glissières latérales, peut résoudre élégamment à la fois le problème de la protection solaire et celui de la protection contre les intrusions, grâce à un choix approprié du matériau et du système de verrouillage.

La même claustra, placée à l'intérieur et dans un matériau différent, pourra en position basse servir également de grille diffusante dans le cas où un convecteur est placé en allège.

L'occultation nocturne pourra être souple et s'enrouler en allège par exemple, ou rigide (volets pleins) et venir se replier contre les joues latérales. L'accent sera mis sur les propriétés thermiques et esthétiques. Une tringle à rideau intégrée en linteau sera prévue en option. Un cahier du Moniteur (6) fournit un inventaire des fermetures existantes et une norme (7) indique les spécifications auxquelles doit répondre une fermeture

4.6 Eléments de remplissage (EdR)

Il s'agit essentiellement du panneau d'allège. Une directive Européenne (8), fournit un cahier des charges relatif aux propriétés d'un EdR. Un avis technique du CSTB (10) décrit sa mise en oeuvre et un cahier du Moniteur (9) fournit une liste très détaillée des produits existants.

4.7 Panneaux de façade

Un DTU, quoique ancien (11) fournit des prescriptions techniques relatives aux panneaux de façades menuisé s. Divers documents fournissent des éléments sur les défauts rencontrés dans les façades légères, la liste des produits et systèmes d'isolation et d'étanchéité complémentaire agréés, ainsi que des procédés de mise en oeuvre des panneaux-baies visant à réduire les ponts thermiques (10), etc...

4.8 Joints de façade

Deux normes importantes (13 et 14) et un texte du CSTB (15) définissent les règles du jeu et des systèmes-types concernant la liaison entre panneaux de façade et gros-oeuvre et la compatibilité entre panneaux.

Du respect de ces spécifications dépendra le débouché commercial du composant, la compatibilité avec les systèmes existants étant une condition sine qua non de son succès.

Un texte extrait d'une revue professionnelle (16) recense les diverses garnitures d'étanchéité des joints de façade et leur utilisation ; une directive Européenne définit un agrément des mastics pour joint (17).

4.9 Homologation

L'ensemble constitué par :
- la fenêtre équipée de ses vitrages,
- l'allège,
- les joues latérales formant cadre,
- l'occultation nocturne isolante et la protection solaire,

sera testé comme un panneau de façade menuisé , de façon à être en mesure d'obtenir l'Agrément du CSTB, ainsi que la garantie décennale.

5. DESCRIPTION ANALYTIQUE DU BLOC (cf. Figure II)

5.1 Dimensions

Hauteur : 250 cm
Largeur : 140 cm hors tout

Epaisseur : au moins 30 cm entre face interne de la fenêtre et bords intérieurs des jouées.

5.2 Eléments de remplissage

/1/* Ouvrant supérieur pivotant autour d'un axe horizontal
/2/ Ouvrant inférieur coulissant verticalement
/3/ Panneau d'allège

5.3 Encadrement

/4/ et /5/ Jouées latérales (pièces universelles au sens du § 1.7)
/6/ Traverse en linteau
/7/ Socle

5.4 Répartition dans l'épaisseur

/8/ Claustra (protection d'été)
/9/ Volets isolants intérieurs (occultation nocturne)
/10/ Convecteur (éventuellement)

5.5 Fonctionnement des parties mobiles

5.5.1 Idées directrices :

- Eviter les efforts en bras de levier sur la menuiserie, à cause des faibles qualités mécaniques du PVC.
- Concevoir les déplacements des ouvrants et leur verrouillage de sorte que la poussée du vent renforce l'étanchéité par un écrasement uniforme des joints.
- Adopter un verrouillage "élastique" (grâce au contre-poids du châssis coulissant par exemple), rattrapant les éventuels jeux d'usure (àla différence d'un verrouillage classique par crémone).
- Permettre une manoeuvre et un nettoyage aisés des ouvrants.

5.5.2 Description

- Le châssis supérieur /1/ s'ouvre en pivotant à 150° autour d'un axe horizontal, la traverse inférieure vers l'intérieur du local. Trois positions de blocage à faible ouverture sont prévues.

- L'ouvrant /2/ coulisse verticalement devant l'allège. Baissé à fond, seuls dépassent les 2/3 de la traverse haute qui constituent un rebord pour la tablette d'appui. Pour nettoyer la face externe des 2 vitrages de /2/, on devra pouvoir basculer vers l'intérieur du local soit la totalité du châssis, soit les 2 vitrages séparément.

- La claustra /8/ coulisse verticalement. Baissée, elle ferme le volume formé par l'allège, les jouées, le socle et la tablette ; levée, elle constitue une protection d'été qui n'empêche pas l'ouverture partielle du châssis supérieur. Toutes les positions intermédiaires sont possibles.

- Les volets /9/ repliés chacun en 3 parties, viennent se loger dans l'épaisseur des jouées. Déployés, ils s'appliquent par des joints souples sur un bourrelet périphérique de /4/, /5/, /6/ et déterminent ainsi un volume d'air étanche isolant.

5.5.3 Manoeuvre et verrouillages

- Ouverture des volets : ils se déverrouillent par pincement d'un dispositif à ressort vertical entre le pouce et l'index, et se replient vers l'intérieur. Un clips les maintient alors contre les jouées.

---

* Les numéros entre / / renvoient à la figure II.

- Ouverture des châssis : ceux-ci sont solidarisés au cours du dernier cm de la manoeuvre de fermeture. Le châssis supérieur /1/ s'ouvre par une courte traction vers le bas (c-c') désolidarisant les 2 ouvrants, puis en tirant vers soi (c'-c''). Le châssis /2/ se déplace alors légèrement vers le haut et l'extérieur, ce qui assure à nouveau son étanchéité.

Pour baisser /2/, on déverrouille par pression la poignée de descente /11/. Le relâchement de la pression bloque l'ouvrant dans la position désirée.

5.6 Dispositifs fixes et détails d'architecture 3 idées directrices :

- Regrouper sur un même site plusieurs fonctions (cf. § 1.2 à 1.5)
- Exploiter un même relief pour divers usages quitte à l'adapter
- Unifier les détails par un dessein d'ensemble.

Exemple : cimaise pour rideaux + cache-néon = liaison formelle plafond/bloc.

5.7 Implantation dans la façade

Des variantes sont étudiées, permettant notamment de retourner le bloc en rez-de-chaussée, afin d'avoir, de l'extérieur vers l'intérieur : la claustra servant de dispositif anti-effraction, les volet puis le plan des vitrages. D'autres variantes sont possibles et seront décrites ultérieurement.

6. COUT

L'objectif est, rappelons-le, de proposer un composant de coût légèrement inférieur à celui de l'ensemble des éléments qui le constituent, pose comprise et pour un service équivalent.

A cet égard, le coût relativement élevé des vitrages isolants se justifie par les exigences thermiques.

Par ailleurs, le transfert en usine d'un certain nombre d'opérations d'assemblage jusqu'ici effectuées sur chantier, devrait permettre :

. Une meilleure garantie des performances annoncées,

. Une économie de main-d'oeuvre, les difficultés de mise en oeuvre classique étant en partie résolues par des astuces techniques de montage, mises au point lors de la phase de conception,

les 2 choses étant interdites par les conditions de travail habituelles de chantier (intempéries, écarts de température, exposition aux salissures-ciment, etc...)

Ces gains de productivité, plutôt que de servir à proposer un produit meilleur marché, devra être plutôt mis à profit pour proposer un produit de prix équivalent aux produits répondant aux mêmes fonctions, mais de meilleure qualité, y compris sur le plan architectural.

7. PLANNING PREVISIONNEL (Outlook for future work)

7.1 Calendrier

- Fin Juin 82 : Remise par la SOREIB du cahier des charges à ROSSIGNOL S.A.

- De Juillet à Décembre 82 : Conception du module par une équipe constituée de :
  . un architecte concepteur (J.P. BAILLON, indépendant)
  . un ingénieur de bureau d'études (M. LE BAIL, ROSSIGNOL)
  . un spécialiste des profilés de menuiserie (M. LEVREL, ROSSIGNOL)
  . un "designer" industriel (L. SALVAIRE, Directeur BE ROSSIGNOL)
  . un ingénieur thermicien (S.SIDOROFF, SOREIB)

- De Décembre 82 à Juillet 83 : Construction et essais du prototype.

- Mi-83 : Démarrage d'une fabrication industrielle en pré-série.

7.2 Phase de conception

. Définition complète du produit,

. Etude sur maquette et choix des systèmes techniques en fonction des contraintes du Cahier des Charges,

. Choix des fabricants des parties sous-traitées,

. Dessins et plans,

. Elaboration d'une gamme de fabrication.

7.3 Réalisation et tests du prototype

. Construction dans l'atelier des prototypes de ROSSIGNOL S.A.,

. Essais dans les installations de Saint-Gobain Vitrages, à Chantereine, mesures sur banc et en extérieur (cellules-tests).

BIBLIOGRAPHIE

(1) P. CHEMILLIER & L. CHABREL : Les évolutions technologiques dans le bâtiment, in cahiers du CSTB, Mars 82.

(2) Travaux de miroiterie et de vitrerie en verre épais. CSTB, Document Technique Unifié n°39.4, Mars 77.

(3) Saint-Gobain Vitrages : Mémento 1982.

(4) Le drainage des feuillures, "Techniques Nouvelles en serrurerie , menuiserie, miroiterie" n° 125, Février 81.

(5) Les fenêtres, Encyclopédie des produits composants du bâtiment, Editions du Moniteur, Documentation Française du Bâtiment (DFB), Mars 79.

(6) Les fermetures, Encyclopédie des produits composants du bâtiment, DFB 79.

(7) Fermetures pour baies extérieures équipées de fenêtres. Méthodes d'essais. Norme AFNOR NF P 25.501, Avril 80.

(8) Directives UEAtc pour l'agrément des EdR. CSTB, cahier 1762, livraison n° 237, Mars 82.

(9) Les façades légères, Encyclopédie des produits composants du bâtiment, DFB, Novembre 78.

(10) Conditions générales de mise en oeuvre des EdR ...,Avis Tehnique. CSTB, cahier 1691, livr. n° 216, Janvier/Février 81.

(11) Panneaux de façade menuisés. Prescriptions Techniques. CSTB, D.T.U.Mai61.

(12) Revue Technique du Bâtiment n° 47, Mars/Avril 75.

(13) Joints entre panneaux de façade à bords minces alignés. Compatibilité des panneaux. Norme AFNOR 85.201, Septembre 75.

(14) Règles du jeu pour les joints en système ouvert. Panneaux de façades à à bords minces insérés entre planchers et refends. Norme AFNOR P 85.202 Septembre 75.

(15) Systèmes types de joints, rattrapages et attaches. CSTB, cahier 692, livr. 80, Juin 66.

(16) Façades légères et cloisons industrialisées, chap. IV : les garnitures d'étanchéité des joints. Revue du CIMUR n° 47, Décembre 71/Janvier 72.

(17) Directives UEAtc pour l'agrément des mastics d'étanchéité utilisés dans les façades des bâtiments. CSTB, cahier 1404, livr. 174.

– Caractéristiques thermiques

Les propositions devront respecter les limites suivantes :

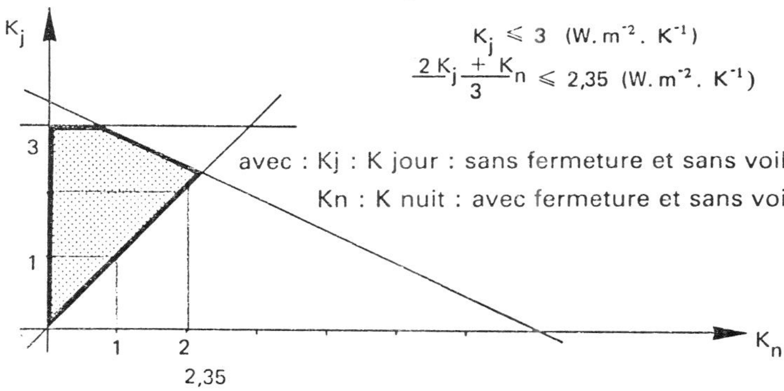

Figure I (Doc. Plan Construction)

Type de vitrage	Epaisseur de lame d'air	Coef. K W/m²/°C
Vitrage simple		5,7
Double vitrage	6 mm	3,4
	8 mm	3,2
	10 mm	3,1
	12 mm	3,0
Vitrage non émissif	8 mm	2,1

Tableau I (Doc. St-Gobain)

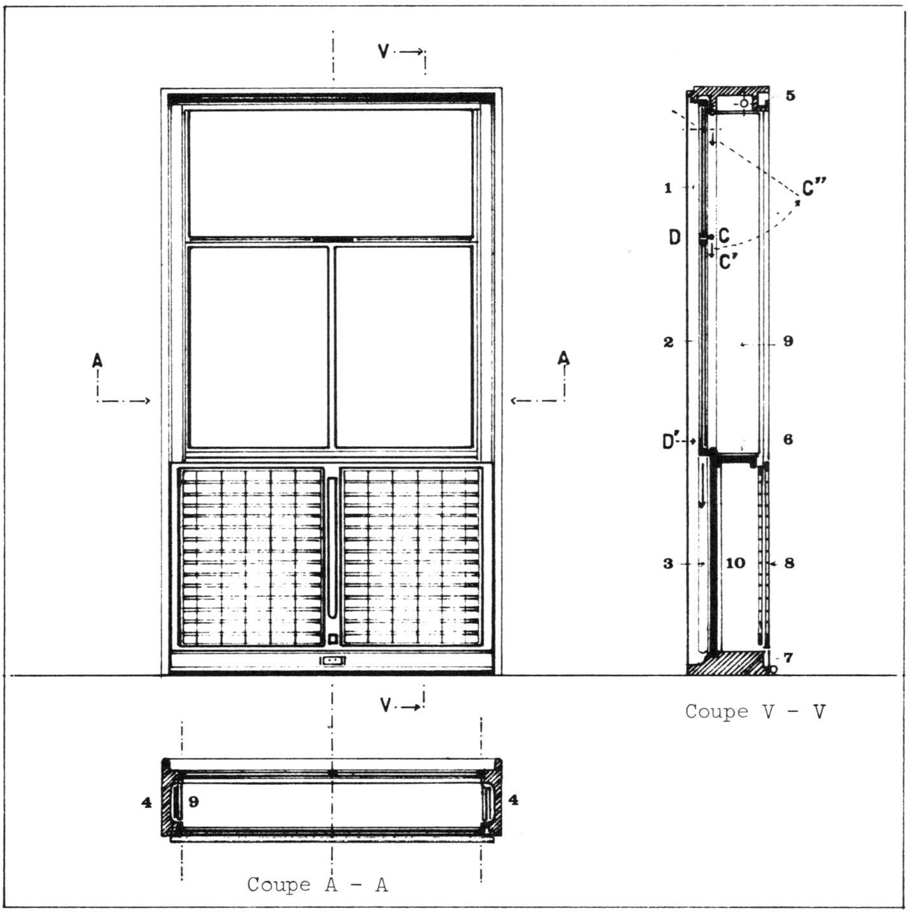

Figure II

TEMPORARY THERMAL INSULATION IN THE PASSIVE SOLAR UTILISATION

Authors : Dr. W. FUHSE, A. FUHSE

Contract number : ESA-PS-143-D (B)

Duration : 18 months 1 Jan 1982 - 30 Jun 1983

Total budget : DM 285.093,- CEC contribution : DM 130.000,-

Head of project : Dr. W. FUHSE

Contractor : FES Fuhse Energie System GmbH

Adress : FES Fuhse Energie System GmbH
Schulstraße 37
D-7997 IMMENSTAAD

SUMMARY

We are presently developing a rolladen- like strukture for the temporary thermal insulation for passive solar utilisation. The k- value of the thermal insulation device should be about 0.5 W/m^2°K. The design has to take account of the structure to withstand influences like wind, hale, dust, rain, solar radiation ranging from infrared to ultraviolett.

Other design criteria are: low maintenance costs, long life time, good architectural integration of the rolled up device, and easiness of operation.

The work programme covers conceptual work and design of the highly insulating roll- shutters, set up of a testing- and proving stand, performance of the heat transfer measurements, development of control gear, initiation of life tests and proving program.

Our early considerations have shown that the rolladen elements must be designed very carefully. We concentrated on "meander"- like boundaries on the top- and bottom- side of the elements in order to reduce the heat fluxes.

In a finite element computer code, we have evaluated the internal heat flows of the rolladen profiles in order to assess in a first order approach the effectiveness of the thermal insulation.

The considerations have shown that the thermal efficiency depends essentially on the shaping of the elements.

1. INTRODUCTION

The applicability of passive solar engineering in Europe depends essentially on the availability of means for temporary thermal insulation. There are, however, practically no products on the market, which yield a satisfactory thermal insulation; therefore we are presently developing a rolladen- like strukture, which should have a high thermal insulation value.

Our initial assessment showed that the k- value of the thermal insulation device should be about 0.5 $W/m^2 {}^oK$, but of course this design figure depends on basically three factors: the overall costs, the energy saving potential, and the acceptance by the market.

There are essentially three ways of including thermal insulation in the passive solar windows: The device can be put on the outside of the window; thereby the insulation has to withstand all influences of weather like wind, hale, rain, radiation ranging from infrared to ultraviolett; also dust, acidic rain and other effects have to be encountered in the design of the temporary thermal insulation.

The second way of fitting the thermal insulation is to put it on the inside; apparently all the difficulties, we have mentioned for the outside fitted device do not encounter for the inside thermal insulation; one can easily do with light structures, because they do not have to withstand extremes gusts or rain; on the other hand, these thermal insulation means, which have a high insulation value, tend to be quite voluminous; during day time, when the window should not be obscured, the thermal insulation requires a lot of valuable space inside the building. Furthermore the insulation will most likely be deposited near the window, which tends to reduce the area of the direct gain openings. In some cases like the clerestory window, the integration of internal insulation becomes very difficult.

The third method of integration temporary thermal insulation in the building is by putting the insulation inside the window itself. One intruiguing method has been implied in the airport terminal building in Aspen, Colorado, where polysterene powder is blown in the open space between the double glasing of the windows. Allthough the thermal insulation has been proven to be very good, the operating experiences have shown difficulties due to static loading of the foam and adhesion to the glass surfaces. Even the use of antistatic sprays did not turn out to yield the window areas to be unobscured by polysterene powder during daytime.

Allthough the method we have mentioned first (putting the insulation on the outside) apparently showed the greatest difficulties for the development, we believe that it will, in the long term future, be the most effective one.

2. WORK PROGRAMME

In principle, the work concerns the development of a rolladen with a low k- value and a low air- leakage rate. Since we also expect in the future some of the direct gain windows to be not vertical but tilted, even horizontally orientated, we would like the structure to be rigid enough to bear the snow forces etc. for a span of say 3 meters; however, this criterion is not an essential one, since the span can always be reduced to satisfy the stability requirements.

Other design criteria are: low maintenance costs, large life time, good architectural integration of the rolled up device and easiness of operation. For the latter one, we originally intended to prepare a fully automatic control device; however, since the development efford has to be small, we will only include a manual operation in our development work. This has also the advantage that hasards by automatic shut down of the rolladen have not to be included in this stage.

The work programme comprises the following items:

1. The conceptual work and design of the highly insulating roll- shutters including frames and seilings.

2. Set up of a testing- and proving stand.

3. Performance of the heat transfer measurements under various operating conditions.

4. Development of a (manual) control gear.

5. Initiation of life tests and proving program.

6. Evaluation of tests and improving of the design.

3. PRINCIPLE OF TEMPORARY THERMAL INSULATION

The following picture shows the basic principle of the rolladen- like strukture for the temporary thermal insulation in passive solar engineering. The essential feature of the construction is that the insulation thickness should be about 5 cm or even more. The elements of the rolladen have to be filled with an insulating material such as polyurethane- foam; it is not possible to use hollow elements alone, since the thermal convective flows inside the elements increase the heat conductivity to values similar to those of standard rolladen elements. We have, however, calculated that a honeycomb like internal structure can yield a

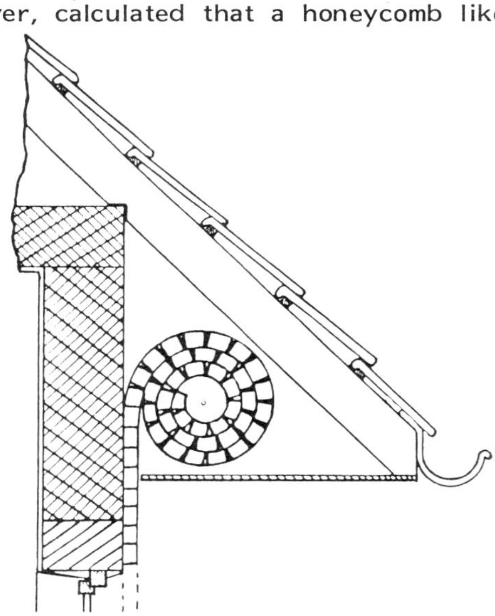

Basic principle of the temporary thermal insulation in the passive solar utilisation.

comparable insulation value, provided the characteristical dimensions of the honeycombs are about 3 mm.

The thickness of the rolladen elements leads to quite a voluminous roll; its diameter might be in the range of 50 to 70 cm depending on the height of the window and the element thickness. Such a big roll can hardly be integrated in the wall; therefore we assume that the rolled up unit is haused outside the wall; this gives the additional advantage that the thermal insulation of the wall is not disturbed above the window (it is well known that the usual rolladen housings are severe thermal leaks; even the best available rolladen housings have effective k-values of about 0.8 $W/m^2 {}^oK$, which is much higher than the k- value of a normal wall).

Our early considerations have shown that the rolladen elements must be designed very carefully. In particular the horizontal connections between the inside and the outside easily tend to be thermal leaks. Therefore we concentrated on "meander"- like boundaries on the top- and bottom- side of the elements in order to reduce the heat fluxes. This principle is indicated in the following drawing.

CROSS SECTIONAL VIEW
OF THE ROLLADEN ELEMENT.

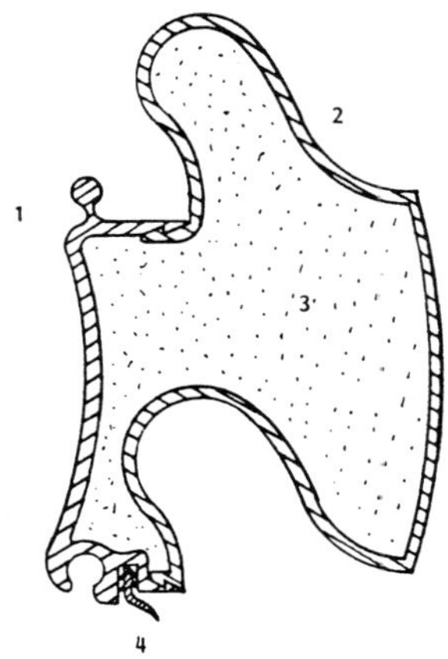

The element consists of basically four parts: 1. The outside facing aluminium extrusion, which should give the elements the strength to withstand the strains and stresses imposed by the weight of the rolladen and the external forces; 2. the inside PVC- extrusion, which protects the internal insulating material; 3. the insulation material, e.g. PUR- foam; 4. a seiling, which should avoid drafts through the rolladen and therfore lead to a small air exchange rate.

4. INVESTIGATIONS ON THE THERMAL BEHAVIOUR OF INSULATION ELEMENTS

The rolladen elements are expected to be by far the most cost effective parts of the whole structure. Therefore we have concentrated our development efforts since the start of the work in January this year on the element lay out.

In a finite element computer code, we have evaluated the internal heat flows of the rolladen profiles in order to assess in a first order approach the effectiveness of the thermal insulation. The calculations assume for simplicity reasons very simple boundary conditions as shown in the schematic drawing: the aluminium extrusions are assumed to be on a constant temperature level; similarly the rear side of the PVC - extrusion is isothermal; additionally the boundary conditions on the upper and lower side of the profile are such, as if the profiles were surrounded by an ideally insulated material.

BOUNDARY CONDITIONS FOR HEAT FLOW CALCULATIONS

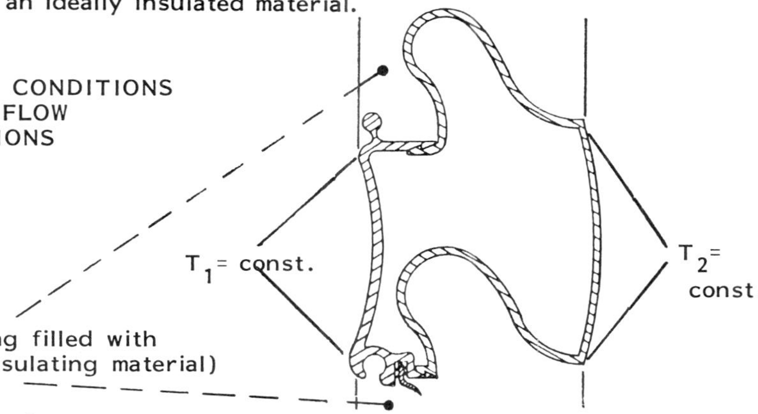

(surrounding filled with perfectly insulating material)

These assumptions are justified insofar, as we can expect the temperature variations of the inner- and outer side of the profiles to be very small as compared to the temperature variations in horizontal direction. On the other hand, the heat flux in vertical direction on the upper and lower side should be small compared to the radial flow (otherwise the profile would not be efficient anyway); this is best modelled by an surrounding material with a thermal conductancy of 0.

We can prove this latter assumption by comparing the local temperatures of the upper and lower side in the results of the computer calculations: if the temperatures on the upper side are equal to the temperatures of the corresponding points of the lower side, then - in fact - the vertical heat flows are small. All calculations, which we have carried out, show the temperatures to be quite similar, thus proving the validity of the assumption.

The following graphs present the results of the calculations for the base-profile and a profile, which does not have the "meander". The drawings show the lines of constant temperatures, whereby the difference between two adjacent lines is one tenth of the total temperature difference.

RESULTS OF HEAT FLOW CALCULATIONS ON ROLLADEN-ELEMENTS:

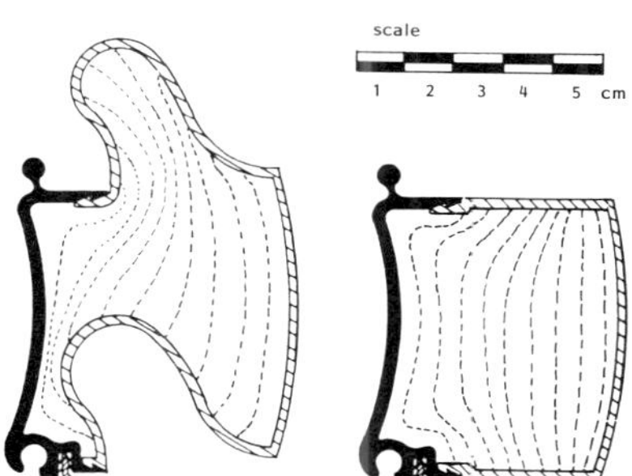

rolladen element with "meander"- type boundary
k= 0.62 W/m²°K

rolladen element with rectangular cross section
k= 0.86 W/m²°K

The drawing also contains the calculated effective k- value of the profiles.

Allthough the lines of constant temperatures show a more homogenous picture for the profile without meander, its k- value is larger by about 50 %.

This result is of utmost importance, since it indicates that profiles with meanders can be more expensive by, say, 30 % than the rectangular profiles. A greater cost difference, however, does not seem justified on the basis of these results.

However, the inhomogenity of the lines of constant temperature in the base profile indicate some room for improvement. It is expected that those profiles are advantageous, where the lines of constant temperature are about parallel to the inner- and outer surface. This can be accomplished by two methods: firstly - Giving the profile a roof- like upper- and lower surface, whereby the roof should be symmetrical and should have a large inclination; secondly - Making the heat-bridges on the upper and lower side as small as possible. The design, on the other hand, has to observe the other criteria as e.g. mechanical stability.

Therefore we have designed several profiles, which seemed us principally suitable for the purpose. The selection of the preferred profile was conducted on the grounds of foreseeable costs, technical feasibility and thermal conductancy. Computer runs have been performed on various profiles, whereby the best profile has a k- value some 30 % smaller than our base profile. This improvement is basically due to the different shaping of the meander like upper and lower part. Since these results have only been obtained recently (and work is still under way), they are not included in this paper.

5. OUTLOOK FOR FURTHER WORK

Despite the continuation of work on the evaluation of profile properties, the work for the second half of this year will concentrate on the following aspects:

1. Fabrication of a test sample of profile elements
2. Set up of a wholly assembled structure (incl. mounting); mechanical tests
3. Considerations on architectural integration
4. Cost evaluation and identification of cost sensitive parts
5. Set up of a test- and qualification stand

The initiation of heat conductivity measurements and life tests is scheduled for the end of this year.

6. CONCLUSION

The considerations described in this paper have shown that the thermal efficiency of rolladen like means for temporary thermal insulation essentailly depends on the shape and the structure of the rolladen elements. A meander shaping of the elements can drastically increase the insulation effect.

DEVELOPMENT OF AN INDEPENDENT SHUTTER SYSTEM FOR PASSIVE TEMPERATURE CONTROL IN BUILDINGS

Authors : J. Schmid, H.-P. Preuß

Contract number : ESA-PS-145-D

Duration : 18 months, 1 January 1982 - 30 June 1983

Total budget : DM 395.016,CEC contribution: 50%

Head of project : Prof. Dr. A. Goetzberger/Dr. J. Schmid

Contractor : Fraunhofer-Institut für Solare Energiesysteme (ISE)

Address : Oltmannsstraße 22
D-7800 Freiburg

Summary

This is a first report on the current activities of ISE concerning the development of self-contained automatic shutter systems that influence the radiant flux through windows.

Combining the principle of roller shades with the spectral filtering nature of various coated films, a dc drive with solar-electric power supply, controlled by an electronic circuit, provides the actually appropriate type of film into action. Details of purpose, principle, function and efficiency of the new universal element for passive temperature control are presented.

1. PROBLEM FORMULATION

1.1 Heat losses and solar radiation gain through windows

It is generally known that windows are responsible for a considerable part of the heat losses in buildings. In the case of a single one-family house about 20 to 30% of the total heating load depending on the insulation quality is due to the windows (1). For example the heat transfer coefficient k (W/m^2K), representing an index to quantify thermal losses through parts of a building, becomes $k \approx 5.8$ for single glazed windows, $k \approx 3$ for double glazed ones and $k \approx 1.8$ for high-quality infrared reflecting windows, respectively - still several times greater than the k-values of walls and roofs.

Thus having been discredited as thermal leakages windows came into the centre of a discussion, at times contradictory, when the energy problem extended to a global crisis (2). As a result it turned out that windows are not only energy sinks but that they also can act as cheap and efficient solar collectors due to their transparent nature relative to the solar spectrum (3 - 6), if the energy flux is controlled in a correct manner. The passive solar radiative gain is described by the total energy transmission index g (3, 4). Considering an arriving radiant flux (W/m^2) the gain term gI reduces the thermal flux $Q = k(T_i - T_a) - gI$
through a window, which is normally a function of k and the room and outdoor temperatures T_i, T_a, respectively. The graph of the above equation shown in Fig. 1, leads to the noteworthy result that a double glazed window stops causing heat losses ($Q \geq 0$) as soon as the solar radiation reaches approximately 100 W/m^2 - even under average winter temperatures.

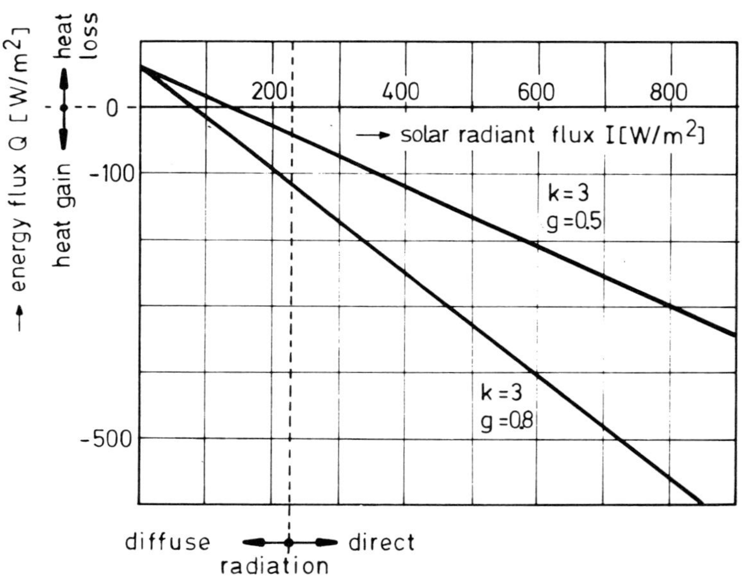

Fig. 1: Energy flux through a double glazed window with k = 3 and g = 0.8 (g = 0.5) in dependence on solar radiation for $T_i - T_a = 20$ K (with reference to (7)).

However large windows alone won't do. On the contrary they often need a temporary sun shield, maybe even in winter. Moreover one has to provide for an additional thermal insulation at night (I = 0), in order to reduce the increasing losses $Q = k(T_i - T_a)$ as much as possible. The role of radiation is of special importance and often underestimated (1). This is why heating costs could be minimized if the energy flux through window glazing is automatically controlled.

1.2 Conventional types of energy conserving windows

The traditional method to conserve heating energy at night is to close the shutters. If they are open during the day there is a maximal utilization of the sun's radiant energy, whereas when closed, they can ensure maximal thermal insulation at night. Conventional shutters are not optimized for radiation reflection and normally they are manipulated by hand, which is an unpleasantness for the occupants. Above all, this is certainly a non-optimal operation, because the shutters are probably not regularly closed every nicht. Although this drawback can be overcome by electrically driven shutters, a power supply line is necessary for each single window, which is expensive or nearly impossible in existing buildings.

Whereas the possibility of reducing the k-values of windows by installing multiple-glazing is limited because of various practical reasons the application of sputtered infrared reflecting heat mirror coatings or self-adhesive metallized polyester-films is said to be a highly effective method of influencing radiative loss through window glazing. Such types of coatings indeed cut the heating consumption or reduce the danger of solar overheating. However as a direct consequence they also impair the desired cooling effect of the thermal emission through windows during summer nights and they restrict the solar thermal gain in winter. Evidently the mentioned devices are mainly permanently mounted with no change in their transmittance properties possible.

A decisive improvement can be achieved with the help of an automatic multi-functional shutter system as described in the following.

2. MULTI-PURPOSE SHUTTER SYSTEM WITH PHOTOVOLTAIC POWER SUPPLY AND AUTOMATIC OPERATION

2.1 Principles of influencing radiant flux

The major attention of the FhG-shutter system research and development is directed to the automatic control of the radiant portion of the energy flowing through the window. This means selecting film materials with different spectral transparency and reflectance properties with regard to the visible (0.38 - 0.77μm) and the near-infrared (0.77 - 2μm) spectrum coming from the sun and the far-infrared (2 - 100μm) thermal blackbody radiation out of the warm room. The filtering of a specified range of the spectrum is favoured by the significant separation between the solar and the blackbody spectrum shown in Fig. 2.

2.2 Objectives and general function description

Shutter elements have been selected which support heating and air-conditioning in an optimal manner by influencing the energy flux through windows under all climatic conditions and which are self-contained because of a power supply based on solar cells.

There are three primary goals:

- Improvement of the classic shutter for thermal insulation and radiant reflectance.

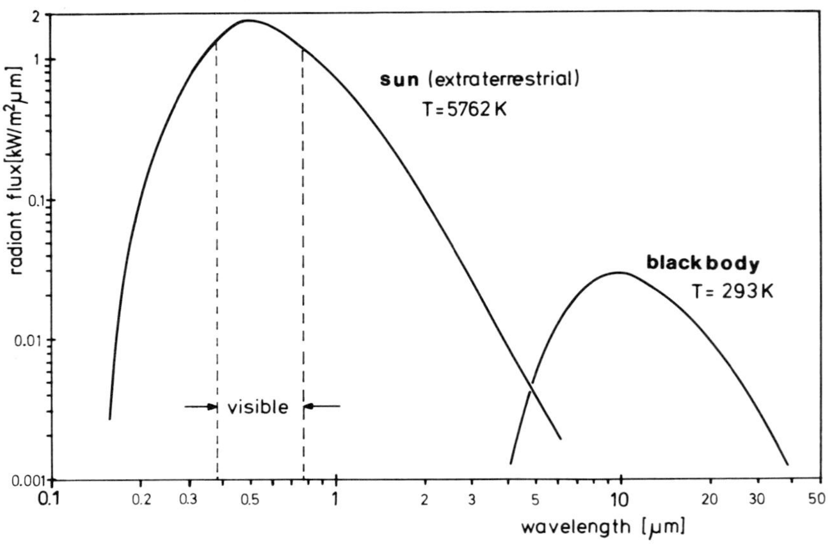

Fig. 2: Solar radiant spectrum and blackbody spectrum

- Development of an electric dc drive with an independent power supply consisting of rechargeable NiCd batteries, solar cells and a voltage regulation unit. An intelligent electronic controller, if necessary including a time reference module together with sensors should guarantee optimum operation according to the actual requirements. Manual operation remains possible.
- Development of a flexible multi-film roller shade with self-adjusting transparency according to different heat and radiation requirements in summer and winter, at day and night, which is equipped with a photovoltaic power supply.

The operation and effect of such a system will be described next. First the intended mode of action is demonstrated guided by the 4 typical scenarios in Fig. 3.

Winter day: maximal utilization of the total solar energy, but reflection of the thermal infrared back into the room,
winter night: complete reflection of all radiant portions improves insulation and ensures optical opacity,
summer day: reflection of all solar infrared radiation reduces the expenditure of active air-conditioning, at the same time daylight is transmitted,
summer night: emission of the far-infrared energy towards the cooler night atmosphere, but no transmission of visible light.

To meet all these requirements one has to design a multi-film roller shade analogous to the destination displays on publish transport vehicles, consisting of 2 rollers and a width of film between. Due to the 4 climatic scenarios one width has 4 different sections with distinct transparency properties. The basic principle of action is shown in Fig. 4. To move the desired section into position, the built-in electronic device is controlling the dc gear motor. To keep a certain tension over the film, a torsional spring is integrated into the second roller.

As for the power supply solar cells with rechargeable storage batteries are ideally suited because of the very low requirement for energy. The final result is an universal roller shade, that represents a compact and autonomous unit.

2 versions of mounting are sketched in Fig. 4.

2.3 Advantages of the new device

The following arrangement of points reflects some outstanding attributes of the new device:

- reduction of the costs for heating and cooling,
- better efficiency than single-film shades,
- automatic operation, optimal utilization of solar and heating energy and high comfort of living,
- manifold control criteria possible, also adjustable to manual operation,
- autonomous photovoltaic power supply,
- all components are integrated into the case,
- low voltages, no large-scale safety measures necessary,
- well suited for mass production.

The possibility of influencing energy flux in both directions (both heating and cooling effect) makes the new multi-film roller shade a fundamental element of passive interior climate control. Not only windows but also greenhouses in particular could profit by solar roller systems such as the one presented.

2.4 Efficiency and economical aspects

Although high efficacy of the multi-film roller shade may be expected the economical efficiency is still not quite clear, for the price is closely connected with the number of pieces in the case of a series production.

Nevertheless a rather rough estimation reveals, considering a single one-family house, a reduction of annual energy costs up to DM 800. Taking an average price of DM 500 per window an investment of DM 5.000 for 10 windows would become profitable after 6 years. This is without considering the increased comfort of living! It is worth mentioning that a hand-driven version could reach the break-even point much faster.

3. STATUS OF THE WORK

Various development activities have been started. The first effort was the integration of an appropriate dc drive into the case of a classic shutter and the dimensioning of the solar electricity supply. Guided by a minicomputer (Apple II) and a data acquisition and control unit (HP 3497A) several situations can be simulated and control criteria can be tested (Fig. 5), in order to enable the design of ideal electronic control circuitry.

At the same time several versions of multi-film roller shades were constructed. The prototypes are in the initial stages.

Optical measurements of most commercially available films have been carried out.

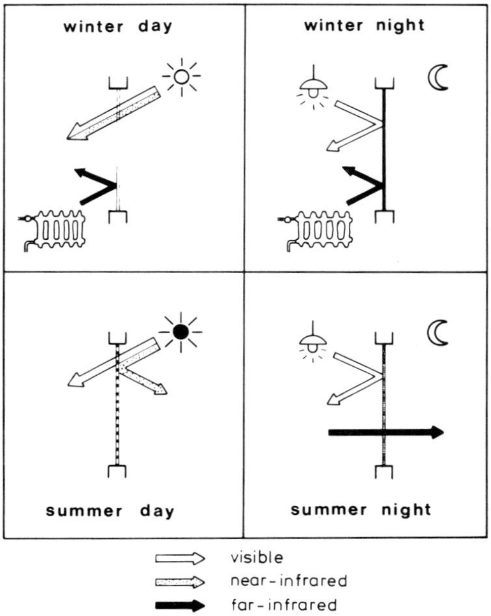

Fig. 3: 4 scenarios to demonstrate the effect of a multi-film roller-shade

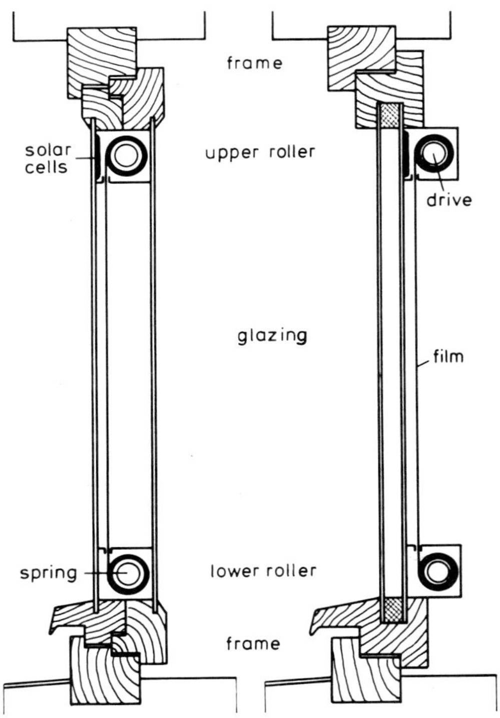

Fig. 4: 2 versions of mounting the roller system to windows (sectional view)

Fig. 5: Operator's position for the computer-aided control of a shutter

4. FUTURE PROGRAMME

A wealth of problems still needs solution. After having finished with the mechanical parts of the prototypes they have to be combined with the drive and the solar power supply. Extensive functional tests have to be made. New types of coatings to be developed have to be identified in co-operation with special manufacturers.

The most time will be necessary for the realization of the complete controller. The applicability of the fluorescent collector (8) shall be examined. While one prototype is to work as a long-time test another one shall be used to determine the efficiency under realistic conditions. This seems to be the only way to get reliable performance data concerning function, stability and, last but not least, the amount of energy actually saved.

The serious interest of the industry is an encouragement to intensify the investigations.

References

(1) H. Hörster: Wege zum energiesparenden Wohnhaus. Philips, Hamburg 1982.

(2) K. Gertis, G. Hauser: Energieeinsparung infolge Sonneneinstrahlung durch Fenster. KI 7(1979), 3/107 - 111.

(3) H. Künzel: Das Fenster als Sonnenkollektor. VDI-Berichte Nr. 316, 1978.

(4) H. Künzel, C. Snatzke: Wärmeverlust und Wärmegewinn durch Fenster. Glasforum 1 (1979).

(5) E. H. Krüger, U. Schuh: Keine Angst vor großen Fenstern. Sonnenenergie 4 (1979), 1/25 - 26.

(6) H. Werner: Wärmedurchgang durch Fenster und Wand unter Berücksichtigung der Sonneneinstrahlung. Haustechnik - Bauphysik - Umwelttechnik 3 (1981), 121 - 126.

(7) H. Künzel, C. Snatzke: Das Fenster als Sonnenkollektor. IBP-Mitteilung 7, Fraunhofer-Institut für Bauphysik, Stuttgart 1979.

(8) A. Goetzberger, W. Greubel: Solar energy conversion with fluorescent collectors. Appl. Phys. 14 (1977), 123 - 139.

DEVELOPMENT OF A THERMAL STORE AND A SOLAR BLIND FOR USE IN CONSERVATORIES AND GLAZED ROOF SPACES

Authors	:	D. STEWART - Fulmer Research Institute Limited I. MEAD - Calor Group Limited D.M.TATLOW - Perma Blinds Limited

Contract Number	:	ESA-PS-147-UK
Duration	:	18 months 1 January 1981 to 30 June 1983
Total Budget	:	£ 81 648 CEC Contribution: £ 30 000 (37%)

Head of project	:	Duncan Stewart, Fulmer Research Institute Ltd.
Contractor	:	Fulmer Research Laboratories Ltd.
Address	:	Hollybush Hill, Stoke Poges, Slough, SL2 4QD, Buckinghamshire, U.K.

Summary

A programme to evaluate the performance of a glazed roof-space fitted with a solar blind and a latent heat store is described. The latent heat store is manufactured from phase change material, PCM, contained in individual tubes 39 m.m. diameter. The solar blind is developed from established current roller-blind technology. The test programme will be carried out in the Calor solar laboratory and performance monitored over a full heating season.

1. Introduction

The current trend in house building to produce low energy houses has been clearly demonstrated at the U.K. Home World Exhibition (May 1981) at Milton Keynes. A particular feature with many designs was the use of glazed apertures, with patent glazing systems. These glazed apertures demonstrated the desire of architects to increase the passive solar gain of housing as well as adding to the architectural features.

The solar gain achieved by using these designs however can be significantly increased if a solar blind is used to prevent the escape of energy (the thermal diode or intelligent window principle). Estimates in the amount of energy saved vary according to the design of the house and the operation of the thermal diode. One estimate (1) for the energy saving of a poorly insulated house in the UK using a single glazed collector amounts to 180/kWhr/year/m^2.

A further saving in energy can be provided by adding thermal storage materials to the glazed roof aperture. A phase change material (PCM) store can be positioned directly in a glazed aperture for new house construction or as a retrofit package to existing houses. The contribution to energy savings made by positioning the PCM in the glazed apertures has been modelled by Carter (2) and Lee and Oberdick (3). Carter concludes that PCM with a high phase change temperature can overheat a building and hence may increase the heat loss. According to Carter, materials with a phase change temperature close to the desired temperature can provide 10 days complete heat in a superinsulated building and 4 to 5 days partial heat in a standard building. The use of PCM in this application however is for purely passive heating without fan assistance to circulate the air. By using fan assistance higher temperature PCM stores can be used more efficiently when combined with a solar blind.

The pro ramme to be described will combine the use of a solar blind with a PCM store in a glazed roof space to determine the benefits that are likely to be achieved.

2. Aims of the project

The aims of the project are to determine:

1. the total useable energy gain using a closed glazed roof-space collector.

2. the total useable energy gain using a closed glazed roof-space collector with a PCM store and an automatic roller blind.

3. the total useable energy gain using a closed glazed roof-space collector with PCM store but without a roller blind.

4. the effect of roof pitch using the above test conditions (1 to 3).

5. the effect of using a vertical glazed conservatory window and the above test conditions.

3. The test programme

The test programme is designed to produce comparative data using the Calor laboratory at Fulmer, 30 km west of London (51°33'N 0°34'W). This laboratory is purpose built with a steel frame and contains a glass/patent glazing bar roof, with 3 different pitches, 20, 30 and 40° to the horizontal. Also, the south facing wall is fitted with glass. The laboratory is shown in figure 1 and the position of the test walls in the laboratory in

Figure 1. Calor Solar Laboratory at Fulmer.

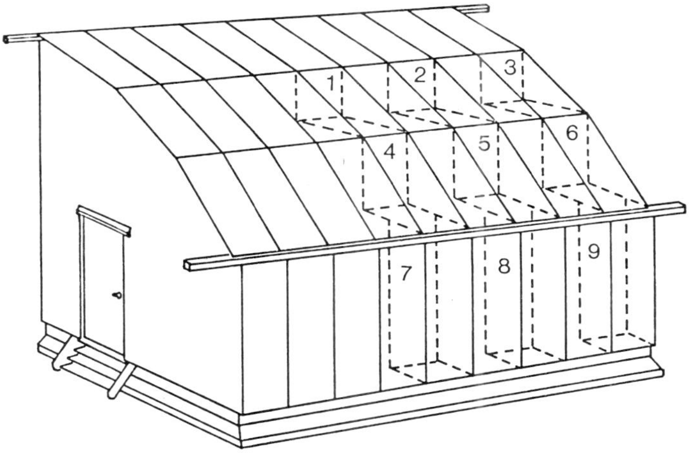

Figure 2. Solar laboratory showing position of individual test cells.

figure 2. These test cells will be fitted with various combinations of blind and PCM store at the roof pitches shown in figure 2. The design of an individual test cell is shown in figure 3.

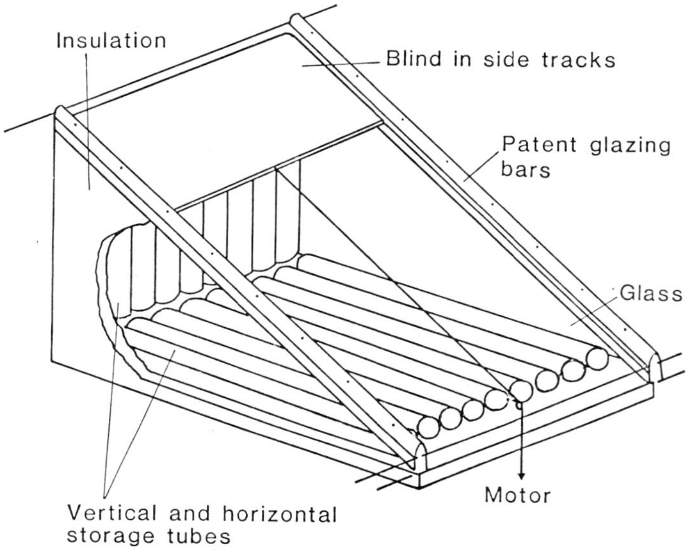

Figure 3. Lay-out of test cell showing position of PCM storage tubes and solar blind.

The energy collected in each cell will be extracted using an air fan and the air temperature increase and flow rate monitored.

Temperature control will be by a matched pair of thermistors adjusted to a preset level of ΔT and hysteresis level.

The programme allows for the monitoring to be made during one heating season.

3.1. PCM Materials

The PCM materials used for the programme will be based upon the stabilised Glauber's Salt ($Na_2SO_4 10H_2O$) having a melting temperature of approximately 32°C. This salt is stabilised using an acrylamide polymer to prevent separation during temperature cycling (4,5). The salt is contained in plastic tubing of various lengths. The equipment for filling the tubes is shown in figure 4.

Initial experiments have been made using the 53 m.m. diameter 1200 m.m long, PCM tube to establish an optimum air flow rate across the tube. These initial experiments have shown that the heat transfer, radially through the tube, is too slow for the current application. Black plastic tubes 39 m.m. (1½ inches) outside diameter with a wall thickness of 1 m.m. are now being evaluated to determine the air flow/heat transfer characteristics. These smaller tubes have been improved by adding 2% carbon black to the copolymer (polyethylene-polypropylene) to increase their solar absorption.

Figure 4. Mixing and injection equipment for filling storage tubes with P.C.M.

Figure 5. PCM storage tubes prepared prior to temperature cycling.

Temperature cycling of the PCM is continuing on full scale tubes 1200 m.m. long mounted in a water circulating system to speed up the test sequence. These are shown in figure 5. The circulating system to temperature-cycle the tubes is shown in figure 6. 230 charge/discharge cycles have been completed prior to Jan '82.

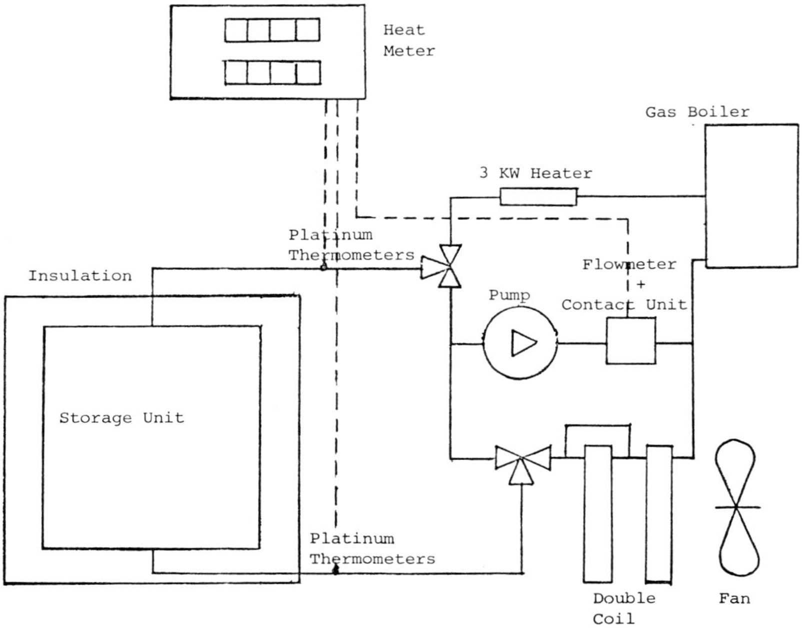

Figure 6. Temperature cycling rig for testing of PCM storage tubes.

One complete charge-discharge cycle is accomplished every 24 hours between 19 and 65°C.

Alternative PCM materials having a melting temperature approximately 25°C may appear to be more suitable for the programme proposed but the long term stability and reliability of these materials has not yet been established.

3.2. Solar Blind

The use of a solar blind to act as a 'shutter' or an intelligent window appears attractive; the blind would open when the insolation was high and close to prevent heat loss when the insolation was low particularly at night. However, there are a number of problems that need to be considered to ensure that the long term reliability of the roller blind is maintained. These include:

- the number of times the blind will be asked to operate,
- the extremes of temperature that the blind will encounter and any long term creep that may occur to the blind material,

- the environmental factors, such as condensation from the glass, interfering with the operation of the mechanism,
- the ingress of dirt or insects into the rolled fabric causing non-uniform rolling.

These problems can be reduced if the blind is not controlled by the insolation arriving on the glazed roof-space but only twice per day; at dawn and dusk. Some loss in total energy gain will result when operating in this mode but the long term reliability of the blind will be improved.

The materials to be used for the blinds are being evaluated to ascertain if they can be rolled repeatedly and uniformly. These properties are as equally important to the commercial companies as their thermal transmittance properties. Guides to seal the edges of the blinds where they meet the glazing bars will reduce convection losses from the test cells. The winding mechanisms for operating the blinds are standard types already developed for other applications. Longer term reliability can be achieved by using these standard, proven components wherever possible.

4. Future Work

The immediate work programme is to complete the construction of the test cells and commence the data acquisition since some delays have occurred due to the change in the dimensions of the PCM tubes and rolling problems associated with the solar blinds.

References

1. Passive Solar Housing in the UK, published by Energy Conscious Design, Earlham Street, London, 1980.

2. Phase Change Storage in Passive Solar House Heating, C. Carter, Solar World Forum, Brighton, England, August 1981.

3. Development of a Passive Solar Simulation Technique Using Small-Scale Models, K.S. Lee, W.A. Oberdick, Solar World Forum, Brighton, England, August 1981.

4. European Patents No. 99 and 793 023 70-6.

5. Thermal Storage Materials and Components for Solar Heating, J.K.R. Page, R.E.H. Swayne, I.K. Mead, C. Hayman, Solar World Forum, Brighton, England, August 1981.

DEVELOPMENT AND TESTING OF ARCHITECTURAL COMPONENTS FOR PASSIVE SOLAR ENERGY USE IN TRANSPARENT BUILDING FACADES

Authors	:	Dr. R. Baumann, Prof. Dr.Th. Herzog, Dr. W.Jaensch
Contract number	:	ESA-PS-149-D (B)
Duration	:	18 months (1.1.1982 - 30.6.1983)
Total budget	:	DM 231.800,-- CEC contribution: DM 115.900,--
Head of project	:	Prof. Dr. Th. Herzog, Gesamthochschule Kassel Universität
Contractor	:	Gesamthochschule Kassel, Fachbereich 12 - Architektur
Address	:	Gesamthochschule Kassel, Fachbereich 12 - Architektur D-3500 Kassel, Menzelstraße 13

Summary:

The goal of this research project is the development of interior-situated, translucent moveable elements in the area of transparent building openings, which are applicable as measures for temporary insulation as well as for protection from the sun, and achieve the required effects simultaneously.
The development project is concentrated on two main points - the development of an efficient support and transport system for the incorporation of lamellae and the development of the lamellae themselves.
Every component system should be perfected according to respective requirements. With changing requirements, it should be possible to either exchange the lamellae or alter their structure.
A product of this type should be achieved which is capable of incorporating new developments over a long time period, and which can accommodate changing demands.
It should be so adaptable that it allows for broad areas of possible application-installation on already existing structures as well as integration into new planning.

Problemstellung

Bei transparenten Gebäudeoberflächen sind in der Regel die Transmissionswärmeverluste im Winter wesentlich größer als bei nichttransparenten.
Durch temporäre Wärmeschutzmaßnahmen kann der Wärmeverlust wesentlich reduziert werden.
Konventionelle Wärmeschutzsysteme haben aber häufig den Nachteil, daß sie weitgehend verdunkeln und Sichtbeziehungen zwischen Innenraum und Außenraum unterbinden, das Gebäude wird zum Außenraum hin isoliert (auch psychologischer Effekt).
Eine Benutzung bei Tag ist in Aufenthaltsräumen ohne künstliche Beleuchtung kaum möglich.
Geschlossene Wärmeschutzsysteme verhindern auch die somit beträchtlichen Wärmegewinne durch transparente Flächen.
Konstruktionen des Sonnenschutzes sind bislang nicht für den Zweck des temporären Wärmeschutzes optimiert, sondern dienen primär der Verschattung (Beispiel Jalousetten, Markisen), Wärmeschutzfunktionen erfüllen sie nur am Rande.
Starre Sonnenschutzelemente reduzieren das Tageslichtangebot in den Räumen und sind nicht für beliebige Orientierungen der Gebäude und verschiedene Neigungen der Gebäudeoberflächen gleichermaßen geeignet.
Sollen Räume vor übermäßiger Sonneneinstrahlung geschützt werden, haben sich bewegliche, den jeweiligen Erfordernissen anpaßbare Systeme als die am geeignetsten erwiesen.

Allgemeine Zielvorstellungen

Derzeit wird in allen westlichen Industrieländern an Lösungen für den temporären Wärmeschutz gearbeitet. Gut funktionierende, kostengünstige und universell einsetzbare Systeme fehlen jedoch bislang.
Ziel dieses hier vorzustellenden Forschungsvorhabens ist die Entwicklung innenliegender beweglicher Elemente im Bereich transparenter Gebäudeöffnungen, welche sowohl als Maßnahme zum temporären Wärmeschutz als auch zum Sonnenschutz geeignet sind und die erforderlichen Wirkungen gleichzeitig erbringen.
Die Elemente sollten es ermöglichen, den Eintritt der Sonnenenergie durch transparente Gebäudeöffnungen entsprechend den jeweiligen Anforderungen, unabhängig von der Orientierung und der Neigung der Gebäudeoberflächen, zu kontrollieren und gleichzeitig in der kalten Jahreszeit den Verlust von

thermischer Energie während eines möglichst großen Zeitraumes erheblich zu reduzieren.
Es soll eine technisch und gestalterisch anspruchsvolle kostengünstige Lösung gefunden werden, die beide Qualitäten integriert.
Das Element sollte so anpassungsfähig sein, daß sich breitgefächerte Anwendungsmöglichkeiten sowohl bei der Nachinstallation in vorhandene Bausubstanz als auch bei der Integration in Neuplanungen ergeben.
Die Elemente sollten geeignet sein, die Effizienz bekannter Systeme der passiven Solarenergienutzung zu steigern bzw. neue Möglichkeiten zu eröffnen. Die Motivation zur Anwendung derartiger Elemente sollte durch große Benutzerfreundlichkeit gegenüber herkömmlichen Elementen gesteigert werden.
Die Elemente sollten auf der Rauminnenseite angebracht werden, gegenüber außenseitiger Anordnung ergeben sich dabei folgende Vorteile:

- geringer Verschleiß
- kleinere Korrosionsgefahr
- keine Störanfälligkeit durch Bewitterung
- keine Windgeräusche
- einfache Wartung
- einfachere technische Ausführung in Material und Detail
- Bedienung von innen
- das Aussehen der Gebäude wird nicht so stark verändert.

Die vorgeschlagenen Komponenten sollten sich speziell für mitteleuropäische Klimabedingungen eignen.
Es wird angestrebt, durch die Art der Materialwahl und der Konstruktion, die Herstellung vielerorts - mit geringem technischen bzw. Invenstitionsaufwand - zu ermöglichen. Ebenfalls sollte Massenproduktion möglich sein. Um Chancen für den Export zu eröffnen, sollte davon ausgegangen werden, daß der Transport mit geringem Verpackungsvolumen und möglichst geringem Gewicht möglich sein muß.

Zum geplanten Arbeitsablauf

Die Entwicklungsarbeit konzentriert sich auf zwei Schwerpunkte, auf die Entwicklung eines leistungsfähigen Trag- und Bewegungssystems für die Aufnahme und Bewegung von Lamellen und auf die Entwicklung der Lamellen selbst. Jedes Teilsystem sollte entsprechend dem jeweiligen Anforderungsbereich (z.B. Nutzung, Betrieb, gestalterische Anforderungen usw.) optimiert werden können.

In einem dritten Arbeitsabschnitt soll die Wirkung des Systems in einer speziellen Meßstation überprüft werden.

Zur Entwicklung der Lamellen

Hierbei werden einmal der mögliche konstruktive Aufbau und die Anschlußbedingungen an das Bewegungssystem untersucht, zum anderen werden mögliche Materialien nach ihrer Verwendbarkeit überprüft.
Um die bauphysikalischen und energietechnisch relevanten Kenngrößen zumindest überschlägig zu ermitteln, wird in Anlehnung an DIN 4108 und DIN 67507 eine Meßanlage für Zwillingsmessungen gebaut, in welcher die für eine wärmetechnische Beurteilung der möglichen Materialien wesentliche bauphysikalische Größe ermittelt und gegenübergestellt werden (z.B. Abminderungsfaktor für Sonnenschutzvorrichtungen, Gesamtenergiedurchlaßgrad, Wärmedurchgangskoeffizient usw.).

- Verschiedene Materialien und Materialkombinationen werden im Hinblick auf die angestrebte Wirkung vergleichend beurteilt.
- Materialien und Materialkombinationen werden danach auf unterschiedliche Anwendungsbereiche hin optimiert.

Ausgehend von einer konstruktiven Grundlösung sollten in Aufbau und Material unterschiedliche Lamellen entsprechend den jeweiligen Anforderungen alternativ eingesetzt werden können.
Die verwendeten Baumaterialien für die Lamellen sollten eine hohe Lichtdurchlässigkeit aufweisen, wodurch bei Tage die Lichtverteilung und Intensität im Rauminneren nicht wesentlich verändert wird, so daß die Wirkung als Wärmeschutzsystem über große Zeiträume des Tages hinweg zum Tragen kommen kann. Direkte Strahlungsanteile sollten beim Durchgang durch das System in diffuse umgewandelt werden.
Das Tageslichtangebot sollte auch in geschlossenem Zustand weitgehend erhalten bleiben. Energiegewinn durch Einstrahlung verbessert den äquivalenten Wärmedurchgangskoeffizienten "$k_{äq}$".

Zum Test der entwickelten Konstruktion an einer Meßstation unter weitgehend realen Nutzungsbedingungen

An einer Anlage für Zwillingsmessungen können die für das thermische Verhalten von Gebäuden oder Bauteilen wesentlichen Strahlungs- und Temperaturübertragungsvorgänge erfaßt werden.
Die Versuchsanlage besteht aus zwei Raumeinheiten mit je einer Grundfläche

von 3,60 m x 4,20 m und einer Höhe von 3,00 m, welche in einem Abstand von 3,60 m voneinander angeordnet und mit ihrer Längsachse nach Süd-West ausgerichtet sind. Eine gegenseitige Verschattung der Gebäude ist somit weitgehend ausgeschlossen. Innerhalb dieser beiden Raumeinheiten ist jeweils eine Innenzelle mit einer Grundfläche von 2,40 x 3,00 m und einer Höhe von 2,40 m so eingebaut, daß sie die Süd-West-Fassaden mit den Außenzellen gemeinsam haben. Die Seitenwände und Decken stehen zu den Außenzellen auf 60 bzw. 120 cm Abstand und bilden somit eine thermische Pufferzone.
Anhand einer Versuchsreihe soll geklärt werden, welchen Einfluß die Verschattung einer Pufferzone (Wintergarten, Anlehngewächshaus o.ä.) auf das thermische Verhalten von Gebäuden in den verschiedenen Jahreszeiten hat.
Zu diesem Zwecke wurden die beiden Testräume mit vierflügeligen Isolierglasfenstern versehen.
Vor einen der Räume wurde zusätzlich ein Wintergarten gebaut. Mit Hilfe der nachstehend beschriebenen Meßanordnung können die wesentlichen, für eine Beurteilung notwendigen Strahlungs- und Temperaturverläufe, registriert werden.

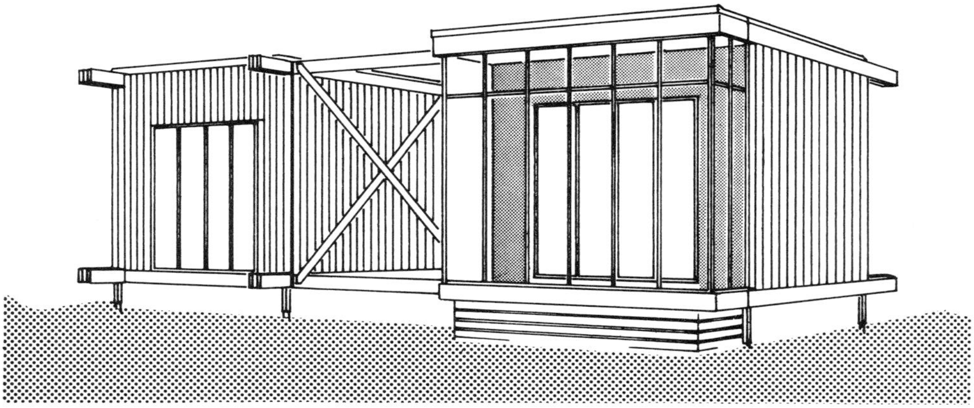

Darstellung der einzelnen Untersuchungsphasen:

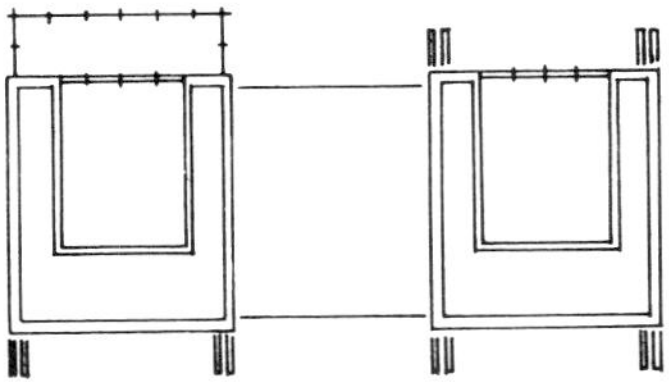

1

Temperaturverläufe während der Heizperiode ohne Heizmaßnahmen

Das System wird sich dabei vollkommen selbst überlassen. Aufheiz- und Abkühlungsvorgänge der beiden Räume werden verglichen

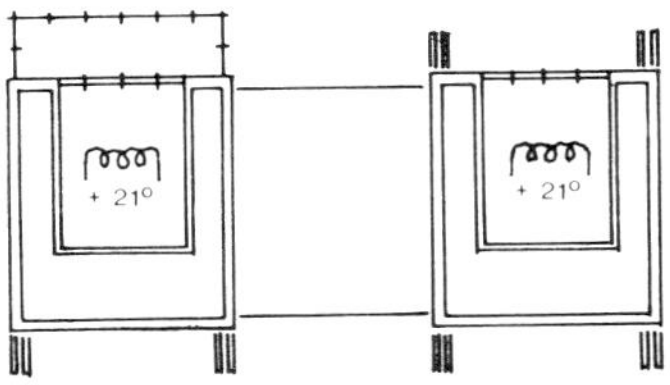

2

Temperaturverläufe während der Heizperiode mit Heizung

Beide Räume werden auf gleichem Temperaturniveau (21° C) gehalten und der dazu notwendige Energiebedarf gemessen (Winter 81/82)

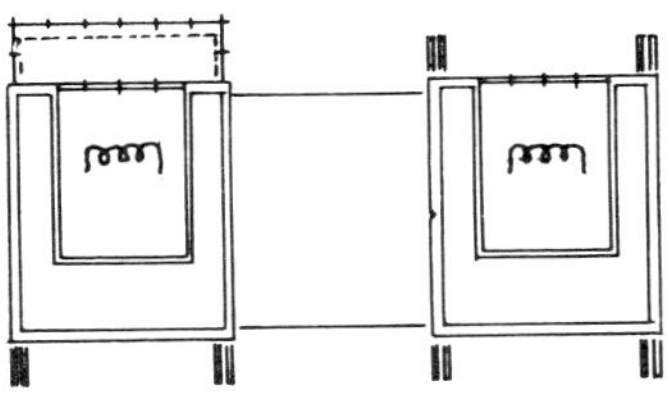

3

Temperaturverlauf während der Heizperiode mit temporärem Wärmeschutz

In dieser Phase werden die entwickelten Sonnen-Wärmeschutzelemente angebracht und mit Phase 2 verglichen (Winter 82/83).

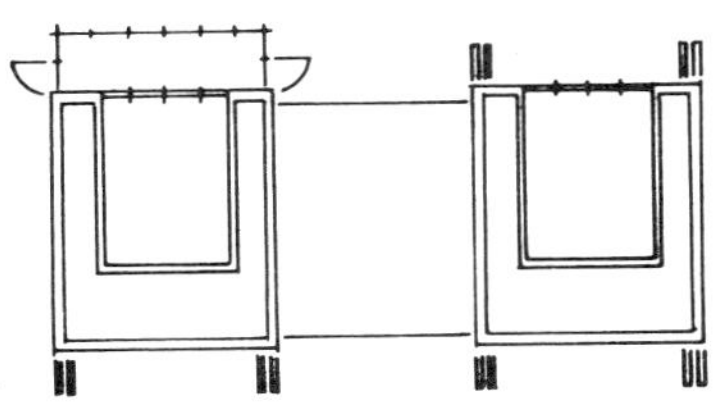

4

Temperaturverlauf im Frühjahr und Sommer ohne Pflanzen - mit und ohne Lüftung

Hierbei wird untersucht wieweit durch Querlüftung übermäßiger Temperaturanstieg in der Pufferzone verhindert werden kann.

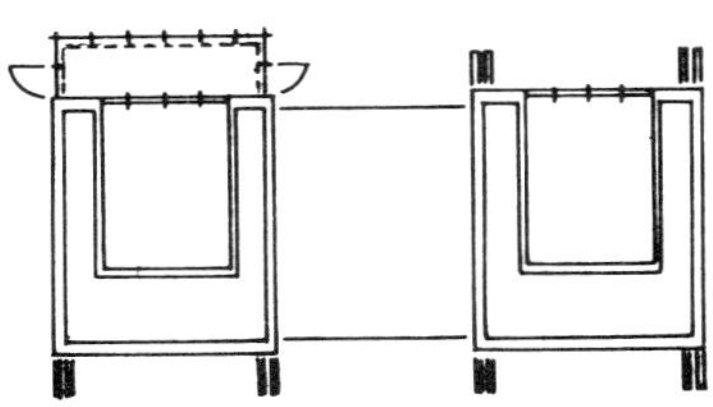

5

Temperaturverlauf im Frühjahr und Sommer mit Sonnenschutz

In dieser Phase wird die Wirksamkeit des entwickelten Systems unter sommerlichen Bedingungen untersucht und mit Phase 4 verglichen

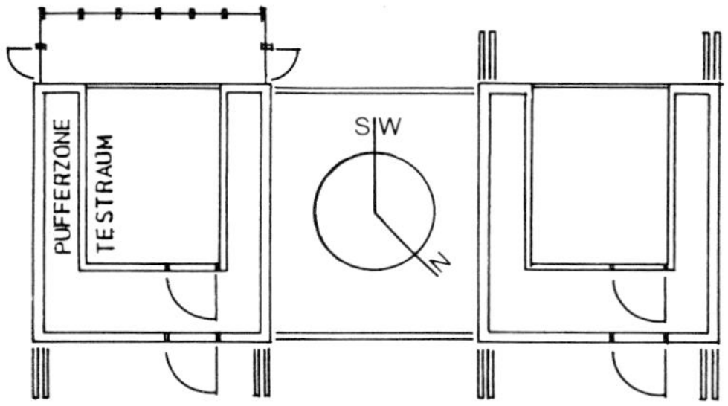

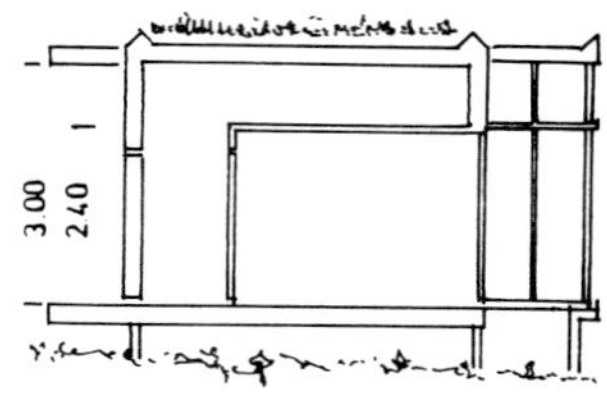

Schematische Darstellung des Testgebäudes
Grundriss und Schnitt M 1 : 100

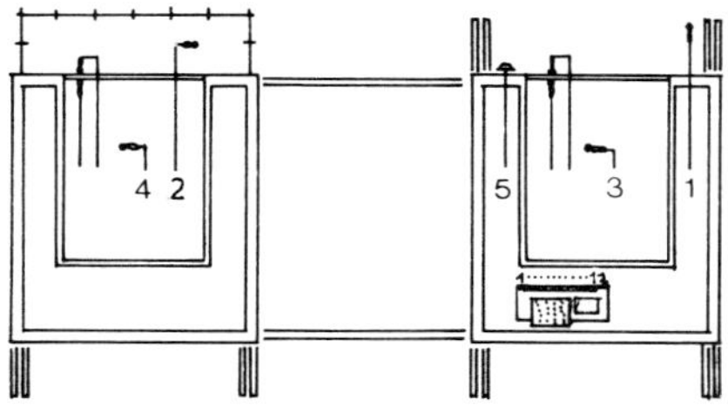

Meßstelle

1 Außen Lufttemperatur
2 Lufttemperatur Wintergarten
3 Lufttemp., Testraum ohne Wintergarten
4 Lufttemp., Testraum mit Wintergarten
5 Globalstrahlung

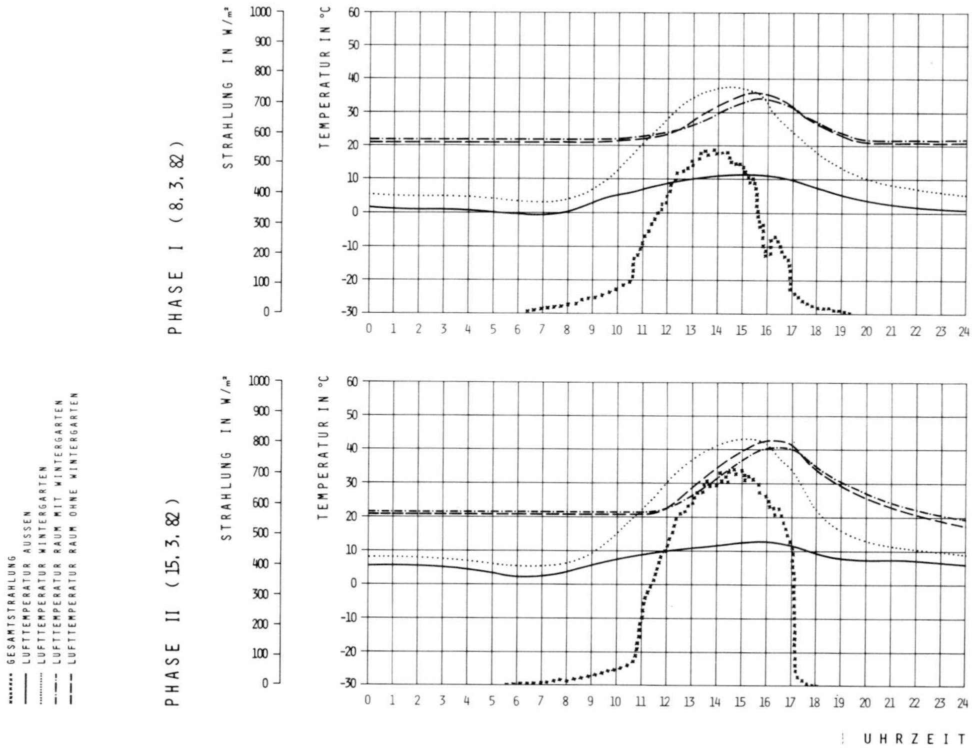

Wetterbedingungen/ Jahreszeit	Gebrauchsstellung	Effekt
Sommerhalbjahr/Tag/ Sonnenschein	Elemente ausgefahren und in reflektierender Stellung oder Elemente gerafft	Verschattung gegen Aufheizung, geringfügige Sichtbehinderung. Geringe Verschlechterung des Tageslichtkoeffizienten.
Sommerhalbjahr/Tag/ bedeckt	Elemente gerafft oder ausgefahren in geöffnetem Zustand	Kein Sonnenschutz, geringfügige Sichtbehinderung, Lichtumlenkung nach innen möglich
Sommerhalbjahr/Abend (flachstehende Sonne)	Elemente ausgefahren und auf reflektierende Stellung gebracht	Blendschutz, hohe Durchlässigkeit für sichtbare Strahlung, reduzierte Sichtverbindung
Sommerhalbjahr/Nacht	Elemente gerafft oder ausgefahren in geöffnetem Zustand	Nächtliche Durchlüftung möglich

Darstellung der wichtigsten Wirkungsphasen

	Wetterbedingungen/ Jahreszeit	Gebrauchsstellung	Effekt
	Winterhalbjahr/Tag/ Sonnenschein	Elemente ausgefahren, aber geöffnet oder gerafft (abhängig von Himmelsrichtung und Sonnenstand)	Einstrahlungsgewinn. Wirkung als Blendschutz möglich durch schrägstehende Lamellen. Geringe Verschlechterung des Tageslichtkoeffizienten

	Winterhalbjahr/Tag/ bedeckt	Elemente gerafft	Einstrahlungsgewinn vorhanden
	Winterhalbjahr/Abend	Elemente ausgefahren und geschlossen	Fenster wirken von außen nicht "tot" zum Straßenraum hin. Beleuchtung innen ist in Straßenräumen sichtbar. Thermische Wirkung als temporärer Wärmeschutz.
	Winterhalbjahr/Nacht	Elemente ausgefahren und geschlossen	Wirkung als temporärer Wärmeschutz

Darstellung der wichtigsten Wirkungsphasen

DEVELOPMENT OF A RANGE OF DOUBLE WINDOWS INCLUDING A SOLAR CONTROL BLIND AND INTERNAL SHUTTERS FOR USE WITH PASSIVE SOLAR HOUSES

Author : W. Houghton-Evans

Contract number : ESA-PS-142-UK

Duration : 15 months

Total budget : £16,419 CEC contribution : £6,568

Head of project : W. Houghton-Evans, Civil Engineering Department, The University of Leeds

Contractor : Stephen George and Partners, Architects and Town Planners

Address : 170 London Road
Leicester
LE2 1ND

Summary

The contractor is collaborating with a window manufacturer and the University of Leeds in the design and production of a range of modular window assemblies which are intended to improve the control of heat and light transmission through apertures facing the sun.

1.1 Introduction

The window is a most significant element in any building. It is required simultaneously to do many things. Among these are:

to illuminate;
ventilate;
decorate;
provide a view;
control the transmission of heat, sound and moisture;
afford adequate security.

In a solar-heated building the window's significance increases. Some larger windows are likely to be used to increase solar gains. But without improved control, these can result in over-heating during the day and excessive heat-loss at night - especially in northern Europe, where the nights are much longer than the days in winter.

There are already available various devices for improving control. These include double-glazing, blinds and shutters. What, however, is not so far available, is a window assembly which combines the most suitable devices in a manner designed to produce the best results, and marketed in a form and at a price which will ensure widespread use in inexpensive housing.

1.2 Description

In this project we propose to develop a range of modular window assemblies illustrated diagrammatically in Fig.1. The components will be designed in such a way that it will be possible to fit the window initially with only one of the sashes, to which the blinds, the other sash and the shutters may later be fitted as they can be afforded or are found to be wanted.

2.1 Design

The design of any window should have regard to long-established practise and tradition if it is to succeed. Our first task was to assemble information about practices and products already available and assess their suitability. It is already apparent that much may be learnt from a comparison of different national traditions and current practices, and this in itself is likely to yield interesting results. In the diagram we show only side-hung sashes, one pair opening inwards, the other outwards, thus already combining British with other European practice. Other arrangements - notably those employing sliding sashes - will also be investigated.

2.2 Solar Control

Of the many types available, our initial choice is the 'Venetian' blind (as it combines control of heat and light while permitting ventilation) but this will be confirmed by further investigation. The location chosen for this blind within the assembly we consider to be of the first importance. An external blind will reject unwanted solar gains the most effectively (with a solar-gain factor less than 0.15, compared with the values for an internal blind of approximately 0.5). But an external blind is exposed to the weather and is known to deteriorate. An internal blind may be used with an open window, but this may admit unwanted cold air. A compromise is to place the blinds between the panes in double-glazing, which gives a solar gain factor of about 0.25. The arrangement proposed in the diagram allows the blind to be external in fine weather (when the outer sashes may be opened) and protected at other times. It is also readily accessible for adjustment and cleaning via the openable inner sashes.

2.3 Shutters

These again have a long tradition, and are sometimes placed externally, sometimes internally. Their use pre-dates the widespread availability of glass, and has probably had at least as much to do with security as heat control. Increased security is in our proposal an incidental gain. Products of the type we illustrate are already marketed, and the energy saving from them can be shown to be around 0.3GJ/m^2 during an 8-months heating season in the UK. With heat costing about 4 £/GJ, a saving of approximately 1.2 £/m^2 year is achievable, and shutters costing 10£/m^2 will be repaid in about 8 years. Credit should also be given for increased comfort and security, and for curtaining not used.

2.4 Ventilation

During daylight controlled ventilation is assured through normal use of opening lights. The illustrated arrangement makes no provision for night ventilation, and our present intention is to incorporate a manually-controlled ventilator within the head or cill of the assembly.

3.1 Performance Evaluation

With the aid of our associated manufacturers, we intend to test assemblies for air and rain penetration, using standard rigs available in the UK, and on this basis we shall select the most appropriate materials and details. Air leakage is likely to be especially significant. The modular system of assembly and the incorporation of manually-operated moving parts results in numerous joints and matching surfaces through which air movement must be kept to a desirable level. Draughtiness is likely to be less of a problem than heat loss and condensation. We do not expect the double-glazing we propose to come up to the insulation level achieved with sealed units, but condensation control is likely to be less troublesome, as good accessibility allows water to be collected and wiped or drained away. Again, we may eventually look to tradition for a guide.

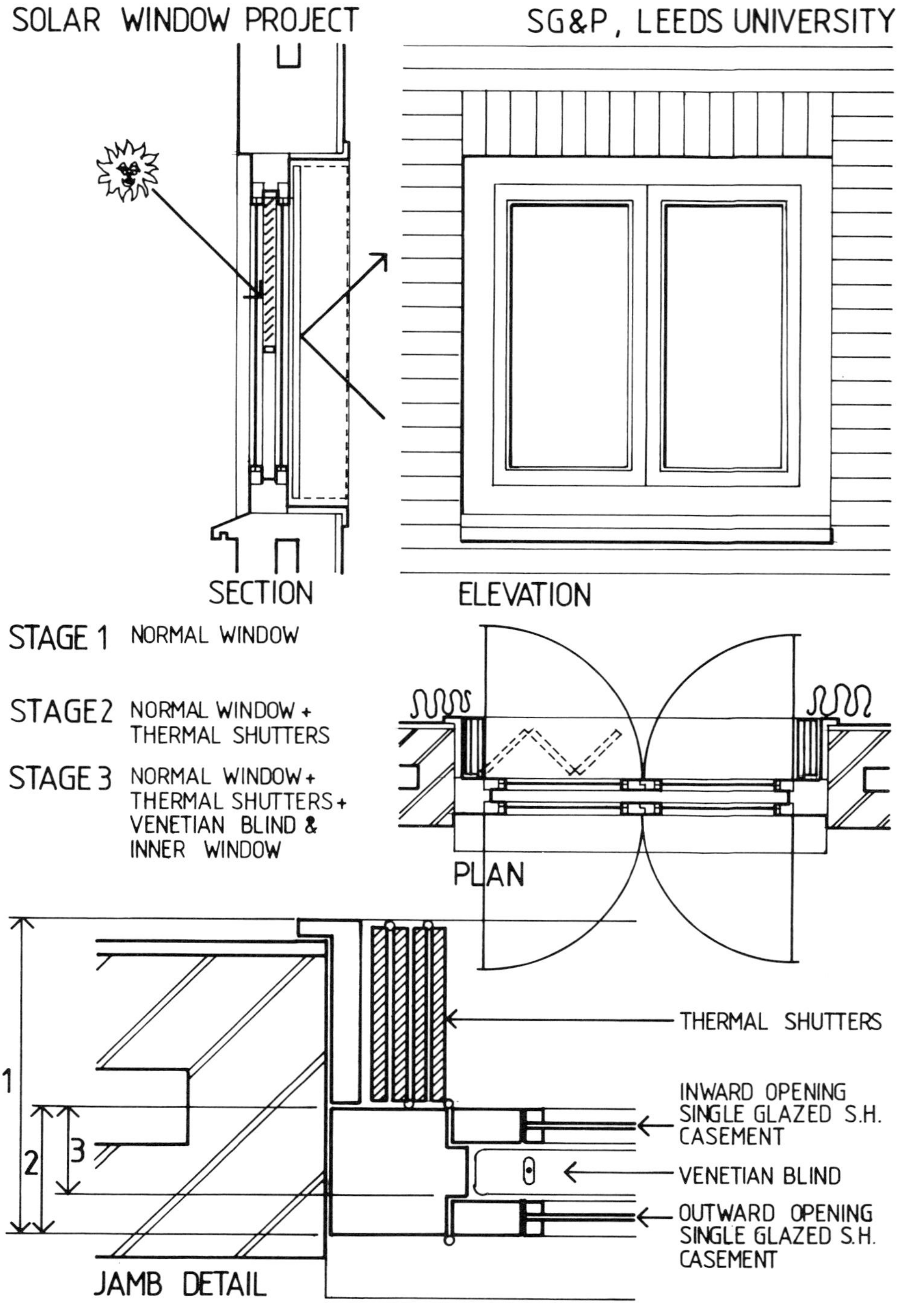
SOLAR WINDOW PROJECT
SG&P, LEEDS UNIVERSITY
SECTION
ELEVATION
STAGE 1 NORMAL WINDOW
STAGE 2 NORMAL WINDOW + THERMAL SHUTTERS
STAGE 3 NORMAL WINDOW + THERMAL SHUTTERS + VENETIAN BLIND & INNER WINDOW
PLAN
THERMAL SHUTTERS
INWARD OPENING SINGLE GLAZED S.H. CASEMENT
VENETIAN BLIND
OUTWARD OPENING SINGLE GLAZED S.H. CASEMENT
1
2
3
JAMB DETAIL

USE IN PASSIVE SOLAR ARCHITECTURE OF SOLAR DOUSERS WITH A SEASONAL EFFECT

Authors : U. FLISI, F. SCAMONI, U. MONACO, G.V. NARDINI
Contract number : ESA - PS - 138 - I (s)
Duration : 18 months 1 January - 30 June 1983
Total budget: Lit. 106.000.000 CEC contribution : Lit. 53.000.000
Head of project: Dr. U. FLISI, Montepolimeri CSI
Contractor: MONTEPOLIMERI CSI
Address: MONTEPOLIMERI CSI
Viale Lombardia, 20
20021 BOLLATE (MI)
ITALY

Summary

The main objective of this research is that of verifying the ability of PMMA plates, grooved on one surface in the shape of straight prisms to act as sunlight dousers in civil architecture.

When used in glazing south-facing walls, said dousers will allow sunlight to enter these walls during the Fall and Winter seasons, whereas, owing to the different astronomical path of the Sun, they will stop and reject direct sunlight by means of the optical effect called Total Internal Reflection during the central period of the Spring-Summer seasons.

The performance of these devices has been already calculated theoretically and will now be investigated experimentally firstly on reduced scale models and subsequently on real scale ones.

This paper describes the research project and the work done in the first months of its period, in order to prepare the samples and to assemble the apparatuses to test them.

1 INTRODUCTION

The idea of controlling sunlight transmission through transparent plates in accordance with the position of the sun appeared several years ago in some patents (1 - 7) and publications (8 - 10). However, the means generally used for this purpose were different from the one proposed here. In most cases, a totally "passive" effect was not deemed effective enough, so that a liquid optical medium was introduced to modify the optical properties of the system when desired. In those cases which concern entirely "passive" system, the shape of these is quite different from the present proposal, which can be shown to be both simpler and more efficient than all of them.

Glass products having shape and properties very close to those of the present proposal were produced and marketed in Italy by St. Gobain around 1955 and its powerful "mirror effect", experienced by one of the authors, induced the studies which led to the invention of the device described here.

2 DESCRIPTION OF THE PROJECT

2.1 Theoretical background

The invention is based on the phenomenon of "total internal reflection" that will take place whenever a light beam reaches, from within, the surface of a transparent medium having an index of refraction higher than 1.0, at an angle greater than the critical angle of the medium (all said angles must be measured between the direction of the beam and the normal to the surface at the point of incidence). Use is made of said phenomenon in connection with the fact that the apparent height of the sun in the sky varies throughout the year, and is higher during the summer months than during the winter months.

By an appropriate choice of angle 1 (see fig. 1a), and by taking into account the index of refraction of the medium, one can obtain a convenient value for the critical height of the sun that will set the limit between the summer behaviour and the winter behaviour.

For example, by using ordinary window glass with an index of refraction of 1.52, and by choosing angle 1 equal to 20 degrees, one can obtain the critical height of 53°, which the sun reaches at noon on April 1st and on September 15th, at the latitude of 40° North.

If the glass sheet having this profile is installed in a vertical position facing South, it will reflect sunlight away for the whole length of the day during the summer period between said dates, according to the scheme of fig. 1a), and it will admit it during the winter period according to the scheme of fig. 1b).

Actually the passage between the winter and the summer period is not abrupt but there will be two transition periods of about one month each, respectively before April 1st and after Sept. 15th, in which sunlight will be progressively stopped and admitted. Before April 1st sunlight will be admitted during a progressively shorter time interval around noon each day, while being reflected away during the early and late hours of the day. After Sept 15th sunlight will be transmitted for progressively longer intervals around noon, until when it will pass through for the whole daytime.

A glass sheet made of a sequence of prisms like those reported in fig. 1a) shows two undesiderable side effects, the correction of which requires a rearrangement of the internal angles of the prisms resulting in a profile like that represented in fig. 1c). In this rearrangement the lower prism faces f have an inclination equal to the critical height of the sun, while angle 2 is equal to the sum of the critical angle of the medium plus 90°. Given the refractive index of window glass, equal to 1.52, the condition of fig. 1c) yields a critical height of 56° 40' that is still satisfactory, since the sun will reach this critical height at a latitude of 40° North, on April 8th and September 5th at noontime.

A final remark concerns the fact that one single profile can satisfy the requirements of a great variety of locations, since, the lower the latitude the longer becomes the period in which the summer behaviour sets in and the shorter the winter period. Accordingly, dousers having the profile shown by fig. 1c) will reflect sunlight away:

. From April 8th to Sept. 5th, at a latitude of 40° N.
. from April 20th to August 24th, at 44° 30' North
. from April 1st to Sept. 15th at a latitude of 37° N.
. from October 5th to March 10th at a latitude of 35° South (this is the latitude of such big cities in the Southern hemisphere as Buenos Aires, Argentina, and Sydney, Australia).

2.2 Computerized elaboration

The theoretical performance of this prismatic glass, having a profile like that of fig. 1c), has been calculated with a computer in comparison with that of an overhang supposedly installed at the top of the South-facing wall of a building and having three depth/height ratios.

The results of such an exercise are reported in fig. 2 as percentages of incident light versus the date. Solid lines I, II and III are calculated for overhangs whose depth is in a ratio of, respectively, 15 - 30 - 45 percent with the total height of the wall. Dashed lines instead report the performance of a prismatic plate with refraction index (n) = 1.52 and front faces inclined at

15.5°.

The performance of the prismatic glass appears to be much better than that of the overhang, since the glass will admit considerably more light during winter and much less during summer than any overhang, whatever the depth-to-height ratio chosen for the latter. Substantially the same results are obtained when calculations are made for other latitudes. The advantage of prismatic glass also holds when the walls are oriented 10° or 20° offset from full South, but in such cases the time interval of total reflection is reduced compared to that of full South.

2.3 Experimental part

The main objective of this research is that of controlling with experiments the results of our calculations and therefore of verifying the ability of these prismatic plates to act as sunlight dousers.

Calculations have been made for a glass of refractive index equal to 1.52, but in these experiments PMMA will be used for various reasons:

- it is easier for us to shape in the desired form
- it is lighter and safer to handle
- its refractive index of 1.49 is very close to 1.52.

The sunlight dousers will be placed on physical models and tested in two phases.

In the first phase the model consists of a light integrating device internally coated with a high diffusivity high reflectivity coating (11) mounted on an automatic solar tracker, as described in fig. 3a. The mounting is polar (fig. 3b), that is with the pivoting axis placed parallel to the earth's axis.

Inside this box, a sensor is placed in the center of the opening, facing the interior, in order to measure the radiation passing through the PMMA prismatic plates. In this way all the radiation coming to the opening is detected and the diffused part of it can be determined by shielding the integrator from all direct light. By subtracting the diffused part from the total, the direct radiation can be obtained.

In the second phase the PMMA prismatic plates will be placed on the south wall of a real-size test cell similar to that described in fig. 4, which is now used in a project concerning passive solar architecture. The south-facing opening of this test cell will be closed with a Trombe wall, consisting of a water-filled drum-wall with no air circulation, fitting the description made by Balcomb et al in (12), where is reported as case 2. The dousers of the presently proposed design will be used as the transparent cover of said Trombe wall; as an alternative, they are replaced by plain transparent sheets, with or without shading overhangs.

The testing method involves writing the thermal balance of each system, as exposed to the sun in the open air, according to its known thermal properties (trasmittance and capacity), and taking into account the measured variables, such as the direct and diffuse radiation, the temperatures of the outside and inside air and of the water wall, possibly the wind velocity and the radiative temperature of the sky; the unknown properties will thus be worked out by means of statistical regression methods, with special attention to the degree of transparency of the cover sheet or douser for direct radiation, as a function of the angle of incidence, as well as for diffuse radiation.

In operating terms, all the above mentioned variables will be monitored by dedicated sensors; the signals of the latters will be scanned at regular intervals by a specially programmed data logger, then recorded on a support, such as a magnetic tape. The latter will be transferred in turn to a computer carrying the instructions required for working out the unknown properties from the available data.

3. PRESENT STATUS OF THE RESEARCH

The research project has started at the beginning of 1982 and the situation in May 1982 is as follows:

- A mold has been constructed for shaping PMMA sheets in the form of fig. 1c, and some samples of prismatic plates have already been obtained.
- The light integrating device will be ready before the end of June and the solar tracker will be built in a few months, hopefully before September
- The test cell with the Trombe wall, will be built before December 1982, whereas experience is now being made on a similar cell in order to get acquaintance with instrumentation, testing and calculation of the same type as required by this project.

4. REFERENCES

1) O.H. Brandi - German Patent No. 831,449 (1952)
2) Belgian Patent No 569,237 to St. Gobain (1959)
3) F.C. Henriques - U.S. Patent No 4,056,090 (1977)
4) N.S. Kapany - U.S. Patent No. 4,078,548 (1978)
5) E. Herbert - U.S. Patent No. 4,148,563 (1979)
6) M. McClintock - PCT Application WO - A 79 00276 (1979)
7) A.G. Eltreva - European Patent Application No 0 005 259 (1979)
8) J. Deltour, A. Nisen - Bull. Rech. Agron. Gembloux 5 (1-2), 232 (1970)

9) A. Nisen - ibid 6 (1- 2), 152 (1971)
10) A. Nisen, J. Deltour - ibid 6 (1-2), 182 (1971)
11) C.S. Herrick - Solar Energy 28 (1), 5 (1982)
12) J.D. Balcomb, J.C. Hedstrom, R.D. Mc Farland ibid 19, 277 (1977).

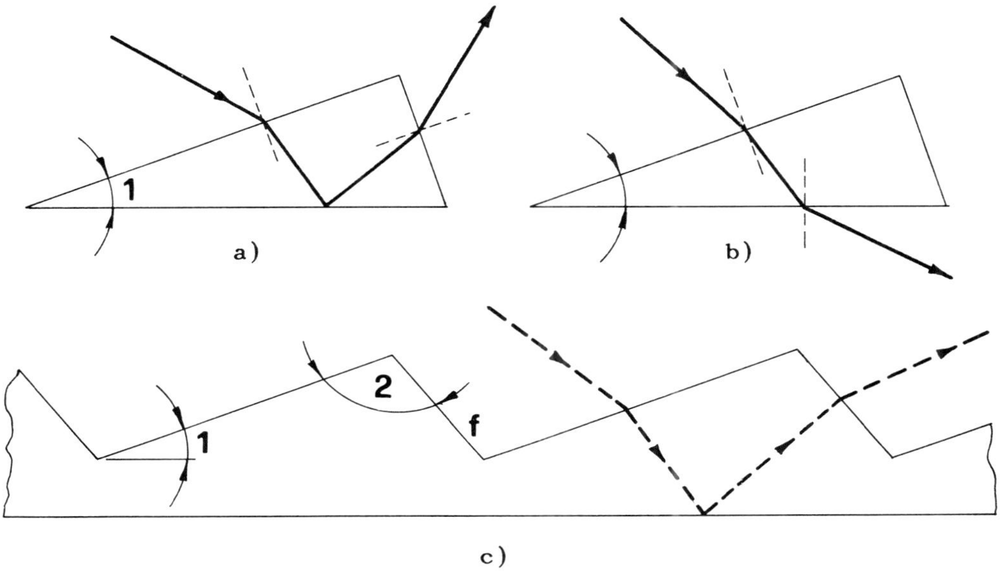

Fig. 1 Mechanism of total internal reflection (a) and trans_ mission (b) of radiation through a prism; profile of the glass sheet (c).

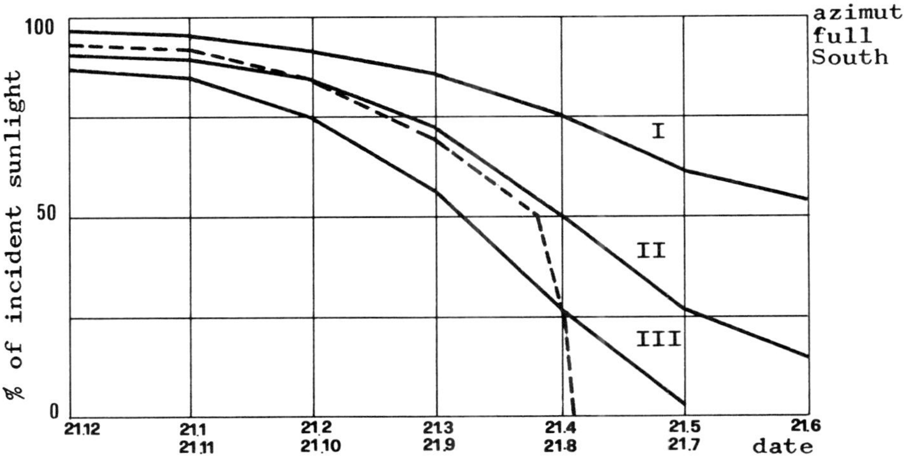

Fig. 2 Percentage of direct sunlight reaching a vertical wall covered by a prismatic plate (dashed line) or equipped with an overhang (solid lines) with the following depth/height ratios : I = 0.15 II = 0.30 III = 0.45 Values for Milan (Italy) - Lat. 45.4° N

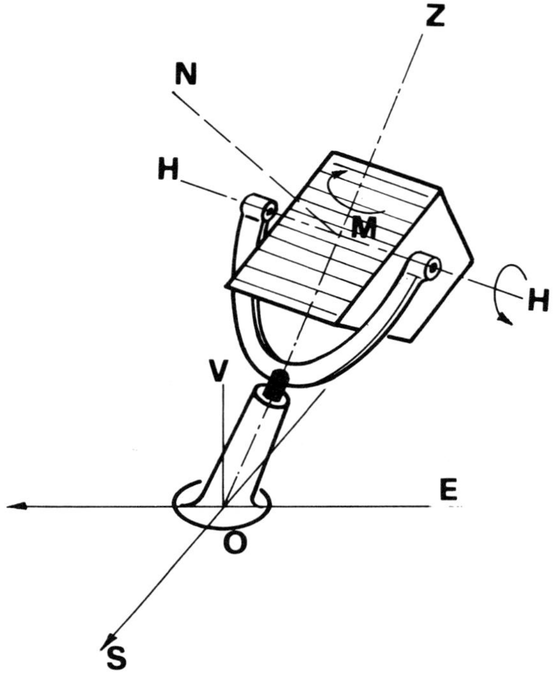

a)

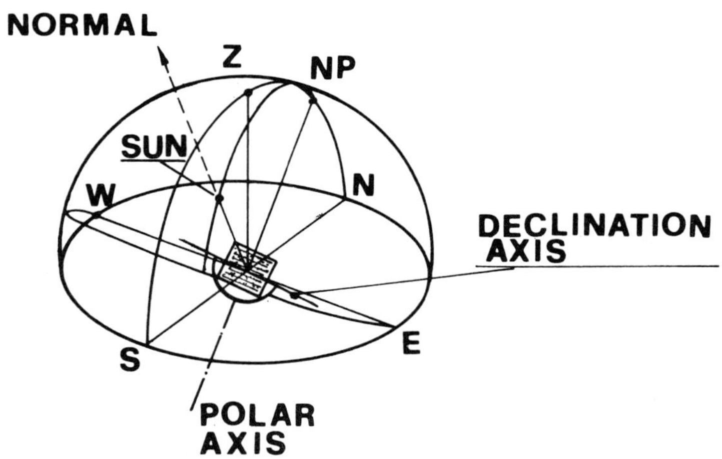

b)

Fig. 3 Light integrator on solar tracker (a) and its orientation (b).

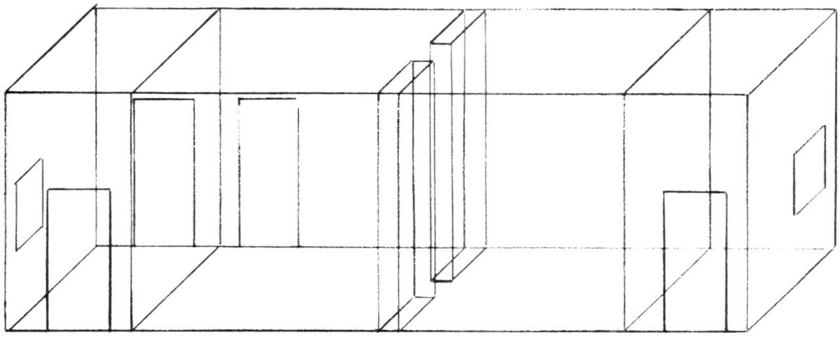

a)

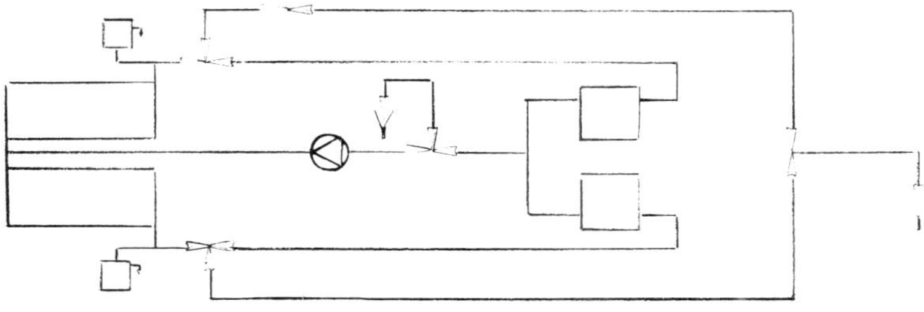

b)

c)

Fig.4) Test cell for the real scale tests:a)scheme- b)liquid circuit- c)photograph of the present cell.

DEVELOPMENT OF AN AUTOMATIC TEMPERATURE SENSITIVE AIR CONTROL DEVICE FOR USE WITH PASSIVE SOLAR GAIN COLLECTORS

Author	:	A.C. LYALL
Contract Number	:	ESA-PS-155-UK
Duration	:	7 months (ending December 1982)
Total Budget	:	£17,087 CEC contribution £7260
Head of Project	:	A.C. Lyall
Contractor	:	Waterloo Grille Company Limited 14 Parsons Road Benfleet Essex England

Summary

The energy benefits of advanced passive solar gain collectors such as Trombe Walls and conservatories are known to be improved where controlled air movement is provided. The aim of this research is to develop a low technology component which is simple, inexpensive and reliable and which is capable of sensing when energy benefits are obtainable and carry out action to secure those benefits. Control devices currently available are sophisticated, relatively expensive and have attendant installation and maintenance problems which could be solved by alternative self contained controls requiring no external power supply.

The project programme will commence with an appraisal of possible solutions to achieve the objectives mentioned above from which a prototype or prototypes will be developed, tested and rated on a "simulation" test apparatus. The final stage includes refining of the product to ensure its commercial and practical viability. During all stages, the performance can be compared with that of a local solar project utilising similar prototypes already developed by Waterloo Grille Company and the Essex County Council.

Introduction

With advanced passive solar gain collectors such as Trombe Walls and conservatories, the close control of energy collection and use is essential if maximum efficiency is to be obtained. The aim of this project is to develop a simple, inexpensive and reliable temperature sensitive air control device which can control the flow of warmed air from the collection area to be occupied zone. A typical arrangement for a solar wall system is shown in Fig. I. This forms the basis for testing of developed control devices.

Performance Criteria

In order to achieve the function described above, the following design objectives have been established involving development of:-

(a) a basically standard device capable of controlling air flow according to building demands and solar gain/collection availability.

(b) a long life, low maintenance device of acceptable architectural form that can be easily installed in new or existing passive solar projects.

(c) an automatic device that requires no external power supply or manual assistance.

Project Stages

1. Discussion of design objectives and physical design parameters in relation to all possible solutions. Appraisal of these solutions in order to eliminate impractical and expensive devices/systems.
2. Development - a logical progression from Stage 1 which will include the design and construction of one or more possible solutions which satisfy the design criteria.
3. Testing and rating of the device(s) resulting from Stage 2 using a simulation collector apparatus.
4. Commercial development - modification and refining of the test device(s) to ensure a marketable commodity.

Timetable

Work commences May/June 1982.

Stage 1: ½ month
Stage 2: 2½ months
Stage 3: 3½ months
Stage 4: ½ month

Stage 1 : Discussion of design objectives and appraisal of possible solutions

The current project file was studied by all group members to familiarise them with the project aims, design objectives and project programme. During many group meetings a design strategy was evolved to satisfy the objectives and an appraisal of possible solutions was made. The results are summarised below in the form of direct and related parameters.

A. Sensing temperature

A simplified model is given in Fig. II which describes the probable application of the temperature sensitive air control device.

Collector void air temperature indicates when solar energy is available while room temperature indicates when solar energy is required. Sensing both temperatures is desirable to maximise energy benefits.

The ideal collector sensing location could not be specified but would be identified during the testing stage. Sensing collector surface or radiant temperatures do not appear to offer practical advantages over sensing dry bulb temperatures. An extreme collector temperature range of 0-80°C was identified with a calibration setting range of 18°C-30°C. Provision for manual overide and overtemperature protection seems desirable.

B. Air Control Device

The obvious solutions include : multi-shutter, opposed blade, parallel blade, gate or flap dampers. The device would have a reasonably solid interior finish to provide good architectural integration combined with a functional air control assembly operated by a local or remote actuator. Size : width and height should if possible approximate to a brick for easy installation; and depth should be minimised to facilitate installation and operation within a narrow panel/collector area. Probable orientation would be vertical, although the device may be installed in a horizontal surface for special applications.

C. Actuator

The design objectives rule out the use of electrical, electronic, pneumatic and hydraulic powered devices using a control supply. Passive assemblies are also outside the scope of this work. The mode of actuation best suited to the application could not be specified and may be on/off, stopping or modulating. Response speeds of around two minutes were identified as reasonable at this stage but would be investigated during testing.

The following table appraises the relative merits of possible solutions identified during discussions.

ACTUATING PRINCIPLE	SOLUTION	FORCE DEVT.	STROKE	RELATIVE COST	COMM- ERCIAL AVAIL - ABILITY	DIS- ADVANTAGES
GAS EXPANSION	GAS IN PISTON OR BELLOWS	LOW	-	HIGH	-	LEAKAGE
LIQUID EXPANSION	LIQUID IN PISTON OR BELLOWS	-	-	MODERATE	-	LEAKAGE
SOLID EXPANSION	WAX FILLED PISTON	HIGH	HIGH	LOW	GOOD	LOW RETURN STROKE FORCE
	SOLID	HIGH	LOW	LOW	GOOD	
PHASE CHANGE	SOLID/LIQUID GAS IN PISTON OR BELLOWS	MODERATE	-	MODERATE	-	LEAKAGE
DEFORMATION	BIMETAL	LOW	MODERATE	LOW	GOOD	REQUIRES LOW FORCE AIR CONTROL DEVICE
	MEMORY METAL	HIGH	HIGH	LOW	GOOD	

D. Related Parameters

It was noted that the physical parameters of the collector, collector void and internal partition wall would influence the performance of the evolved device during test and operation. However, as the shape, size and construction of these elements could not be clearly defined, it was decided that these factors should be considered again after development of the air control device. The method of testing and rating would probably indicate the sensitivity of these items in relation to the performance of the device.

Conclusions from Stage 1

As the time available for development was limited it was decided that one or more of the commercially available actuating devices should be used in the design of prototypes. The memory metal and wax filled piston devices appear to offer practical and economic advantages over the other solutions and appear to have proven reliability.

The air control device should be relatively small and modular to provide simple installation with the facility for ganging or interlocking a number of units to satisfy air flow requirements. Leakage and the possibility of back flow or air should be eliminated at the design/development stage.

The assembly should be completed with the addition of a local or remote sensing device linked to the air control damper. It is envisaged that the sensor and actuator may be integrated components.

FUTURE WORK

Having established a conceptual model of the air control device and its typical application, the development work can commence. Suitable actuators will be developed to suit the selected type of air control damper. Testing and rating will be carried out on a simulation solar wall apparatus which is shown in Fig. III.

When a reliable and practical device has been developed, the sensitivity of various performance parameters will be tested, eg. room and collector sensor location, air control device location etc.

Finally, the design team will refine the device to make it a commercially viable product for use with the more common air type passive solar gain collector systems.

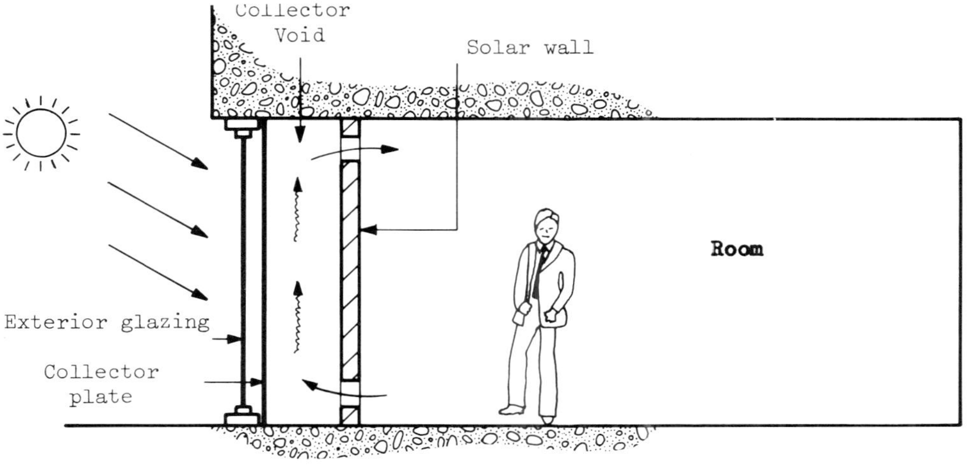

Fig. I - Typical solar wall system

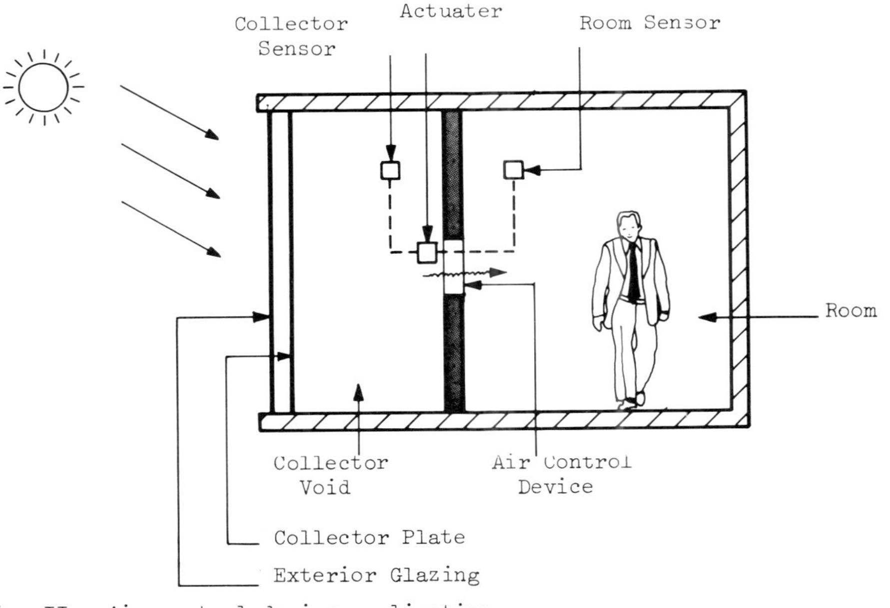

Fig. II - Air control device application

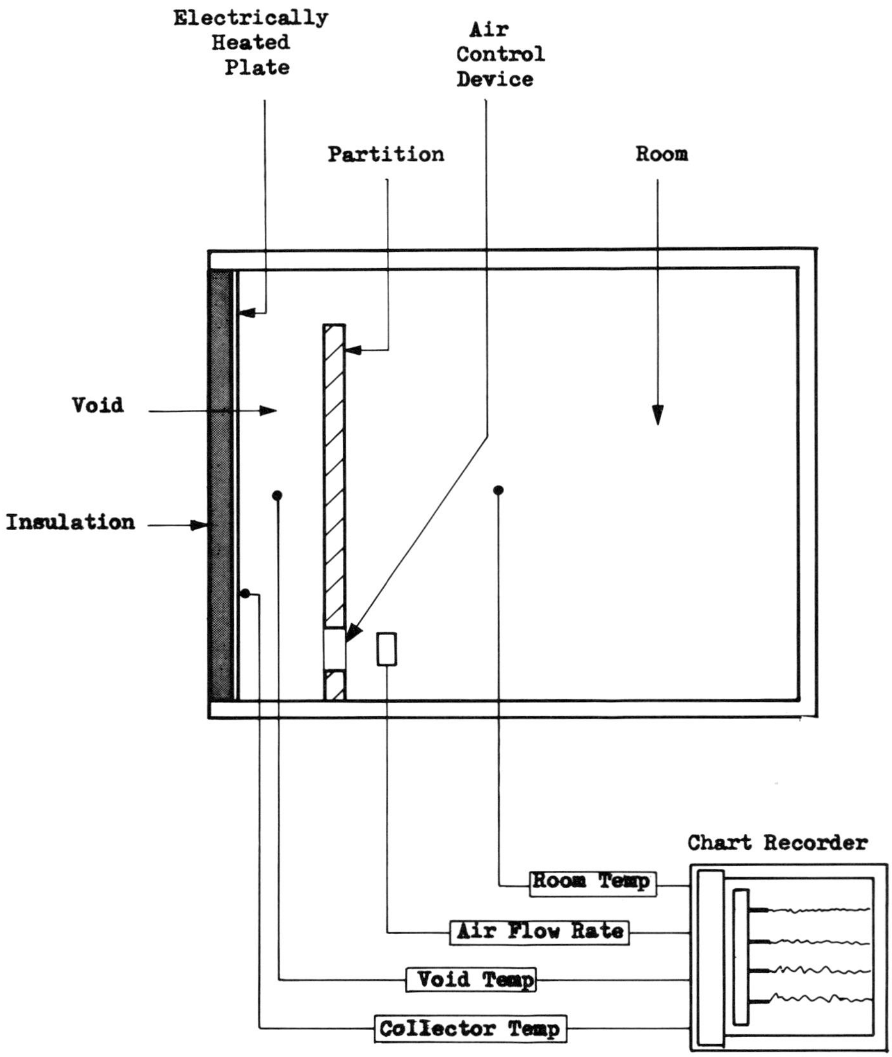

Fig. III - Proposed test arrangement

SELF-ACTING POWER CYLINDERS FOR CONTROLS IN PASSIVE SOLAR BUILDINGS

Author : D. FITZGERALD

Contract number : ESA-PS-154-UK

Duration : 18 months 1st June 1982 - 30 November 1983

Total budget : £15 009 CEC contribution : £6 004

Head of project : Dr. D. Fitzgerald, Civil Engineering Department, The University of Leeds

Contractor : Stephen George and Partners, Architects and Town planners

Address : 170 London Road
Leicester
LE2 1ND

Summary

Passive solar houses may overheat during long periods of sunshine. This could be prevented by using equipment now on the market to deal with the same problem in greenhouses. This contract is to survey the market , buy suitable devices and test them, to see whether they might be suitable for passive solar houses, or whether modifications are needed.

1. The Problem

A passive solar building will be designed to make the most of partial winter sunshine. When the sun shines brightly for a long time there is a danger of overheating. This danger is greatest in Spring and Autumn, becomes less important in Winter, because of lesser average possible insolation, and more cloud cover, and less important in Summer, when the average intensity on the vertical during a sunny day is less than two thirds of what it is in the Spring and Autumn. This is shown in figure 1, for 52°N.

2. A Solution

Overheating could be prevented by venting collectors to the outside using ordinary electrical power cylinders but this violates the philosophy of the passive solar approach, and is likely to be an expensive and complicated solution to the problem.

A solution is offered by the automatic greenhouse window opener, a product now on the market, to solve the same problem in greenhouses, without the use of external energy. A cylinder, fitted with some organic mixture usually based on paraffin wax and petroleum jelly, expands when heated, and provides a force to push a window open.

3. The Contract

This contract is for surveying the market in the product, both within and without the Community, buying promising cylinders, and testing them, to determine whether any are suitable for use in solar houses, or might be modified to be so.

4. Tests

The following tests are proposed.

4.1 The No Load test, to determine the movement of the cylinder, with both rising and falling temperature, when the cylinder is allowed to move freely. This determines the coefficient of volume expansion of the working material.

4.2 The No Movement test, to determine the force generated, as a function of rising and falling temperature, when the cylinder is constrained not to move. It is shown in the appendix that this determines the product of the bulk modulus and the coefficient of volume expansion of the working fluid.

4.3 The Practical test, in which the cylinder is offered a fixed load, and and the movement / temperature characteristic is observed, at both rising and then falling temperature. Each cylinder will be tested at a number of loads, in the region of the maker's nominal load.

The tests will be carried out in a small air oven, in which the temperature is controlled by a temperature programme controller.

5. The ideal cylinder will behave reproducibly. There will be no movement until the operating temperature is reached, and then there will be relatively rapid movement, to provide a significant damper or window movement for a moderate further rise of temperature. When the temperature falls, the cylinder should show little hysteresis. The operating temperature should be simply adjustable. The cylinder should not be damaged if its thermal movement is prevented.

6. Points of application are

venting to the outside of conservatories and other air heaters, to

prevent overheating (see 1 in fig.2)
dampers in conservatories, to allow warmer air to enter a building, and to allow cool air from a building to enter the conservatory (see 2 in fig.2)
dampers with similar purposes in air heaters (see 3 in fig.3)
dampers to allow warm air in a warm southfacing room to reach other parts of the building (similar to 2 in figure 2)
dampers to suppress parasitic air movement by natural convection in system using a fan, when the fan is off.

APPENDIX

Nomenclature

a	coefficient of volume expansion of working material	K^{-1}
A	area of cross section of cylinder	m^2
b	fraction of air in cylinder	-
F	force exerted by cylinder	N
K	bulk modulus of working material	Pa
l	movement of cylinder	m
lo	required movement of cylinder	m
L	length of cylinder	m
P	a pressure	Pa
T	rise in temperature of cylinder	K
To	rise in temperature of cylinder causing no increase in length	K
T_1	further rise in temperature to give required movement	K
V	a volume	m^3

If the temperature of the cylinder rises by T, and the expansion is not opposed, the increase in length

$$l = aLT \tag{1}$$

If the expansion is prevented, the force generated

$$F = aAKT \tag{2}$$

If the expansion is opposed by a force $F^1<F$, there will be an increase in length

$$l^1 = L(aT-F^1/AK) \tag{3}$$

If we fix a temperature rise To for which there is to be no movement, when the restraining force is F^1, and we fix a further temperature rise T_1>To for which there is to be a cylinder movement of lo, then it is easily shown that the material in the cylinder has to have its coefficient thermal expansion

$$a = \frac{lo/L}{T_1-To}\ K^{-1}$$

and its bulk modulus

$$K = \frac{F^1L}{Alo}\left(\frac{T_1}{To}-1\right)$$

For a particular cylinder $L = 0.22m$, $A = 3x10^{-4}m^2$ and the observed rate of increase in length with temperature at no load was 2.8mm/K, so that

$$a = 0.0028/0.22 = 1.27x10^{-3}\ K^{-1}$$

The cylinder was said to be able to provide 7kgf at the window, or (by use of levers) 7x3x9.8N at the cylinder, so that taking T_1 = 20K and To = 10K

$$lo = 1.27x10^{-3}(20-10)0.22 = 2.8x10^{-3}m$$ at the cylinder or 3 times this, $8.4x10^{-3}m$ at the window and

$$K = \frac{7 \times 3 \times 9.8 \times 0.22}{3 \times 10^{-4} \times 2.8 \times 10^{-3}}\left(\frac{20}{10}-1\right) = 53.9 \times 10^6\ Pa$$

Most simple organic materials have the product Ka within about 30% of 1.4×10^6 Pa/K whereas the material used in this cylinder has Ka = $53.9 \times 10^6 \times 1.27 \times 10^{-3} \sim 0.007 \times 10^6$ Pa/K less than the expected value by a factor of 20.

A small amount of air in the cylinder could explain this unexpected result. If the proportion of air in the cylinder is b, then equation 1 becomes

$$1 = LT(a(1-b)+b/273)$$

where the expression in brackets is the observed coefficient of thermal expansion. The bulk modulus K = VdP/dV by definition, where an additional pressure dP causes a volume change dV. For the liquid part of the mixture

$$dV_f = dP(1-b)LA/K$$

and for the air

$$dV_a = dPbLA/Po$$

where Po is the absolute atmospheric pressure of 10^5 Pa. Thus we have

$$dV = dV_f + dV_a = LAdP((1-b)/K + b/10^5)$$

where the reciprocal of the expression in brackets is the observed K. Thus

$$a(1-b) + b/273 = 1.27 \times 10^{-3}$$

and

$$(1-b)/K + b/10^5 = 1/(53.9 \times 10^6)$$

If we assume that for the air-free material in the cylinder Ka = 1.4×10^6 Pa/K, these equations can be solved for a, b and K. The results are

$$a = 1.27 \times 10^{-3}\ K^{-1}$$

$$b = 1.8 \times 10^{-3}$$

$$K = 110 \pm 30 \times 10^6\ \text{Pa}$$

Less than 2 parts per 1000 of air leaves the coefficient of thermal expansion unchanged, but doubles the bulk modulus. It is easily shown that the large uncertainty in Ka only affects K.

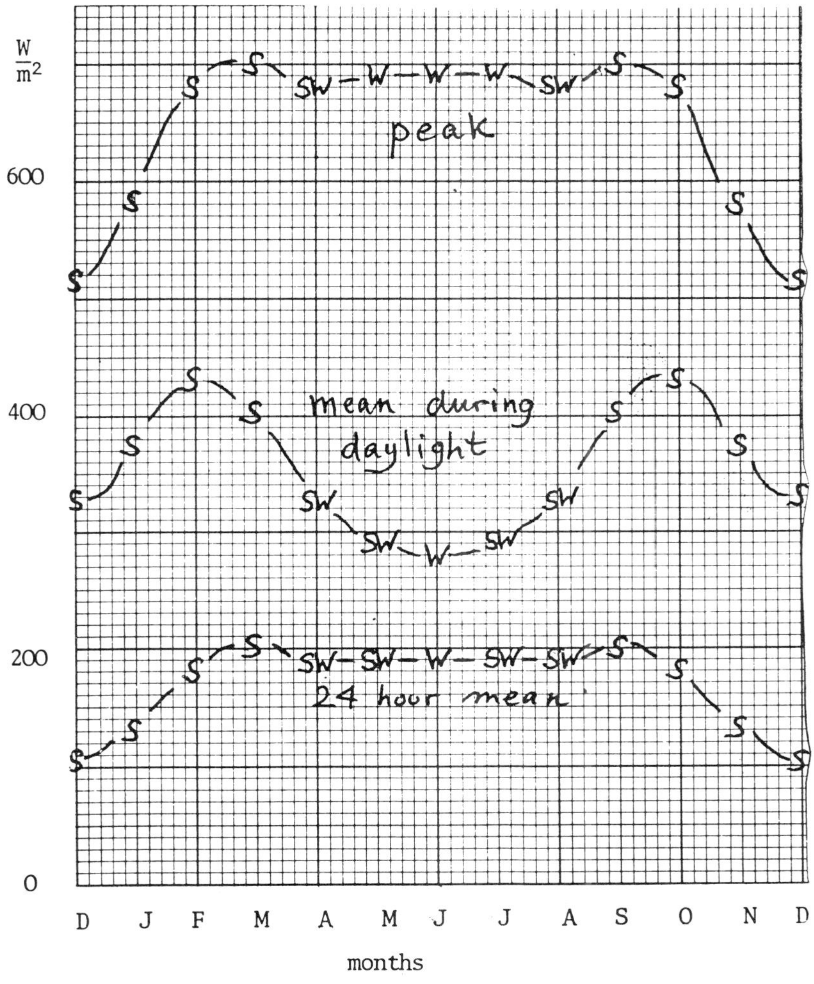

Figure 1. Insolation on the vertical, 51.7°N

The highest values are from the directions shown. Where W or SW is given, the same insolation is also received on the E or SE.

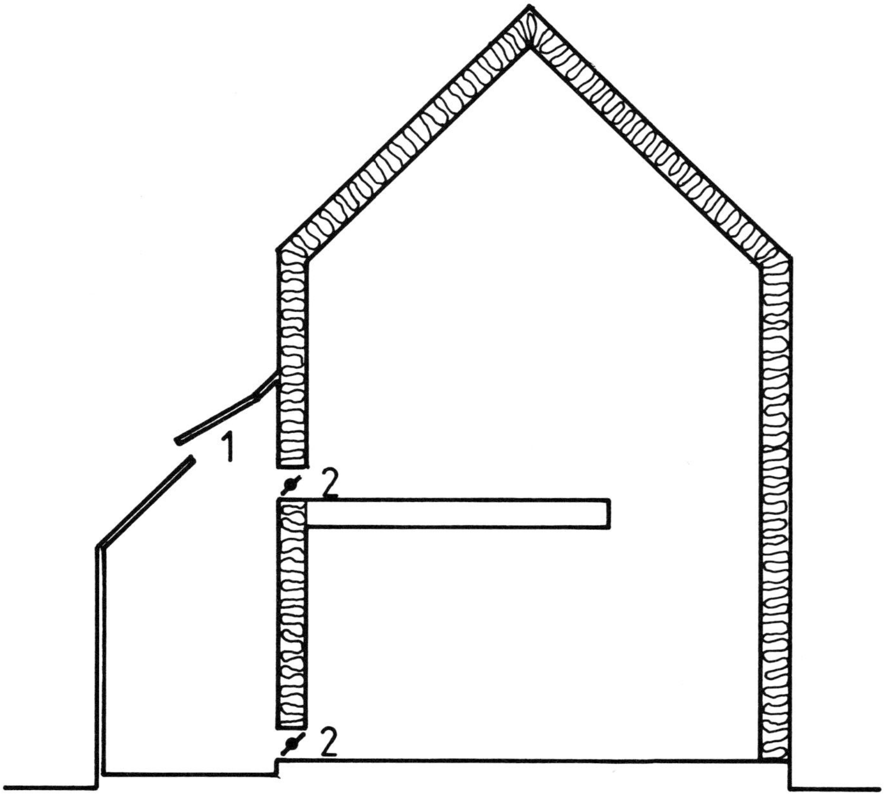

FIGURE 2.

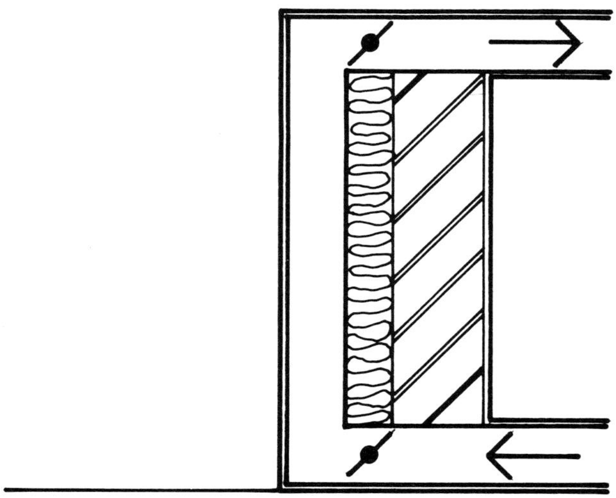

FIGURE 3.

UNIT FOR LOCAL STORAGE OF SOLAR GAIN

Authors : S. Svendsen, L.S. Christensen

Contract number: ESA-PS-150-DK

Duration : 15 months 1 April 1982 - 30 June 1983

Total budget : D.kr. 1,149,500 CEC contribution: 50%

Head of project: S. Svendsen

Contractor : Thermal Insulation Laboratory
Technical University of Denmark

Adress: : Building 118
DK-2800 Lyngby, Denmark

Summary

The aim of the project is to develop a unit for local storage of solar gain. The unit will be based on the principle that excess heat in a room will be stored in a latent heat storage placed in the room and released when needed.

The storage medium is a salt water mixture based on the extra water principle and with a melting temperature close to room temperature.

The project is planned to be carried out in three phases. In phase one - setting up the design criterias - a parameter analysis of heat storage capacity, heat loss coefficients and heat transfer coefficients for the storage unit in a typical room will be carried out. In phase two - design of storage unit - the unit will be designed based on the design criterias found in phase one. In phase three - construction and test of storage unit - the storage unit will be constructed and tested under realistic conditions.

Computer calculations simulating the thermal performance of the storage in a typical room have been carried out. Some of the preliminary results show, that the storage unit saves about 30% of the heating load in the room on a yearly basis.

1.1 Introduction

The background for the project is that in many houses the heat gained from the solar radiation coming through the windows is not used efficiently. It is often seen that a building on a sunny day is overheated in the day time but has to be heated in the night. If the thermal mass of the building could be increased the solar gain could be stored and used during the night. For existing buildings it is practically impossible to increase the thermal mass of the construction. But if a heat storage unit which could work within a narrow temperature interval was developed it might be possible to install it in existing houses and thereby use the solar gain to a greater extent.

The aim of the project is to develop a unit for local storage of solar gain. The unit will be based on the principle that excess heat in a room will be stored in a latent heat storage placed in the room and released when needed.

The project is planned to be carried out in the following three phases: 1) setting up the design criterias, 2) design of storage unit, 3) construction and test of storage unit. The planned work in each phase will be described in the following.

1.2 Setting up the design criterias

Characteristic sets of data for a number of typical rooms will be set up. These data will include window area, thermal capacity and heat loss coefficients. Based on these sets of data a simulation programme will be used to calculate the energy savings obtained by using a unit for local storage of solar gain with different heat storage capacities, heat transfer coefficients and heat loss coefficients. In this way a relation between performance of the unit and its characteristic parameters is established.
Based on practical considerations maximum sizes and weight of a storage unit will be set up.

1.3 Design of storage unit

Some potential salt hydrates will be evaluated with respect to melting temperature interval, storage capacity, practical considerations and price. One of the salt hydrates will be selected to be used in the further work.

A storage unit consisting of container, heat transfer system and insulation will be designed based on the above mentioned design criterias.

The use of natural or forced circulation of the air of the room through the heat transfer system will be investigated with respect to performance and thermal comfort of the room.

1.4 Construction and test of storage unit

A prototype of the designed storage unit will be constructed. This prototype will be tested at the Laboratory in order to find its thermal performance characteristics.

The storage unit will be installed in a test room of an experimental solar house at the campus test field and its performance during a period of some months will be measured. Based on these results a calculation of the energy savings per year will be carried out.

2.1 Calculated energy savings for a room with a storage unit

According to the work plan a number of typical rooms have been selected. Calculations have been carried out for one of these and some preliminary results will be described. The room is the living room of a typical Danish one - family house. The room is located in the south - west corner of the house. The floor area of the room is 35 m^2 and the window area is 10 m^2. All the windows are facing south. The room is well insulated and has a relatively small storage capacity. In a room with the described thermal characteristics one will expect to benefit from a storage unit.

The simulation programme to calculate the energy savings is the BA4-programme, which has been developed at the Thermal Insulation Laboratory. The programme is calculating the room temperature, the heat loss and the heat gain every half hour all the year round. A black box model of a storage unit has been added to the programme. The temperature of the storage and the heat stored in or delivered from the storage is also calculated every half hour. The calculations are based on the Danish Reference Year.

In the following example the wanted room temperature during the day is 21°C and 17°C during the night. The storage medium used is a salt water mixture with 24% Na_2CO_3 and 76% of water based on weight. The storage unit has a volume of 200 l and a heat loss coefficient of 2 W/°C. The heat transfer coefficient is 500 W/°C.

In figure 2.1.1 the solar gain of the room is shown on a monthly basis. The useful part of the solar gain for the room without a storage unit is also shown in figure 2.1.1. It is seen that although the solar gain reduces the heating load with a relatively large amount there is still large quantities of solar gain which are not utilized.

In figure 2.1.2 the heating load of the room is shown without and with the storage unit. It is seen that the storage unit saves about 1200 kWh on a yearly basis or about 30% of the heating load. This result indicates that a storage unit for local storage of solar gain can be a useful device for energy saving.

Calculations like this will be carried out varying the parameters of the storage unit and in this way design criterias for a storage unit will be set up.

2.2 Evaluation of storage medium

The storage medias of interest are based on inorganic salt hydrates. That is: Either eutectics or salt water mixtures with a melting temperature close to room temperature and based on the extra water principle which has been succesfully demonstrated in several projects at the Thermal Insulation Laboratory supported by the European Communities (1), (2), (3).

In order to obtain experiences with different heat storage materials small scale experiments are carried out with 6 different storage materials, 3 salt water mixtures based on the extra water principle and 3 eutectics, see table I.

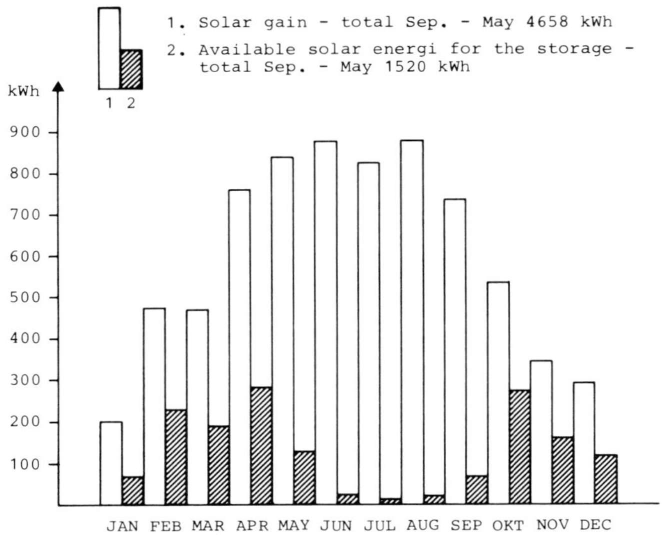

FIG. 2.1.1 Solar gain for the room on monthly basis.

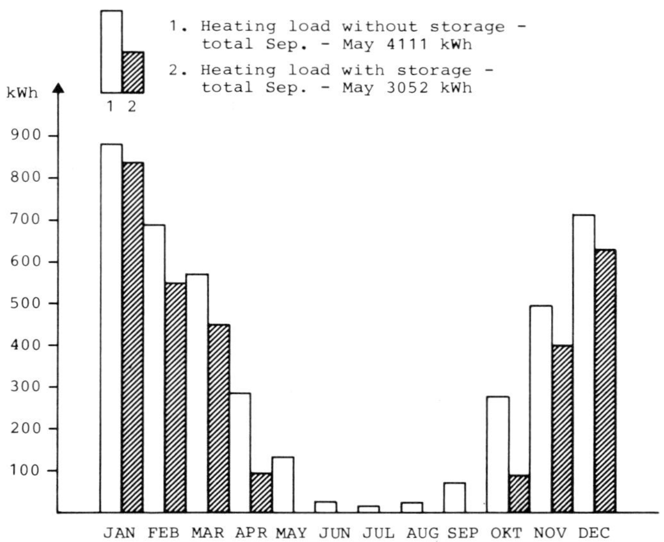

FIG 2.1.2 Heating load of room with and without the storage.

Heat storage material	weight %	Heat content in the temperature interval 20-26°C kWh/m^3	Fraction of solid phase based on weight at 20°C
Na_2SO_4 H_2O	22 % 78 %	27	0.23
Na_2CO_3 H_2O	24 % 76 %	34	0.32
$CaCl_2$ H_2O	46 % 54 %	32	0.37
Na_2CO_3 K_2CO_3 H_2O	21 % 16 % 63 %	information not available	
$CaCl_2$ $MgCl_2$ H_2O	31 % 14 % 55 %	information not available	
Na_2SO_4 NaCl H_2O	11 % 17 % 72 %	information not available	

Table I. Investigated heat storage materials

Melting and crystallization for the three first mentioned salt water mixtures take place in the temperature interval going from 26°C and downwards. The investigations have just been started and the results will therefore not be given here.

2.3 Facilities for realistic testing of the storage unit

A prototype of the designed storage unit will be constructed. This prototype will be tested at the laboratory in order to find its thermal characteristics.

The storage unit will be installed in a test room of the new experimental solar house at the campus test field. During a period of some months the performance of the storage unit will be measured. The results will be compared with the thermal performance of a room of reference. Based on these results the computer model simulating the behaviour of the storage will be validated and the yearly energy savings by using the storage will be calculated.

Litterature references

(1) "Report on heat storage in a solar heating system using salt hydrates" S. Furbo and S. Svendsen, Thermal Insulation Laboratory, Technical University of Denmark, July 1977, revised February 1978.

(2) "Investigations of heat storages with salt hydrate as storage medium based on the extra water principle". S. Furbo, Thermal Insulation Laboratory, Technical University of Denmark, December 1978.

(3) "Heat Storage units using a salt hydrate as storage medium based on the extra water principle" S. Furbo, Thermal Insulation Laboratory, Technical University of Denmark, January 1982.

DESIGN AND DEVELOPMENT OF AN EASY MADE CONVECTIVE TYPE SOLAR AIR COLLECTOR

Authors : A.D. Sferides, M. Papadopoulos

Contract number : ESA - PS - 156 - GR

Duration : 18 months 1 January 1982 - 30 June 1983

Total budget : 2.800.000 DRA CEC contribution : 50%

Head of project : Prof. M. Papadopoulos - University of Thessaloniki- School of Technology - Department of Construction and Building Materials

Contractor : University of Thessaloniki

Address : University of Thessaloniki
School of Technology
Department of Construction and Building Materials
Thessaloniki - GREECE

Summary

The design of a solar air collector to be mounted on walls of existing or new buildings to form a passive solar heating system is investigated theoretically, as the first stage of the design and development project. The collector, operating by loop convection air heating, will be the result of optimization among various factors such as efficiency, price, weight, easiness of transportation, installation and maintenance. So, emphasis is given on simple, inexpensive solutions in relation to careful design to maximize efficiency. While in the stage of theoretical analysis it has been realized that a careful and detailed modelling and simulation of collector operation could distinguish certain solutions as worth-while for experimental study, at the same time rejecting constuction methods and materials having poor performance by calculation. As a result of this fact, the original project timetable has been changed in that collector construction and testing will begin after a good deal of numerical results would be available. Through those results a limited number of solutions would be selected for subsequent experimental investigation. At the same time an extensive search of tne market for commercially available construction materials is performed, to give a complete picture of all possible solutions.

1. Introduction

An easy-made solar air collector operating on the simple principle of free convection, which is to form part of the south-oriented walls of public or private buildings, seems to be a promising solution for passive solar heating systems.

The design and development of such a collector that would combine low cost, efficiency, simple construction, light weight, modular construction, easiness of transportation, installation and maintenance, is a result of a number of successive project stages : First of all an accurate mathematical model should be constructed, which is to be extensively investigated numerically. This would lead to a number of possible solutions which in turn would result to a number of experimental collectors - care taken to conform to standard series-production industrial procedures. The next stage would be the performance measurements of those experimental collectors under real-measurable-conditions for a considerable period in winter. The results of those measurements, together with data on material and construction cost would be processed by a financial study which would finally evaluate the proposed solution(s).

2. Theoretical Modeling

The purpose of theoretical analysis-of which numerical investigation should be considered an integral part-in this project, can be summarized in three major topics :

- To investigate the importance of the various design parameters which enter the calculations. Selection of appropriate values must be decided upon at this stage of the project. This would be realized through evaluation of the influence of those parameters in the overall collector performance. The external dimension of the collector, the direction and path(s) of air stream, the cover-absorbing plate distance, the absorbing plate-back insulation distance etc are some of the design parameters to be investigated.
- To pre-select a number of materials to be used subsequently for the construction of the collectors to be studied experimentally. Certainly material choice is not only a matter of performance but of price, weight, maintenability etc as well. The best compromise among all those factors is to be, of course, the final criterion.
- To simulate the collector operation under real radiation conditions. This stage would predict the long-term collector performance using horizontal total radiation data. Final decisions on the selection of the collector design parameters and construction materials should now be made for the collectors to be constructed for experimental study.

Having thus determined the aims of the theoretical analysis, the necessity of detailed and, as much as possible, error-free calculations is obvious. Indeed, selection of one or another type of design or material, may be a matter of a small percentage fraction. The analysis must be capable of detecting those minor yet important differences.

Because of the required high accuracy level, the analysis is as detailed as possible : Calculations of solar energy, heat losses, convective exchange etc, may be considered "exact", in the sense that simplifying assumptions have been kept to an absolute minimum. Use of such an assumption is made whenever the "exact" solution imposes unjustifiably extensive calculations i.e. a trementous extra calculative work improve-or generally changes -the final result by anegligibly small percentage.

2.1 Assumptions used for the Theoretical Analysis

As mentioned previously, a free convection solar air collector is to be calculated, in the sense that it is to be used in a clearly passive solar heating system.

Calculation is performed for a collector in vertical position, small deviations from this position being only permissible. It is immediately apparent that this position reduces by a considerable amount the effectiveness of incident solar radiation. Yet a proper slope of the collector surface to take advantage of insolation, essentially destroys the boundary-layer free convection which is the fundamental heat exchange procedure for the operation of this type of collector, at the same time increasing convective heat exchange between absorbing plate and cover thus dramatically decreasing the collector efficiency.

For all insolation calculations use is made of horizontal total radiation data, taken from two pyranometers measuring total and diffuse radiation since several months ago in Thessaloniki. Measurements for longer periods already exist for nearby locations.

For the diffuse radiation assumption is made (1) that most of it comes from an origin near the sun and, as a result, scattering of radiation is mostly forward scattering. Under this assumption, properly justified in (1) and (2) especially for the Greek atmospheric conditions, the relation

$$H_T = H_b R_b + H_d$$

will be used for the incident radiation H_T on a (generally) tilted surface where

H_b is the beam radiation
H_d is the diffuse radiation
R_b is the correction factor for the beam radiation (a function of the various angles influencing the beam radiation)

It is obvious that under the above mentioned assumption for the diffuse radiation the corresponding correction factor for this radiation is unity.

For all calculations a steady-state operation is assumed.

Through the cover only one-dimensional heat flow is assumed. Yet the cover may absorb solar energy, may have conduction losses and finally may have a temperature gradient in the flow direction.

Similar assumptions are made for the absorbing plate. It is to be stated here that all those complicating assumptions, in contrast to those prevailing in the majority of collector calculations, are made so that plastic double-layered commercially available plates may be incorporated as cover plates for the experimental collectors, after their performance against reduced losses is theoretically investigated.

For the absorbing plate, use of non-metal materials e.g. asbestos cement, necessitates the assumption on temperature gradient along the flow direction.

The outer metal enclosure of the collector is taken to have the same (ambient) temperature on all parts of it. The same is assumed for the cover outer surface temperature.

Property values of the materials used in the collector construction, as well as those of the flowing air are assumed independent of temperature. An exception to this (3) is the term in the boundary-layer differential equation of velocities expressing the body force, clearly gravitational in free convection. In this case air density is taken to vary with temperature.

Strictly theoretically, collector edges can be treated just like the (non-metal) absorbing plate as far as boundary-layer convection, heat exchange and losses are concerned. But in view of several facts such as :

- Edges are that part of the collector surfaces which is visible from the sun for almost half the useful daytime period of its operation.
- The convection boundary-layer on edges is mixed by the corresponding layers of the absorbing plate and the cover in an extremely complicated way tending to diminish any convective exchange.
- Edge insulation (as recomended by Tabor (4)) is as good as that of the collector back, so losses from edges are a very small fraction of the total collector losses.

the edges temperature is not calculated but is assumed to be the mean air temperature in the collector.

Under this assumption edges do have losses through conduction, exchange radiation with the cover and absorbing plate but have no convection losses.

2.2 Development of Equations

By definition, (2) the collector performance is "the ratio of the useful gain over any time period to the incident solar energy over the same time period".

As is the practice, the collector useful gain is calculated through an energy balance;in this the insolation as well as various losses are taken into consideration.

Among the various parameters of the collector surfaces which enter the calculation of losses, there are two unknown temperatures. Those are the absorbing plate temperature t_p and the temperature of the inner surface of the (transparent) cover t_c. Because of the importance of those temperatures in the overall collector performance, a detailed calculation of them through partial energy balances for the absorbing plate and the cover is performed.

To this, use is made of the boundary-layer differential equation for velocities (5)

$$\rho_s\left(u\frac{\partial u}{\partial x} + \upsilon\frac{\partial u}{\partial y}\right) = \frac{\partial}{\partial y}\left(\mu\frac{\partial u}{\partial y}\right) + (\rho-\rho_s)g_x$$

in which

u,υ	are the air velocity components along and across the collector respectively
μ	is the dynamic viscosity of air
ρ_s	is the air density outside the boundary layer
ρ	is the (variable) air density inside the boundary layer
g_x	is the gravitational acceleration component along x direction (flow direction)
x,y	are the components of the coordinate system as shown in the figure

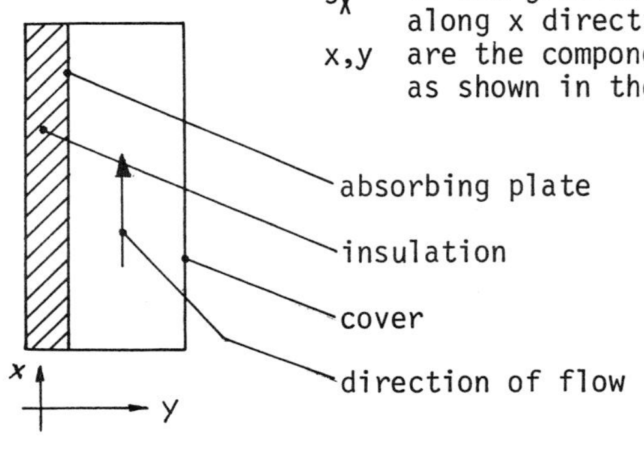

The velocities boundary-layer differential equation is complemented by the equation of continuity

$$\frac{\partial u}{\partial x} + \frac{\partial \upsilon}{\partial y} = 0$$

and the energy (temperature) boundary-layer differential equation (5) according to Pohlhausen

$$u\frac{\partial t}{\partial x} + \upsilon\frac{\partial t}{\partial y} = \alpha\frac{\partial^2 t}{\partial y^2}$$

α being the thermal diffusivity of air ($\alpha = \frac{k}{\rho c_p}$)

t the air temperature inside the boundary layer

The two boundary-layer equations and the continuity equation together with the two equations describing the thermal equilibrium on the absorbing plate and on the cover constitute a system of simultaneous equations which is to be used for the determination of t_p, t_c and at the same time of the velocity profiles along and across the plate and cover.

If surfaces are to be used as plate and cover, which justify the presence of temperature gradient along flow direction, then the simultaneous equations will be modified accordingly to provide for a variable wall temperature.

Solution of the system of equations will-at no sacrifice of accuracy- be clearly numerical or semi-numerical.

Numerical results, in the form of tables and graphs, will be given in the subsequent next project report.

3. Construction and Efficiency Measurements of Experimental Collectors

Shortly after the theoretical analysis for the collector model had began it was realized that parallel progress on the construction of experimental collectors could lead to a problem of choosing a priori materials and parameters without any properly justified arguments. A tremendous variety of materials exist to be used as covers and absorbing plates, not to mention solutions on design parameters.

So, it has been decided to start a systematic search of the market to tabulate and classify as regards cost, properties and dimensions/weight all commercially available materials. In the mean time numerical results from the collector computer model would be available to rely on for the final proper selection of parameters and materials.

For the efficiency measurement, study has already began on the CEC recomended method (draft) for testing solar collectors. Conclusions of this study will specify the constructional problems as well as our needs on measuring instruments and equipment. Provision for acquiring these will soon begin, since their order and shipment takes considerable time.

4. References

1. Beckman W.A.:''The Hottel, Whillier, Bliss Collector Model''. Workshop on Solar Collectors for Heating and Cooling of Buildings, New York N.Y., U.S.A., November 21-23, 1974.
2. Daffie J.A., Beckman W.A.:Solar Energy Thermal Processes, John Wiley & Sons, 1974.

3. Dalbert-Conseil A.M., Peube J.-L., Penot F., Robert J.-F.:"Theoretical and Experimental Investigation of the Flow in a Flat Plate Collector" International Solar Gas Heating Workshop, Perpignan, France, December 17-22, 1979.

4. Tabor H.:" Radiation, Convection and Conduction Coefficients in Solar Collectors" , Bull. Res. Coun. Israel, 6C, 155 (1958)

5. Eckert E.R.G., Drake R.M.: Heat and Mass Transfer, McGraw-Hill-Kogakusha, 1959.

SOLAR WALL WITH TILES COVERED BY A LAYER OF SPECTRAL SELECTIVE ENAMEL

Authors : Pierre DUBOIS, Dominique RAMPAZZO et Robert RAVELET

Contract number : ESA-PS-157 F

Duration : 12 months 1 april 1982 - 1 april 1983

Total budget : 250 000 Francs CEC Contribution : 125 000 F

Head of projet : Pierre DUBOIS , directeur de la Division Energie

Contractor : LABORATOIRES DE MARCOUSSIS
Centre de Recherches de la CGE

Address : Route de Nozay - 91460 MARCOUSSIS (France)

Summary

The reported work relates to solar walls with steel tiles covered by a layer of spectral selective enamel.

- We have started to study the problems involved in fixing the tiles on a concrete or metal wall in order to obtain :
  . good mechanical behaviour of the bonding under thermal cycling,
  . good transfer of the solar heat to the wall.
- We have designed a test bench for a 0,25 m2 element exposed to a sun simulator.
- We shall build and test solar wall elements covered with selective tiles :
  . a 0,25 m2 element on a test bench exposed to a sun simulator,
  . a 1 m2 element set on an experimental house.

0. INTRODUCTION

L'étude porte sur des murs captant et stockant l'énergie solaire qui sont couverts de carreaux émaillés sélectifs pour en améliorer les performances.

Nous présenterons successivement :

- le principe des murs solaires étudiés,
- l'intérêt et les performances des revêtements émaillés sélectifs,
- les résultats obtenus sur la fixation des carreaux sur le mur,
- l'étude du banc pour les essais sous soleil artificiel d'un élément de mur, de 0,25 m2.

Dans le cadre de l'étude, nous poursuivrons les études de fixation des carreaux sur le mur, puis nous réaliserons et nous essaierons :

- un élément de mur de 0,25 m2 sous soleil artificiel,
- un élément de mur de 1 m2 installé sur une maison expérimentale sous ensoleillement naturel.

Les essais permettront de préciser la valeur thermique des revêtements mis en place.

1. PRINCIPE DES MURS SOLAIRES ETUDIES

On voit sur la figure 1 le principe du mur solaire étudié qui comprend, à partir de l'extérieur :

- une vitre,
- une lame d'air avant,
- l'élément de mur proprement dit qui est recouvert de carreaux sélectifs,
- une lame d'air arrière,
- une isolation thermique.

L'énergie solaire est captée par l'élément de mur qui peut être simplement un mur épais en béton ou un mur plus mince contenant un produit stockant la chaleur par changement d'état. On voit sur la figure 2 le principe d'un élément de mur en fibrociment contenant de la paraffine et sur la figure 3 le principe d'un élément de mur métallique contenant de la paraffine.

L'énergie solaire est rétrocédée au logement par réchauffage de l'air neuf fourni au logement, air qui circule dans les deux lames d'air situées de part et d'autre de l'élément stockant. La circulation d'air dans la lame d'air avant permet de récupérer une partie de la chaleur réémise par le mur, cette émission de chaleur étant d'ailleurs limitée par le revêtement sélectif, aussi bien pendant la captation et le stockage que pendant le déstockage.

2. INTERET ET PERFORMANCES DES CARREAUX SELECTIFS

Les modifications des propriétés optiques des revêtements émaillés rendus sélectifs doivent permettre d'augmenter les performances thermiques du mur (cf. figure 4)

Par ailleurs, le comportement dans le temps des revêtements est excellent. On ne note aucune évolution des propriétés optiques après 8 000 heures de vieillissement dans un weather-ometer à 75°C et sous humidité saturante et le comportement des revêtements essayés en brouillard salin est excellent.

3. FIXATION DES CARREAUX SUR UN MUR

31. Les produits essayés

Selon le mur considéré, les carreaux devront être fixés sur du béton ou sur du métal (cf. paragraphe 1 et figures 1, 2 et 3). Nous avons donc fait des essais de fixation sur deux types de supports :

- tôle en aluminium pouvant constituer l'enveloppe d'un stock à paraffine,
- bloc de béton à granulométrie fine pour élément de mur en béton ou en fibrociment avec paraffine.

Nous avons fait des essais avec trois colles industrielles susceptibles d'être appliquées in-situ sur un mur ; ces trois colles sont :

- colle 3M réf. EC 1236 : colle à base de caoutchouc synthétique applicable au pinceau,
- colle BOSTIK 1400 : colle contact néoprène applicable à la racle,
- mastic colle Sapostyrène : dispersion de styrène acrylique en milieu aqueux.

32. Les résultats obtenus

321. Aspects thermiques

On mesure les températures de surface des carreaux émaillés sur les deux faces et la température du support sur lequel ils sont collés, le tout étant soumis à un rayonnement lumineux de l'ordre de 1 000 W/m2.

- Collage sur aluminium

Pour les trois colles essayées, l'écart de température entre les carreaux et la tôle d'aluminium est de 1°C, alors que les plaques sont à 46°C.

- Collage sur béton

On a représenté sur la figure 5 des résultats obtenus avec les trois produits de collage étudiés. L'écart de température entre le carreau et son support est d'environ :

. 1°C pour le carreau collé sur le mastic sapostyrène
. 5°C pour le carreau collé avec la colle EC 1236
. 7°C pour le carreau collé avec la colle B 1400

De plus un écart de température entre le centre et les bords des carreaux est observé pour les deux carreaux collés à la colle EC 1236 et B1400, cet écart étant dû à la non planéité des carreaux.

322. Aspects mécaniques

Des essais de tenue des collages sur supports métalliques et sur supports béton sont en cours :
- tenue aux cyclages thermiques entre 20 et 100°C,
- recherche de la température maximale de non-destruction du collage.

Les résultats actuels sur support béton permettent de classer les colles dans l'ordre suivant : mastic Sapostyrène, colle EC 1236, colle B1400.

4. ETUDE DU PROJET DE BANC D'ESSAIS POUR ELEMENT DE MUR DE 0,25 m2

Le banc d'essais en ensoleillement artificiel pour un élément de mur de 0,25 m2 dont nous avons établi le projet sera placé devant un simulateur solaire existant à Marcoussis.

41. Simulateur solaire (cf. figure 6)

Une lampe au xénon type YBO d'une puissance électrique de 6 500 W produit un rayonnement lumineux représentatif du spectre solaire qui traverse ensuite un condenseur optique type SIMLES à 14 faisceaux élémentaires et est renvoyé en direction horizontale par un miroir. L'ensemble du simulateur est refroidi par un ventilateur d'air, lui-même refroidi par un échangeur à eau. La puissance d'éclairement du simulateur est de 1240 W/m2 pour un diamètre de tache lumineuse de 900 mm avec une uniformité de tache de $\pm$ 1 %. Un variateur d'intensité permet d'ajuster la puissance d'éclairement.

42. Banc d'essais pour un élément de mur solaire de 0,25 m2

Nous avons établi le projet du banc d'essais représenté sur la figure 7. Le support est mobile pour pouvoir faire varier le flux d'éclairement sur l'élément essayé. Il est placé à environ 4 mètres, pour un flux de 800 W/m2. Il permet d'essayer des éléments de mur solaire de 0,25 m2 comprenant un vitrage, un élément de mur proprement dit captant et stockant et une isolation thermique arrière. L'air peut circuler de chaque côté de l'élément stockant et des volets permettent de créer plusieurs organisations de la circulation d'air. Un divergent et un convergent placés de part et d'autre de l'élément de mur assurent une bonne répartition de la circulation d'air dans un plan horizontal. L'air entrant parcourt l'élément de mur de bas en haut, traverse un débitmètre à diaphragme relié à un transmetteur électrique de pression puis un ventilateur d'extraction alimenté par un variateur de puissance. L'énergie solaire incidente est mesurée par un pyranomètre placé dans la zone éclairée relié à un intégrateur. L'ensemble des mesures est enregistré pour intégrer les puissances thermiques au cours du temps.

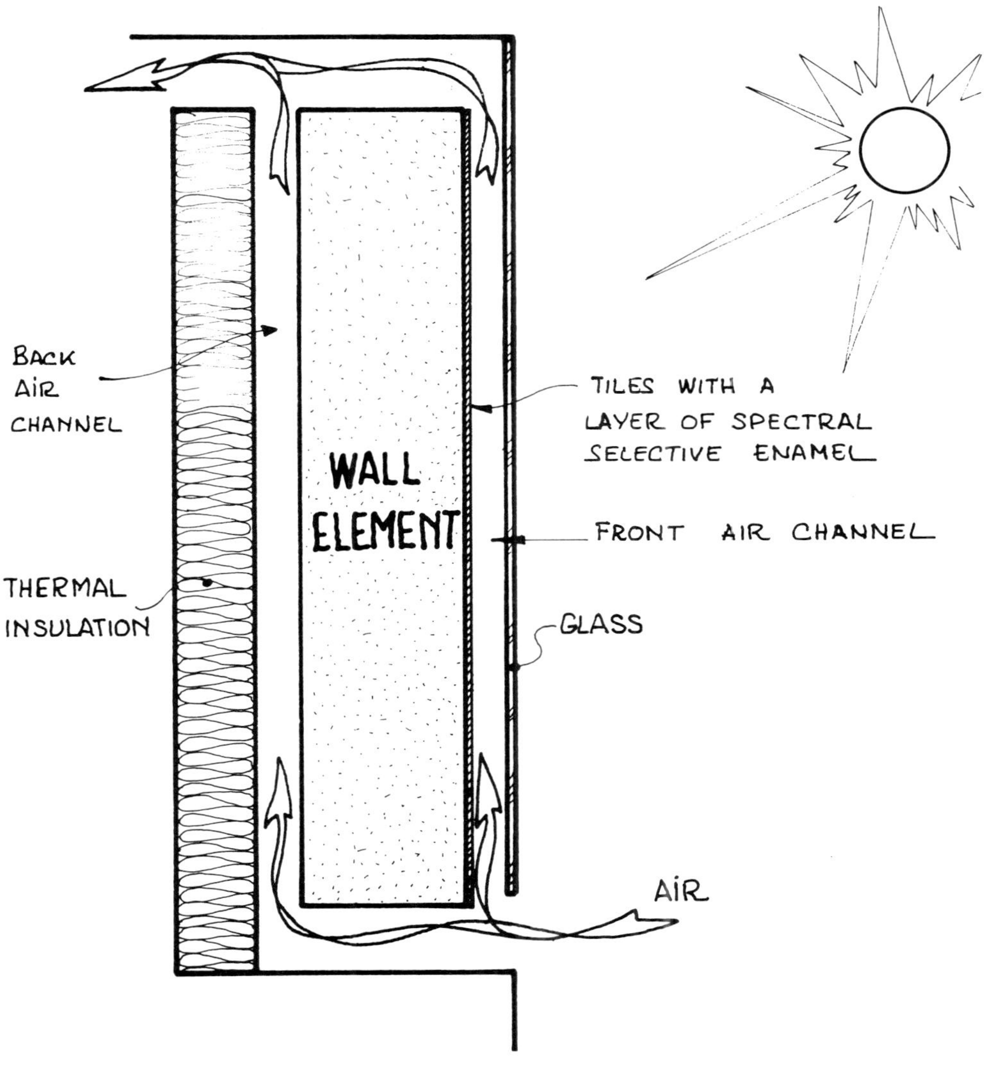

Fig. 1 - Solar wall

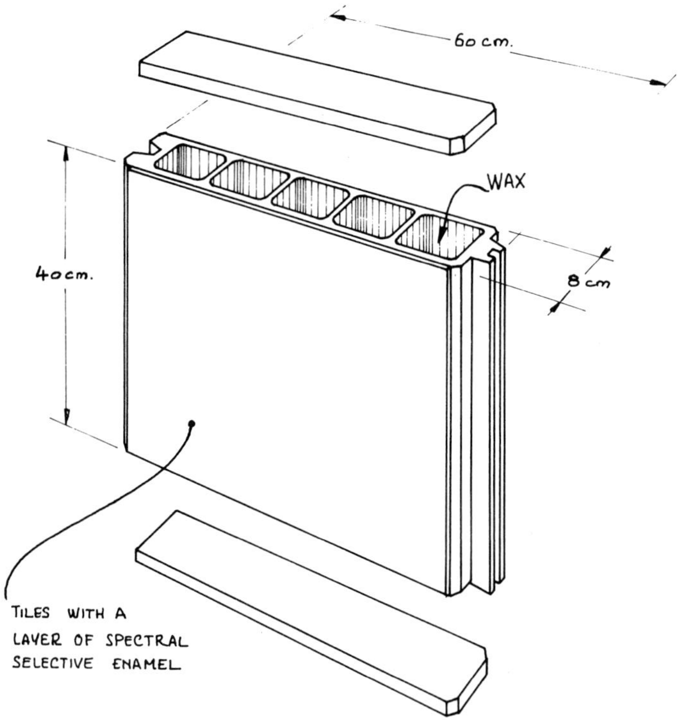

Fig. 2 - Wall element (wax & cement)

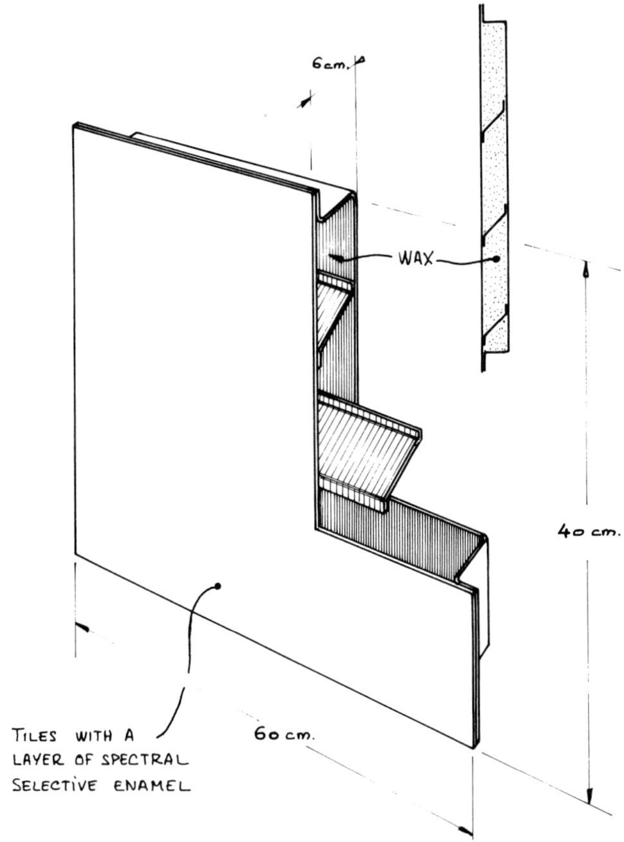

Fig. 3 - Wall element (wax & metal).

Fig. 4 - **TILES OPTICAL CARACTERISTICS**

COLOR	TILES WITHOUT SPECTRAL SELECTIVE COATING		TILES WITH SPECTRAL SELECTIVE COATING	
	Absorption coefficient	Emission coefficient (40°C)	Absorption coefficient	Emission coefficient '40°C)
Yellow	0,43	0,81	0,77	0,30
Orange	0,45	0,81	0,77	0,30
Blue	0,72	0,81	0,84	0,23
Red	0,58	0,69	0,84	0,25
Light green	0,79	0,68	0,86	0,27
Dark blue	0,93	0,81	0,96	0,23

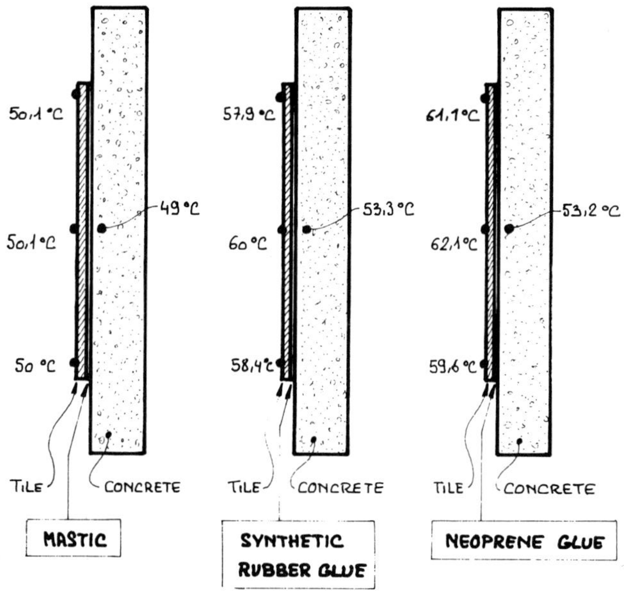

Fig. 5 - Tile-concrete bonding - heat transfer

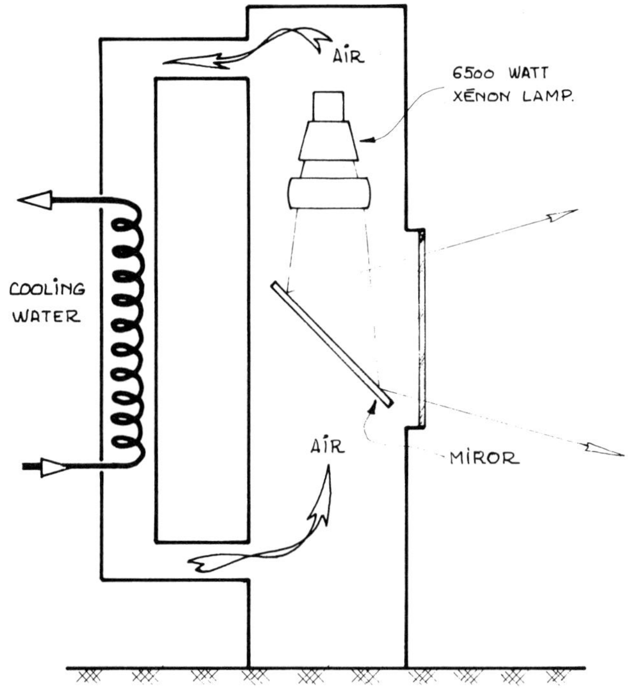

Fig. 6 - Sun simulator

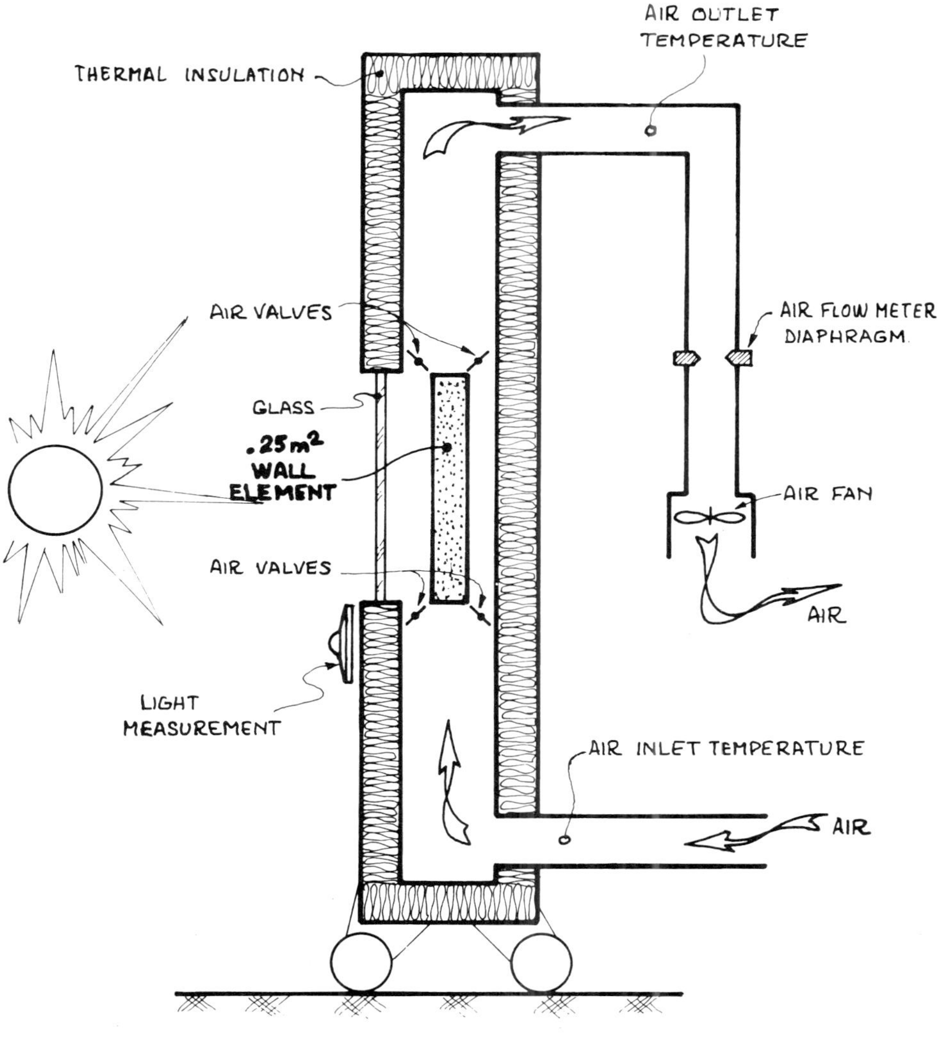

Fig. 7 - Solar wall test bench (for a .25 m^2 element)

SYSTEM FOR WARMING THE NEW AIR USING THE SOLAR ENERGY IN AN INDIVIDUAL HOUSE

Authors : Société LES MAISONS BRUNO-PETIT
Direction Etudes et Projets

Contract number : ESA . PS . 153 . F

Duration : 18 months 1 Januar 1982 - 30 June 1983

Total budget : FF 200 000 CEE contribution : FF 100 000

Head of project : Mr Jean Philippe BOBOT

Contractor : LES MAISONS BRUNO-PETIT

Address : 21 rue des Capucins - F. 92190 MEUDON

Summary

The project designs the capability of decreasing heat losses because of airing in building, by warming the new air using solar energy. The second aim of the study will be define precisely the most performing solution in a given area from the climate datas of each French weather centre. The combination of differents solutions induced to conceive two solar technical solutions :

1. Solar A variant
The renewed air goes into the heat exchanger where it recovers the extracted air calories, with possible air retacking in the greenhouse by a motorized by pass. In winter, the new air taken outside is warmed in the heat exchanger before it is blown in the building. In summer, a manual by pass permits to blown the fresh air from the north frontage.

2. Solar B variant
The airing is made by insufflation of new air in the building. This air passes through a solar energy collector and an accumulating element to delay the solar supply, before it is blown in the building.

3. Complementarity of the greenhouse
The greenhouse is connected to the main system pre-heating the new air and doesn't shorten the performances of this system (exchanger in A variant, solar air collectors in B variant).

1. Introduction

Le projet consiste à étudier la possibilité de réduire les pertes par renouvellement d'air dans la construction en préchauffant l'air neuf introduit par l'utilisation de l'énergie solaire. Pour aboutir à des résultats concrets cette étude nous a conduit à concevoir une maison solaire incorporant tous les dispositifs thermiques, objet de l'étude. La maison solaire est l'aboutissement d'un travail en commun, commencé dès la conception du projet, entre l'architecte et le thermicien. Le cabinet d'architecture EGREGORE a créé pour les maisons BRUNO-PETIT la maison OSIRIS qui prend en compte les impératifs architecturaux des sites et des régions et les impératifs thermiques des zones climatiques correspondantes.

La maison OSIRIS que nous avons dessinée utilise pour sa construction des composants industriels qui s'insèrent dans une technique de construction moderne et performante.

Cette maison a été traitée pour répondre à tous les modes d'urbanisme groupés ou diffus.

Cette maison a pour objectif également de répondre à la demande d'une maison économe en énergie pour un coût d'investissement réduit et un coût d'entretien quasiment inexistant : c'est pour ces raisons que la maison OSIRIS utilise un système solaire passif faisant appel à des composants fiables, connus et éprouvés.

2. Généralités sur l'aspect thermique

L'obtention d'un projet économe en énergie passe avant tout par un niveau d'isolation très élevé.

Le système de construction BRUNO-PETIT utilisé pour réaliser la maison permet d'obtenir des coefficients G correspondant aux normes haute isolation actuelles par suppression des principaux ponts thermiques d'une construction classique.

- continuité de l'isolation mur-plafond,
- absence de refend,
- liaison soignée de la menuiserie en bois à la façade.

L'étude de l'isolation de la maison a permis d'établir pour chaque paroi le rapport $\frac{\Delta C}{\Delta G}$ du surcoût d'isolation envisagé à l'amélioration du G.

Cette étude a mis en valeur l'efficacité :

- de l'isolation en polystyrène sur toute la surface du plancher en terre-plein,
- de fermetures nocturnes thermiquement efficaces (volets isolants) que nous avons étudiées par ailleurs.

L'isolation générale et le renforcement d'isolation en ces points particuliers permettent de réduire la charge thermique du modèle en obtenant un coefficient G moyen (hors préchauffage d'air neuf de renouvellement) de 0,90 $W/m^3{}^\circ C$.

3. Présentation des variantes solaires A et B

La limitation du surcoût solaire et le souci de simplicité et de fiabilité du système thermique ont conduit à opter pour la réduction des pertes par renouvellement d'air par préchauffage de l'air neuf introduit.

Plusieurs solutions peuvent être envisagées pour le préchauffage d'air neuf introduit et en particulier :

- un espace serre : cette solution conduit à une serre de grande dimension (donc de surcoût important) pour obtenir un taux de renouvellement acceptable sans gêner l'occupant,
- un échangeur statique sur l'air extrait : cette solution est classique. Le système est d'autant plus efficace que le climat est rigoureux.

- des capteurs à air : cette solution est intéressante dans la mesure où les capteurs travaillent avec un bon rendement constant et que les surfaces captrices sont relativement réduites.

Ces différentes solutions et leur combinaison possible nous ont conduit à imaginer et à proposer deux solutions techniques solaires en fonction des données climatiques.

3.1 Variante solaire A

Le renouvellement d'air est traité par un système double flux et échangeur statique avec reprise possible sur la serre par une vanne motorisée.

En hiver, l'air neuf pris à l'extérieur est préchauffé sur l'échangeur avant d'être insufflé dans les pièces principales.

Si la température de la serre s'élève au-dessus d'une température de consigne fixée (~ 25°C) la vanne motorisée permet la récupération de cette surchauffe pour en faire profiter les pièces principales.

La hauteur sous plafond importante permet de regrouper en faux-plafond dans la partie dégagement tous les réseaux d'insufflation et de limiter ainsi les pertes par distribution, l'isolation du comble se trouvant au-dessus.

En été un by pass manuel permet de prendre l'air neuf en façade nord et de le souffler directement.

La partie haute de la serre est constituée d'une verrière en double vitrage comprenant 4 trames de 0.90 m x 1.00 m, permettant un gain direct important sur le mur refend accumulateur en fond de séjour. Ce mur traité en maçonnerie lourde permet d'augmenter l'inertie du pavillon.

Le système solaire est complété par un chauffe-eau électro-solaire et 4 trames de 0.90 x 1.00 m de capteur à eau SITRACO.

La proximité du ballon électro-solaire et des différents points d'eau des pièces de services permet de limiter les pertes de distribution.

3.2 Variante solaire B

Le renouvellement d'air est traité par insufflation d'air neuf dans les pièces principales. Cet air transite par les capteurs à air SITRACO (4 trames de 0.90 m x 1.50 m) avant de circuler dans le mur stockeur-déphaseur.

Comme dans la variante A, une reprise sur serre est possible par une vanne motorisée pour récupération des surchauffes.

L'extraction mécanique se fait dans chaque pièce de services.

Les combles perdus permettent un accès facile aux filtres d'entrée des capteurs qui peuvent être facilement nettoyés.

La partie haute de la serre est constituée par les 4 trames de capteurs à air (5.4 m^2).

Le système solaire est complété par un chauffe-eau électro-solaire et 4 trames de 0.90 m x 1.00 m de capteurs à eau SITRACO placés en prolongation des capteurs air.

4. Choix entre les variantes A et B

L'énergie récupérée par l'échangeur de la variante A peut s'écrire :

$$E_A = 0.34 \; Q_{ech} \; \eta_{ech} \; \frac{24 \, DJ}{1000} \quad (Kwh)$$

avec Q_{ech} débit à l'échangeur = 150 m^3/h

η_{ech} rendement à l'échangeur = 0.70

Le préchauffage de l'air neuf par capteurs à air permet de récupérer une énergie annuelle égale à :

$$E_B = \eta c \; Sc \sum_i Ni \; Ei$$

avec ηc rendement des capteurs = 0.42

Sc surface des capteurs = 5.4 m^2

Ni nombre de jours mensuels

Ei énergie solaire globale KWh/m^2.jour

Le calcul des valeurs E_A et E_B à partir des données climatiques de PARIS et de CARPENTRAS conduit aux valeurs suivantes :

PARIS : E_A = 2150 Kwh E_B = 1300 Kwh

CARPENTRAS : E_A = 1680 Kwh E_B = 2050 Kwh

Il semble donc intéressant de pouvoir choisir en fonction des données climatiques de chaque région (ensoleillement et températures extérieures) la solution thermique la mieux adaptée.

D'une façon générale on peut penser que l'on aura le découpage suivant :

Zone à climat froid et ensoleillement moyen ➡ Variante A

Zone à climat tempéré et bon ensoleillement ➡ Variante B

Mais ce choix pourra s'effectuer de façon plus précise pour chaque construction prévue à partir des données climatiques représentatives de son site en étendant le fichier de la méthode de calcul.

Le surcoût voisin des 2 solutions A et B autorise un choix purement thermique pour atteindre la meilleure économie en fonction de la zone climatique du lieu de construction de la maison.

5. Calcul des pertes par renouvellement d'air

Le projet étant axé sur les pertes thermiques par renouvellement d'air, nous avons déterminé spécifiquement ces pertes sur les maisons OSIRIS variantes A et B.

5.1 Perte par renouvellement d'air dans la variante A

Volume habitable pièces principales $Vhpp = 164.1 \; m^3$

$qe = qs \# 150 \; m^3$

$P = \Sigma mAm = 0.3 \; Sv + 1.2 \; S_{PE}$

$$\left.\begin{array}{l} Sv = 21.5 \; m^2 \\ S_{PE} = 2.0 \; m^2 \end{array}\right\} \quad P = 8.85$$

$é = 1.1$ $\Sigma Pé = 9.7$

$D_{RA} = (164.1 + 9.7) \; 0.34 = 59.1$ <u>DRA = 59.1 W/C</u>

Ces pertes correspondent à 24 % des pertes thermiques de jour et à 18 % des pertes thermiques de nuit.

5.2 Perte par renouvellement d'air dans la variante B

Volume habitable pièces principales Vhpp = 154.7 m^3

$qe = qs \# 150 m^3$

$P = \Sigma mAm = 0.3 Sv + 1.25 S_{PE}$

$$\left.\begin{aligned} Sv &= 17.9 m^2 \\ S_{PE} &= 2.0 m^2 \end{aligned}\right\} \quad P = 7.8$$

é = 1.1 ΣPé = 8.6

DRA = (154.7 + 8.6) 0.34 = 55.5 DRA = 55.5 W/°c

Ces pertes correspondent à 25 % des pertes thermiques de jour et à 30 % des pertes thermiques de nuit.

Cela représente (hors apport solaire) environ 10 % du coût énergétique de la maison au lieu de 39 %.

6. Etat d'avancement des travaux

Une maison correspondant à la variante solaire A vient d'être construite et nous allons pouvoir mesurer maintenant les résultats thermiques réels. Cette maison est habitée.

Une maison correspondant à la variante solaire B va être bientôt construite et le chantier sera terminé dans le dernier trimestre 1982. Les mesures pourront donc commencer pendant l'hiver 1982 sur ce modèle.

7. Plans des systèmes de renouvellement d'air

Les solutions pour récupérer les pertes par renouvellement d'air sont définies par les plans joints en annexe. Ce sont ces systèmes qui sont mis en oeuvre sur les deux maisons en construction et que nous allons tester.

Cette présentation étant succincte nous ne donnerons pas ici le descriptif du dispositif et des appareillages utilisés.

ANNEXE I

SCHEMA DE PRINCIPE DES SYSTEMES DE RENOUVELLEMENT D'AIR DANS LES VARIANTES A ET B

VARIANTE **A**

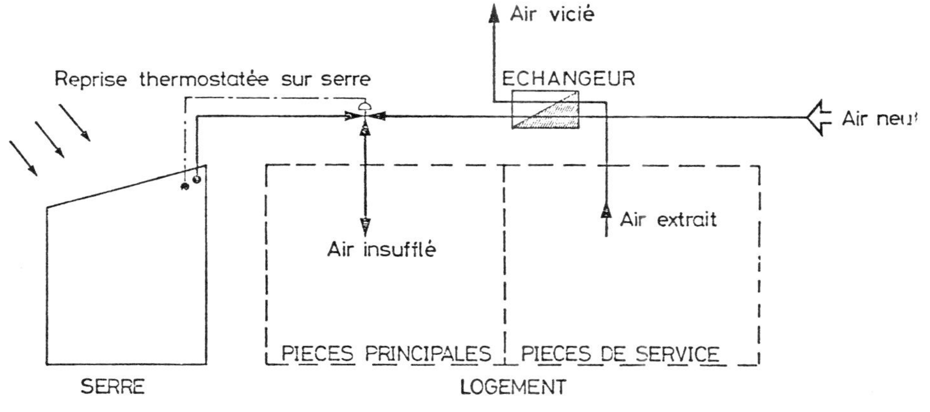

VARIANTE **B**

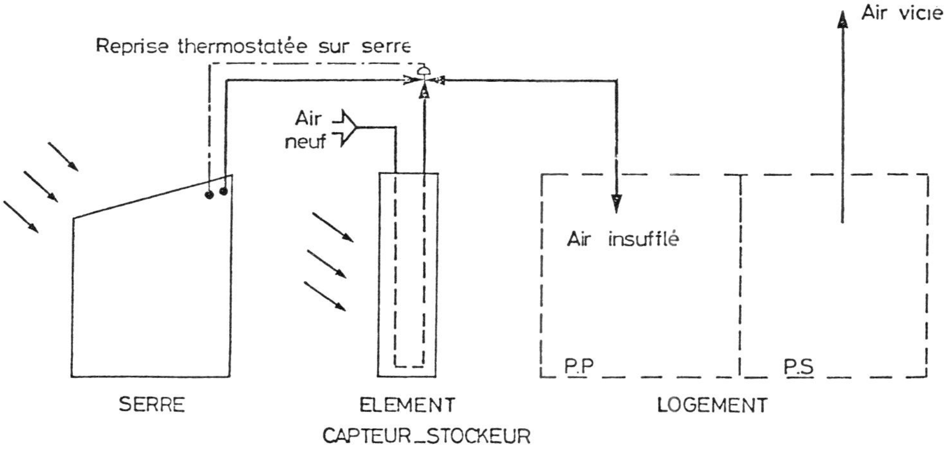

ANNEXE II

PRINCIPE DES MAISONS VARIANTES A ET B

La maison OSIRIS sur laquelle les systèmes de renouvellement d'air sont mis en place

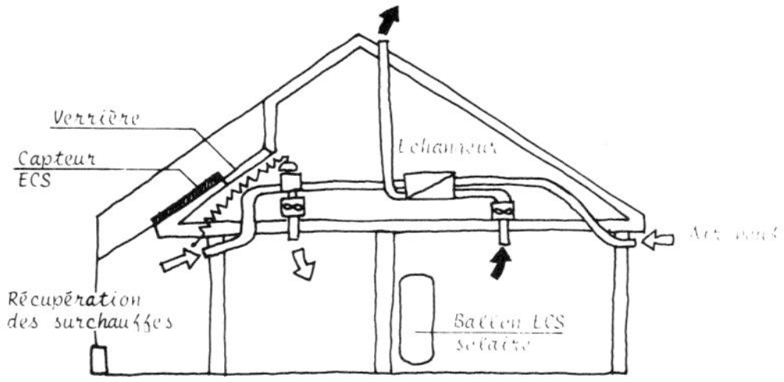

◀ VARIANTE A :

* SYSTEME DE VENTILATION DOUBLE FLUX AVEC ECHANGEUR ET RECUPERATION DES SURCHAUFFES DE LA SERRE.
* GAIN DIRECT IMPORTANT PAR LA VERRIERE SUR SEJOUR.

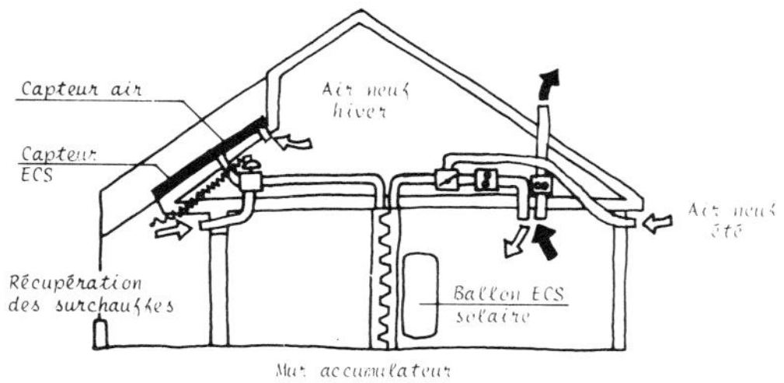

◀ VARIANTE B :

* CAPTEURS AIR ET MUR ACCUMULATEUR POUR DEPHASAGE DES APPORTS SOLAIRES.
* GAIN DIRECT REDUIT POUR LIMITER LES SURCHAUFFES D'ETE.

ANNEXE III

VARIANTE A : PLAN DU DISPOSITIF
SCHEMA DE FONCTIONNEMENT EN HIVER

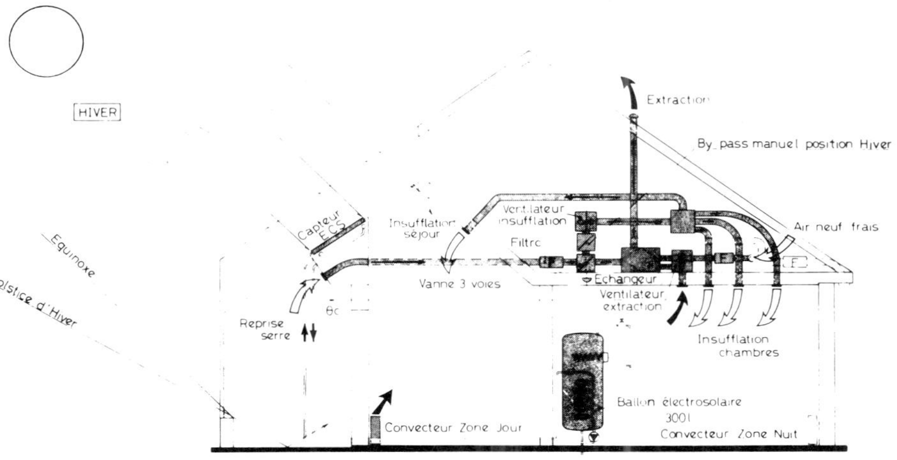

VARIANTE A

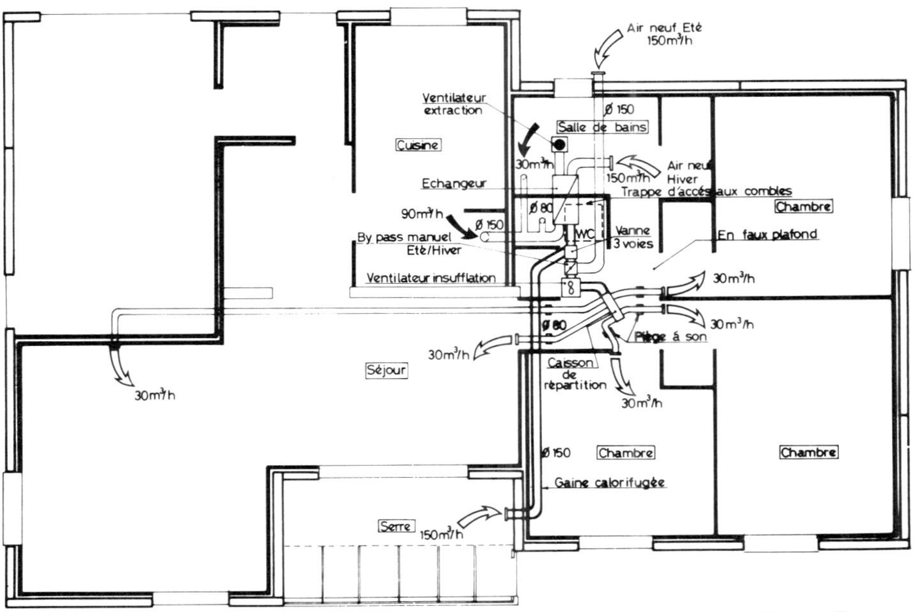

VARIANTE A

ANNEXE IV

VARIANTE B : PLAN DU DISPOSITIF
SCHEMA DE FONCTIONNEMENT EN HIVER

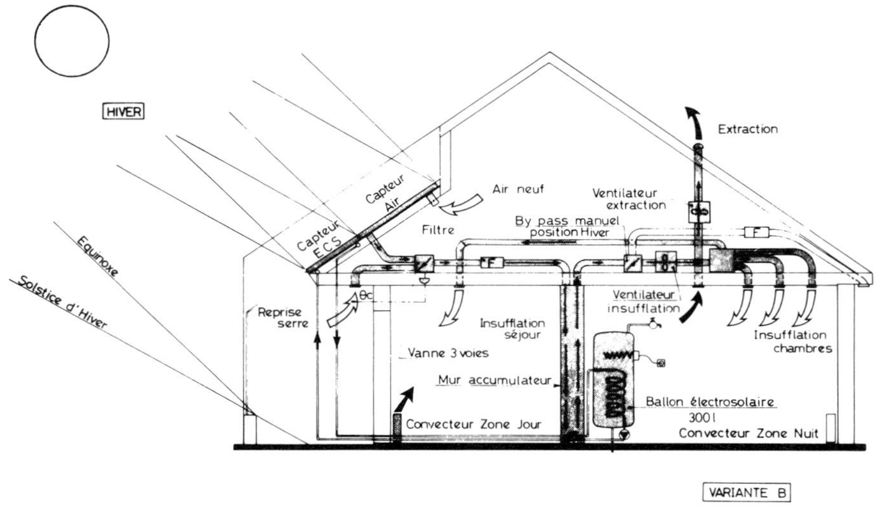

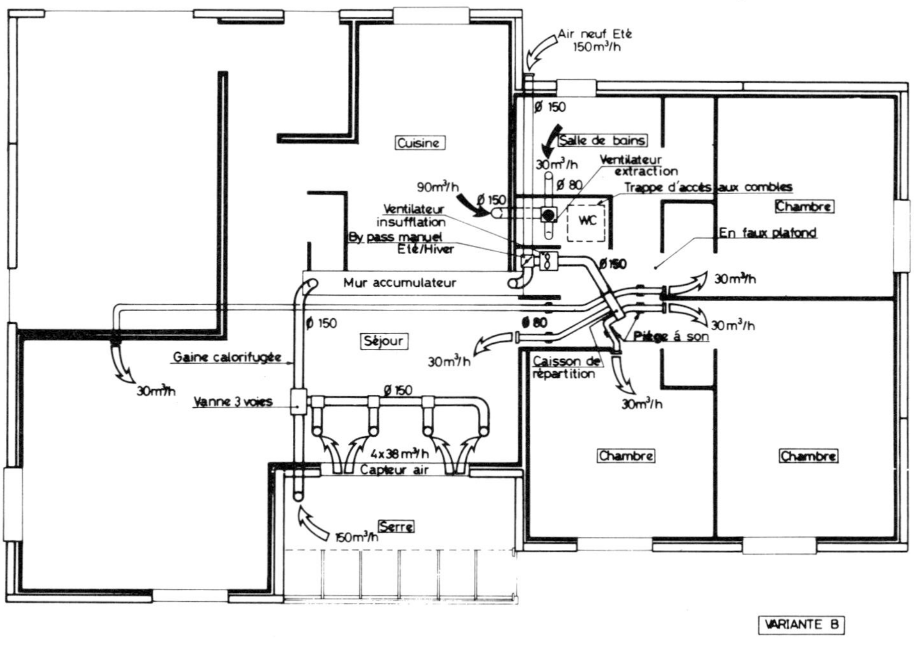

ANNEXE V

VARIANTES A ET B : SCHEMAS DE FONCTIONNEMENT EN ETE

Ces schémas montrent que le système de renouvellement d'air prend en compte la thermique d'été

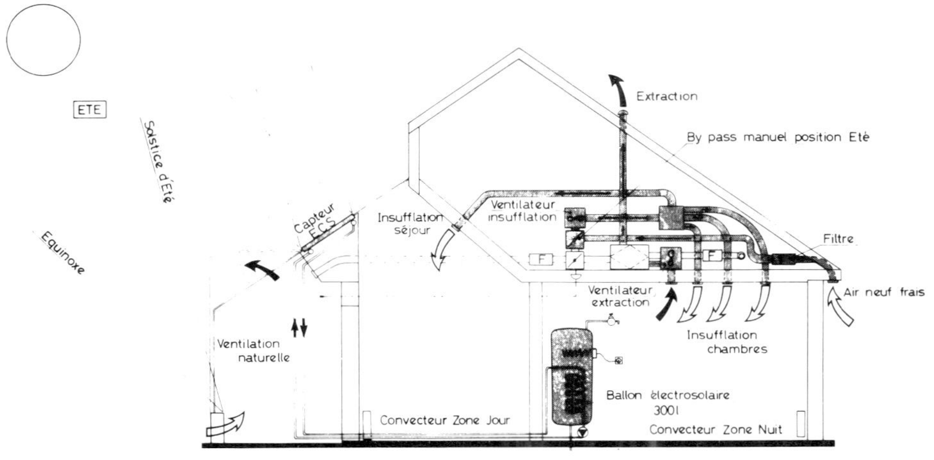

VARIANTE A

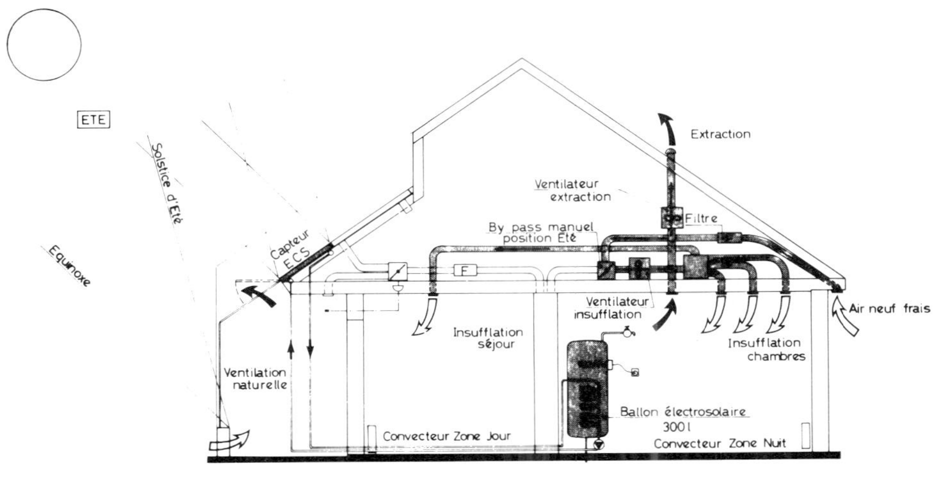

VARIANTE B

SECTION 3 - HIGH PERFORMANCE COLLECTORS

SUMMARY ON HIGH PERFORMANCE COLLECTORS

Author: C. den Ouden

1. Introduction

In the section 'High Performance Collectors' there were four presentations, two of which were devoted to selective coatings, one to a study of a special focussing collector and one to a system specific high performance collector for a solar cooling plant for developing countries.

2. Review of the presentations

The first presentation entitled 'Development of improved solar energy absorber surfaces' was given by Dr. Bannard. He was still working on the theoretical aspects of achieving high values for the absorbtivity and low values for the emissivity and there was some delay with in experimental part of the work.

The second presentation on the 'Development of Cobalt-Oxide based Spectral Selective Coatings for Solar Collectors' given by Mr Boose showed interesting improvement of several types of selective coatings for high temperature applications. Interesting tests of corrosion resistance at high temperatures of several samples have been shown and a new approach for a realistic cyclic humidity test at higher temperatures was discussed. Moreover, a full scale operating electroplating plant had been built for the treatment of surfaces with sizes up to 1 by 2 metres for the most promising types of coating. With this plant it is possible to treat surfaces for a cost of approximately Dfl. 20,--/m^2. He showed the details of these cost, subdivided into the various steps of the production phases.

In the third presentation Prof. Bougard discussed the interim results of the study and construction of a Solar Focussing Collector using a deformable mirror, in the temperature range 100-200 $^\circ$ C.
This interesting focussing collector, uses a mechanism which changes the collector geometry in order to perform seasonal sun tracking without having to adjust or tilt the collector. The system is driven by a single motor which makes both the receiver and the shape of the mirror moving. The entire mechanism is enclosed in the collector casing and protected from atmospheric influences. He showed a realised prototype for outdoor closed loop mid-temperature tests to conform the theoretical and experimental results obtained so far.
The absorber of this prototype has been treated with a selective coating in the full scale electroplating plant discussed in the previous presentation. Experimental testing of this prototype is to be expected during the summer period of 1982.

The final presentation in the section 'High Performance Collectors' dealt with a 'High Performance Collector for deep cooling facilities'. This presentation by Mr. Hause of Dornier discussed the double glazed selective Dornier heat pipe collector as a component in a total system to supply the energy for a cooling machine to cool a food conservation plant. He showed the need for high efficiency collectors to improve the functioning of the cooling machine.

DEVELOPMENT OF IMPROVED SOLAR HEAT-ENERGY ABSORBER SURFACES

Authors : J E Bannard, J Hayden, P O'Malley

Contract number : ESA/C/045/EIR

Duration : 36 months, 1 July, 1980 - 30 June, 1983

Total budget : IR£111,060; CEE contribution 36%

Head of project : Dr J E Bannard

Contractor : National Institute for Higher Education,
Limerick
Ireland.

Summmary

Black selective solar energy absorbers are usually metallic mouldings carrying one or more special coatings. The nature of the composite leads to problems of high IR emissivity, low stability and high cost. The purpose of this project is to solve these problems by showing that a selective surface may be produced simply by controlling the topography of a metallic absorber.

To date the mathematical model has reached an advanced stage and predicts that most of the incident radiation may be adsorbed into a surface if the surface has the appropriate topography. Physical models manufactured by a process of photolithography are being used to test the theory. Additionally, a means is being sought to produce the required topography by electroplating of pure metal.

1. Introduction

Many materials exist for absorbing energy from the solar spectrum, but problems of stability under prolonged use are common. Additionally all of the absorber materials being non-metal exhibit rather high IR emissivity. The ideal selective absorber would be a metal (low IR emissivity) provided it displayed high solar absorptivity and was also stable and corrosion resistant. For concentrated collectors the material should also show high temperature stability.

The purpose of this project is to show that the interaction of incident radiation with the surface of a metal may be made to be non-reflective by ensuring that the surface topography acts as a graded refractive index. Just as coatings are placed on lenses to control reflections, the graded index is a parameter which should lead to total absorption. Because we are dealing with metals, the emissivities in the infra-red region of the spectrum should be low. Thus a mathematical model was to be developed to show that this was correct. At the same time it was thought appropriate to develop a technique to produce physical models to test the proposals. This technique was described in the last report (1). The technique necessarily produces axes of reflection on the surface, and it was therefore, anticipated that a random array of the surface asperities would be necessary in order to achieve 100% absorption. This is being tackled <u>via</u> a study of electroplating; it is proposed that nucleation theory will predict that nuclei can be formed on a surface following pulsing at a certain potential, and that these nuclei may be grown, in the absence of further nucleation, to produce an undulated surface. It was proposed that under the appropriate conditions, a suitable topography will be obtained free from axes of reflection. It is anticipated that both approaches, the physical model approach and the electroplating approach, will lead to the production of cheap selective absorbers. This paper reports progress to date in these two areas.

2. Experimental

2.1. The model for regarding the 'rough' surface of the absorber as a graded refractive index takes layers through the surface parallel to the plane. The mean value of the refractive index of the layer (plane) at the top of the stack is close to that of air, and the value of the bottom layer is that of the absorber bulk material. The preliminary assumptions of the model are:

(1) The light is at normal incidence (this restriction is later eliminated).
(2) There are no absorption losses, and
(3) There are N layers (100 seems sufficient).

The reflection coefficient at the ijth layer interface is

$$r_{ij} = \frac{(n_j - n_i)}{(n_j + n_i)} = \frac{(n_{i+1} - n_i)}{(n_{i+1} + n_i)}$$

where n_i is the refractive index of the ith layer

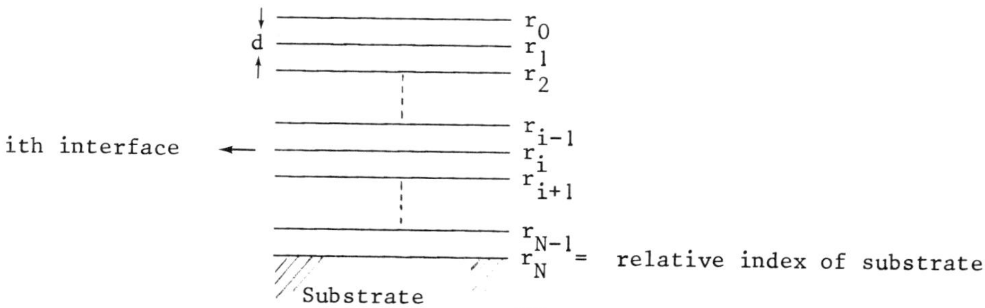

The phase of the wave reflected from r_i relative to the wave from r_{i-1}

$$\Delta\theta = 2\pi . \text{optical path} / \lambda$$

where λ is the wave length of the light ; i.e.

$$\Delta\theta = 2\pi . 2d . n_i / \lambda$$

As all the layers have the same thickness

$$(\Delta\theta)_{i,i-1} = 4\pi d n_i / \lambda .$$

Now the incoming amplitude arriving at the ith interface

$$A_i = A_o (1-r_o)(1-r_1) \ldots\ldots (1-r_{i-1}) \quad \ldots (1)$$

$$= A_o \prod_{j=0}^{i-1} (1-r_i)$$

and the phase of the wave reflcted at r_i relative to the front reflection r_o is given by

$$(\Delta\theta)_{i,o} = \sum_{j=1}^{i} (\Delta\theta)_{j,j-1} = 4\pi d/\lambda \sum_{j=1}^{i} (n_j) \quad \ldots (2)$$

The reflection from the stack is then given by evaluating the total reflected wave at the uppermost interface, i.e.

$$R = \sum R_i$$

where R_i describes the amplitude and phase of the wave reflected from the bottom of layer i when it returns to the top of the stack. Anything transmitted through the last layer is neglected; in a metal no light would be expected to penetrate this far. The amplitude and phase of second reflections are also neglected. The phase of R_i relative to R_o is given in equation (2). The amplitude given in equation (1) must be compensated for a second reflection from each layer as it returns to the top of the stack. So the amplitude returning to the top from the layer

$$R_i = A_o r_i \prod_{j=0}^{i-1} (1-r_i)^2$$

where $(1-r_i)$ is the transmission coefficient at each layer.

The phase of R_i is given by that of the first reflected wave plus the phase lag, $(\Delta\theta)_{i,o}$. For the first reflection the wave may be described as

$$R_o = A_o r_o \cos(\omega t + \theta_o)$$

$$= A_o r_o \cos(\omega t - \theta_o)$$

for the second reflection

$$R_1 = A_o r_1 (1-r_o)^2 \cos(\omega t - (\Delta\theta)_{1,o}).$$

The difference in phase of the wave from layers 1 and 0 is given by $(\Delta\theta)_{1,0}$.

For an arbitrary layer i, the wave reflected from the interface between i and i+1 and returned to the top is given by:

$$R_i = A_0 r_i \prod_{j=0}^{i-1} (1-r_j)^2 \cos\left(\omega t - \sum_{j=1}^{i} (4\pi d n_j/\lambda)\right).$$

For the sum of all such reflected waves

$$R = \sum_{i=0}^{N-1} R_i.$$

A number of different approaches may be made for the reflection coefficients; the transmission coefficient is given by (1-r). We have decided to investigate the polarised components of the incident light i.e. whether the electric vector is parallel or perpendicular to the interface or a combination of the two extremes. Snell's law gives:

$$n_1 \sin\theta_t = n_0 \sin\theta_i$$

where θ_i is the incident angle, and θ_t the transmitted angle, from which we may derive the following:

$$r_\perp = \frac{\left(\frac{n_1}{n_0}\right)^2 \cos\theta_0 - \sqrt{\left(\frac{n_1}{n_0}\right)^2 - \sin^2\theta_0}}{\left(\frac{n_1}{n_0}\right)^2 \cos\theta_0 + \sqrt{\left(\frac{n_1}{n_0}\right)^2 - \sin^2\theta_0}}$$

and

$$r_{\parallel} = \frac{\cos\theta_0 - \sqrt{\left(\frac{n_1}{n_0}\right)^2 - \sin^2\theta_0}}{\cos\theta_0 + \sqrt{\left(\frac{n_1}{n_0}\right)^2 - \sin^2\theta_0}}$$

where $\perp$ is the electric vector perpendicular to the interface, and $\parallel$ is parallel to the interface.

Using the refractive index of the individual layers, the preceding shows us that it is possible to calculate the angles through the layers, which the light will follow and the reflection from, and transmission through, each layer. It is then easy to incorporate an initial angle of incidence and angle of polarisation, and the reflectivity is plotted against a depth/wave length (d/λ) value. The programme is described in Figure 1. It is hoped to extend the programme to account for the imaginary component of the refractive index which is brought about by absorption by a metal.

At the moment only an arbitrary value of refactive index of the substrate (6.0) is used to determine the affects of a) index profile, b) angle of incidence, c) number of layers, and d) angle of polarisation, all as a function of the depth/wavelength ratio.

Figure 2 shows the affect of polarisation as a function of d/λ and reflectivity. The hatched band represents angles of polarisation from 0 to 0.5 radians for four angles of incidence 0 to 0.75 radians. The lower lines represent the advancing band front for angles up to 1.57 radians (90°). The upper band front always remains within the hatched area. The reflectivity of a particular angle of incidence is seen to be affected by polarisation.

Figure 3 shows the affect of angle of incidence at one value of polarisation (perpendicular). It is expected that the effect will be reduced in practice with multipolarised (ordinary) light.

Figure 4 shows the affect of change in profile and Figure 5 the gradation of refractive index through 100 layers for the Sin^2 Sin^2 and inverse ($1-Sin^2$ Sin^2) profiles. The profile is generated using a Sin^2 Sin^2 model of a surface where we relate the area of substrate material or of air at any depth to the refractive index of an imaginary homogeneous layer at that depth by:

$$n_{eff.} = n_{substrate}(Asg/A) + n_{air}(1 - Asg/A)$$

where A = area of segment, Asg = area of substrate element at a particular depth, and (1-Asg) = area of air at that element depth. Figure 6 shows the affect of change in number of layers; there is clearly no benefit in going above 100 layers.

A number of 'moth eye' surfaces have been made as described in the previous report (1). Problems have arisen with the chromium deposit on a number of the samples: tensile stresses in the coating have caused the formation of microcracks. The problem seems to be restricted to the specimens of small amplitude, whereas those with greater depths do not seem to suffer. The initial impact of the computer model suggests that the optimum topography will have a fairly small depth/wave length ratio (≈ 0.25). Taking into account also the possible replication process (1), we are anxious to make studies on specimens with amplitudes in this troubled region. We are therefore, experimenting with coating other metals. Cr was chosen initially only because it was the brightest metal with which to prove the theory. The topography of the 'motheye' is thought to be that of Sin^2 + Sin^2 (like an egg tray). It is proving difficult to prove this by microscopy and we still haven't successfully measured these amplitudes. Spectroscopic studies have shown a dependence on amplitude and to extract further information we need to have a measure of this value.

2.2 The pre-pulse electroplating work proceeds but continues to meet with problems. Potential - current density plots and potentiostatic current-time transients have been studied for a number of electrolyte systems and copper substrate. The former study has given us values of potential range available prior to a limiting current being produced; the latter gives us a minimum time (usually $< 8 \mu s$) for the double-layer charging to be completed and the Faradaic current to reach its equilibrium value.

Theory would predict that a pulse of appropriate hight and duration will produce nuclei above a critical value required for growth, and at a lower potential these nuclei could be made to grow. For three-dimensional nucleation; the electrochemical free energy of formation may be written,

$$\Delta G = ze\eta q + Hg^{2/3}\sigma \qquad \ldots\ldots(1)$$

where g is the culster size Z the number electrons transferred per atom or molecule, e the electronic charge, η the overpotential, σ the surface free energy, and H a shape factor (2). For spherical clusters

$$H = 4\pi \left(\frac{3}{4\pi} \cdot \frac{M}{N\rho}\right)^{\frac{1}{2}}$$

where N is Avogadros number, M the molecular weight and ρ the density.

From equation (1) it follows that ΔG first increases to a maximum,

$$\Delta G^* = \frac{4H^3\sigma^3}{27z^2e^2\eta^2}$$

at the value of g* given by

$$g^* = \frac{8H^3\sigma^3}{27z^3e^3\eta^3}$$

which is the size of the critical nucleus. Relative free energy plotted against relative cluster size, shown in Figure 7, shows that the larger the cathodic overpotential the smaller is the critical nucleus size and the critical free energy change. For g $>$ g* spontaneous growth takes place; ΔG^* acts as an energy of activation in forming three-dimensional nuclei. The distribution of clusters N(g) therefore describes a range of intermediates in the formation of nuclei; nucleation consists of the propagation of clusters according to this distribution. The rate governing three-dimensional nucleation A (nuclei cm^{-2} S^{-1}) is

$$A = Nk_oHN_{(g)}\left(\frac{H\sigma}{9\pi kT}\right)^{\frac{1}{2}} \exp. -\left(\frac{4H^3\sigma^3}{27z^2e^2\eta^2kT}\right).$$

The nucleation rate varies much more rapidly with overpotential at low values of η than the rate of phase growth which has the usual form

$$R = k_o\left\{\exp. -\left(\frac{\alpha z\eta e}{kT}\right) - \exp. -\left(\frac{(\alpha - 1)z\eta e}{kT}\right)\right\}$$

where R and ko are in mol cm^{-2} s^{-1}. It is therefore, possible to form a small number of nuclei by applying a short duration pulse of large overpotential and growing these nuclei at a lower overpotential where little or no further nucleation will take place.

It was hoped that appropriate control of the nucleation and growth conditions would lead to a separation of nuclei of the order of the dimensions of the 'motheye' surfaces. However, a number of problems have arisen. Prepulse plating of nickel onto vitreous carbon was first investigated as it was thought appropriate to eliminate as many uncertainties as possible in the initial stages. Vitreous carbon is thought to have perfectly homogeneous, amorphous surface. On this material a range of nucleation and growth potentials have been identified, e.g. at a nucleation potential of ~ 2000 mV for ~ 10 millisecs and growth potential of ~ 620mV for ~ 200s, individual two-dimensional nuclei can be observed. If the growth time is increased to 600s the two-dimensional nuclei coalesce to form a thin layer of nickel, but still two-dimensional. At present we are investigating the reasons why the growth mode fails to be three-dimensional.

We have attempted to nucleate nickel onto copper but with little success. We are experimenting with different nucleation/growth potentials and with different electrolytes, etc. One difficulty lies in the fact that it is difficult to differentiate on the SEM between the Ni and the Cu substrate. It seems that a thin layer of the Ni rapidly forms before the nuclei have been identified. This is probably because a poly-crystallalline substrate will exhibit large numbers of emergent dislocations, kinks, edges and other lattice sites capable of attracting nucleation. We are investigating the possibility of lowering E nucleation from the theoretical value to account for this higher surface free energy (3).

An area which we are simultaneously investigating is that of controlled-morphology growth. Once a deposit is free from substrate epitaxial influence, it has been found that under controlled conditions of pH and current density, certain crystallographic planes can be made to preferentially grow (4). These zones are shown in Figure 8 for one such system. Under appropriate conditions the $\langle 110 \rangle$ plane can be grown to produce pyramids of defined hight and number density. One drawback is that growth rate is extremely low and may limit the application of the principle, however we are investigating the possibility of speeding up the growth in the $\langle 110 \rangle$ direction by modification of the bath.

3. Future Work

The model will soon be in a state to predict the optimum characteristics of the absorber surface, we will then manufacture a specimen with those characteristics and make measurements to test the model. At the same time the work to produce a replica and to define the replication conditions is proceeding. The nucleation theory has thrown up a number of problems that will be investigated in the next phase of the project, why we have failed to identify nuclei on copper substrate and why we seem to get two-dimensional growth rather than the required outward growth.

4. References

1. Bannard, J.E. Hayden, J & O'Malley, P. Proc EC Contractors, Solar Energy R & D Project A, Athens, 11-13 November 1981.
2. Bindra, P. Fleischmann, M. Oldfield, J.W. and Singleton, D. Faraday Disc. 56 (1973).
3. Fleischmann, M and Thirsk, H.R. Advances in Electrochem & Electrochem Engineering. Vol 3, Wiley, 1963.
4. Amblard, J. et al. J. de Physique, Colloque C1, 42 (1981) 147.

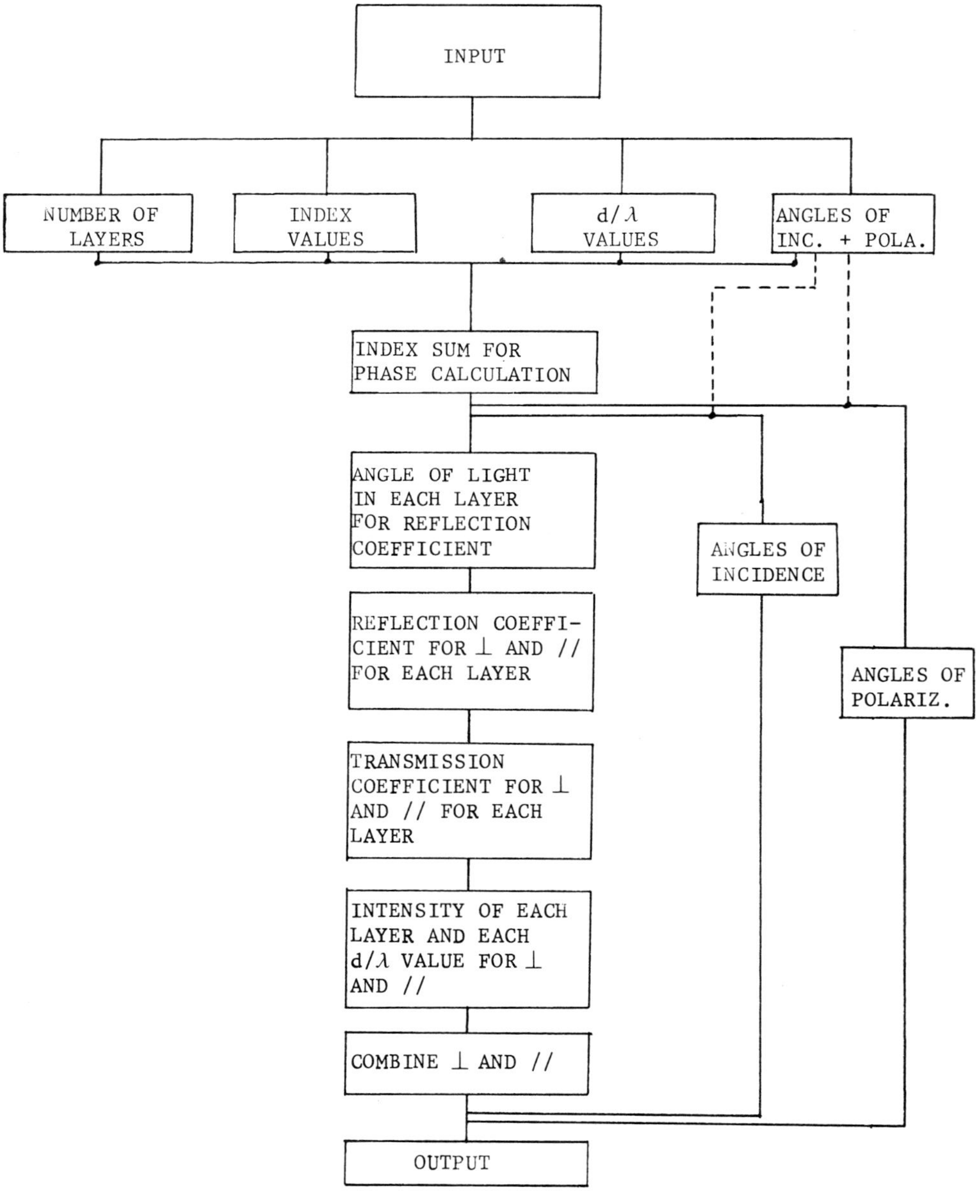

FIGURE 1 - Block diagram of the program.

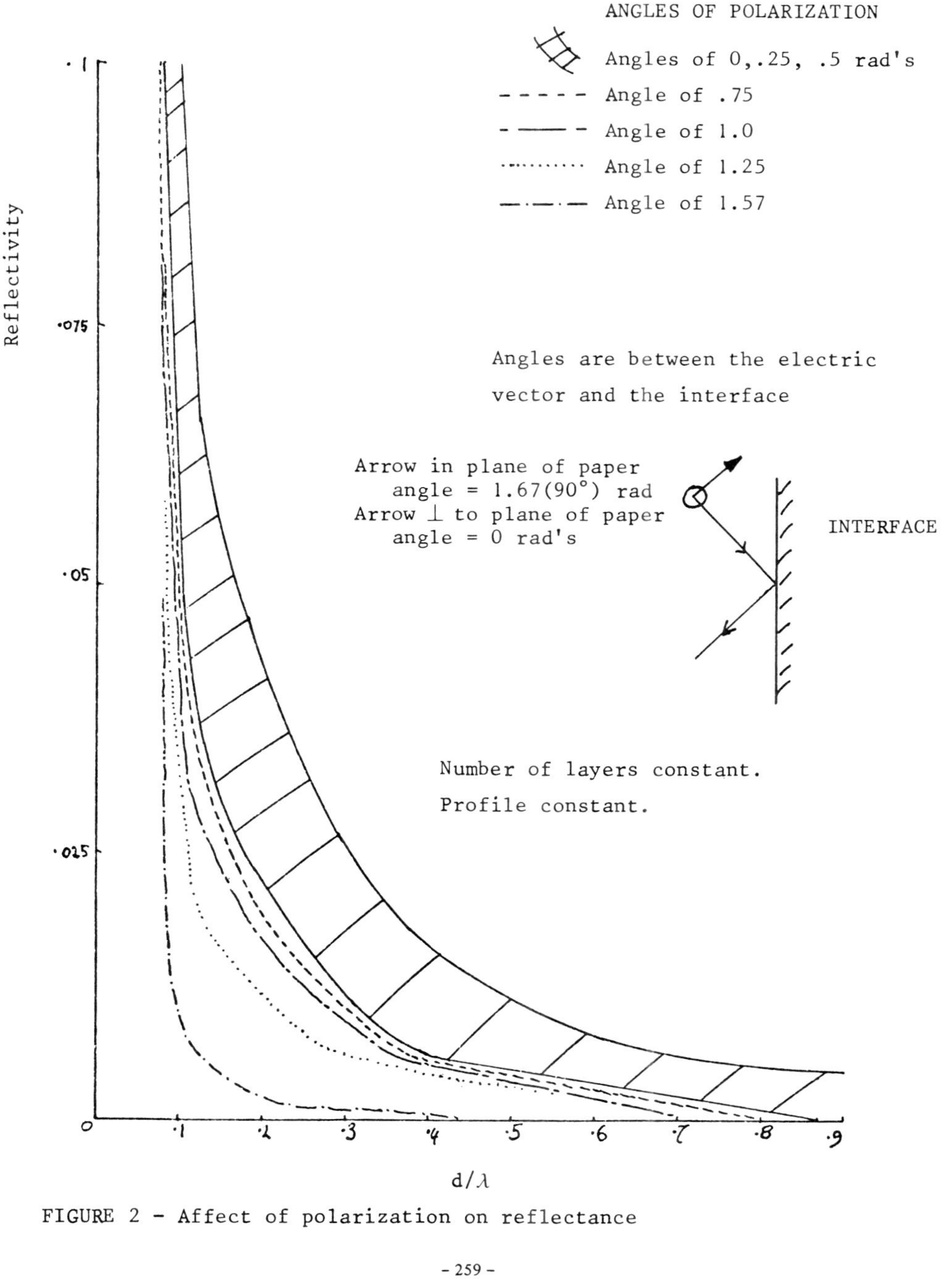

FIGURE 2 - Affect of polarization on reflectance

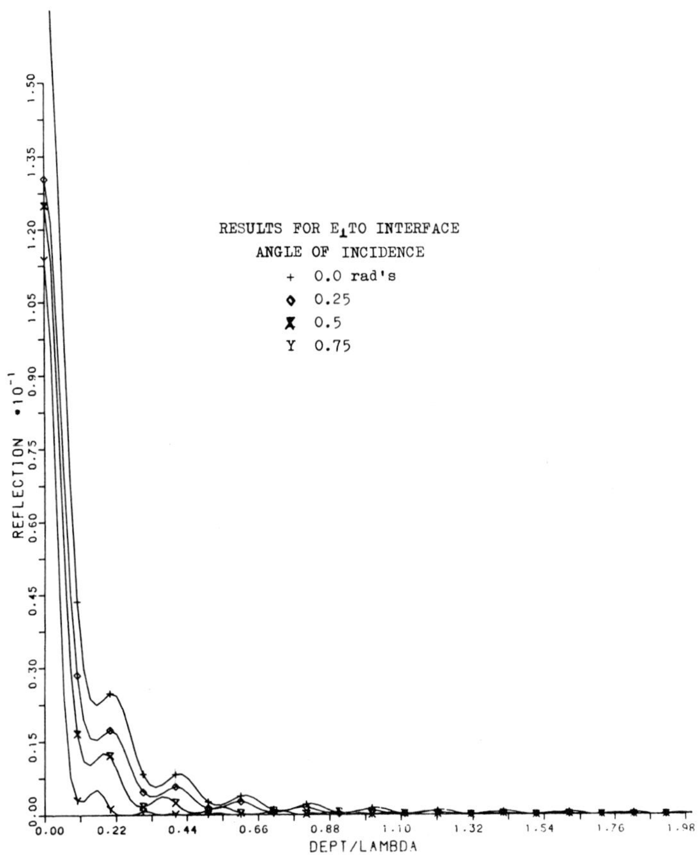

FIGURE 3 - Affect of angle of incidence

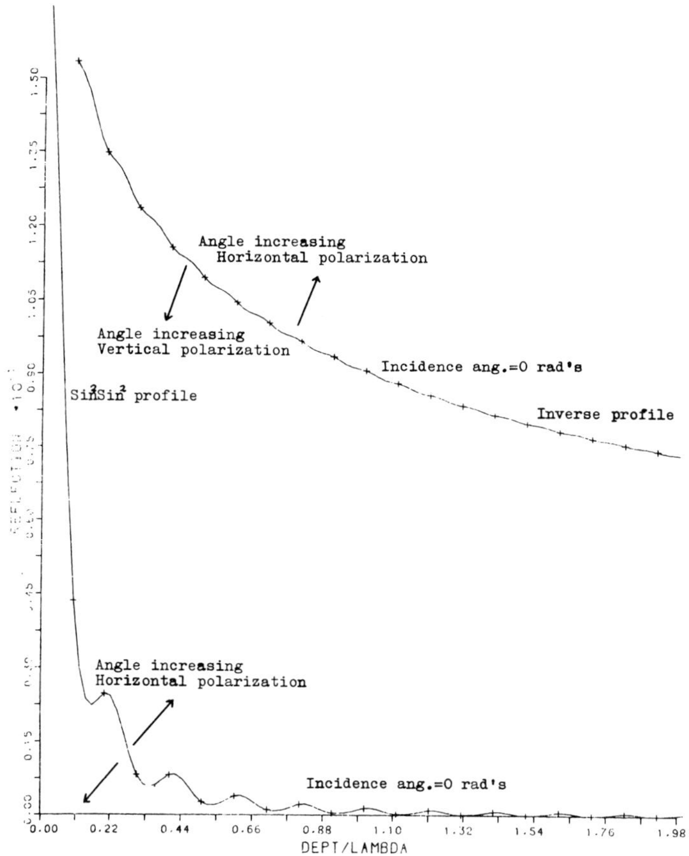

FIGURE 4 - Affect of change in profile

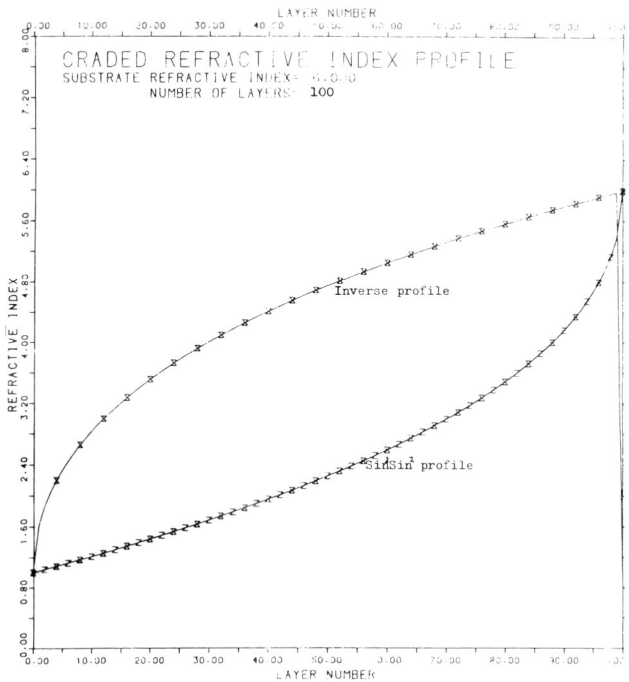

FIGURE 5 - Change of n through 100 layers for 2 profiles

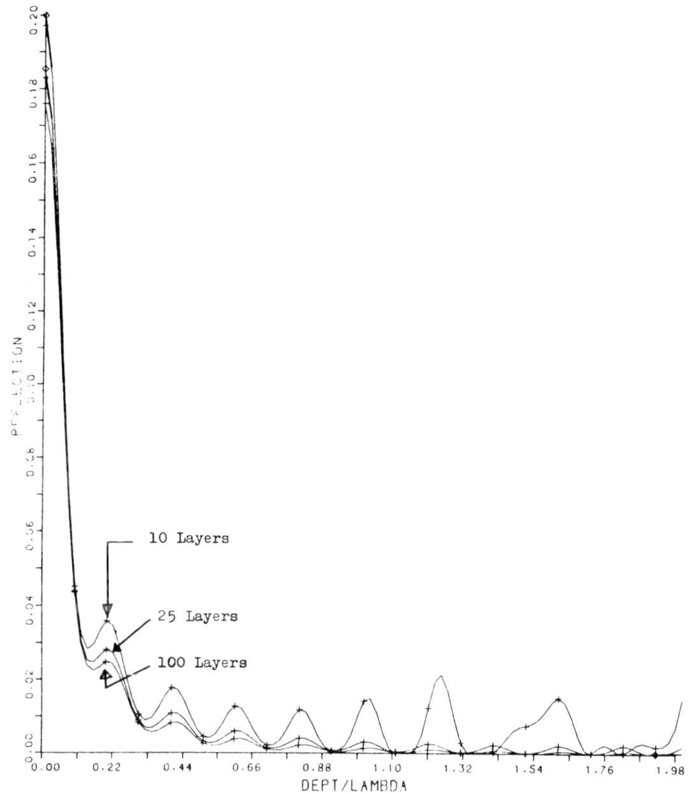

FIGURE 6 - Affect of number of layers used in the model

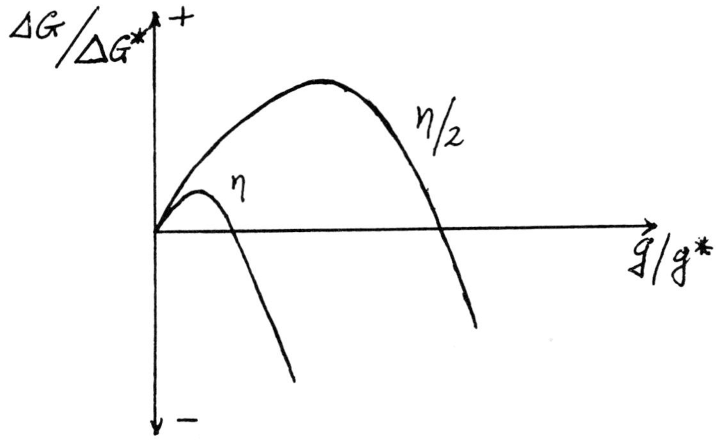

FIGURE 7 - elative free energy change against nucleus size

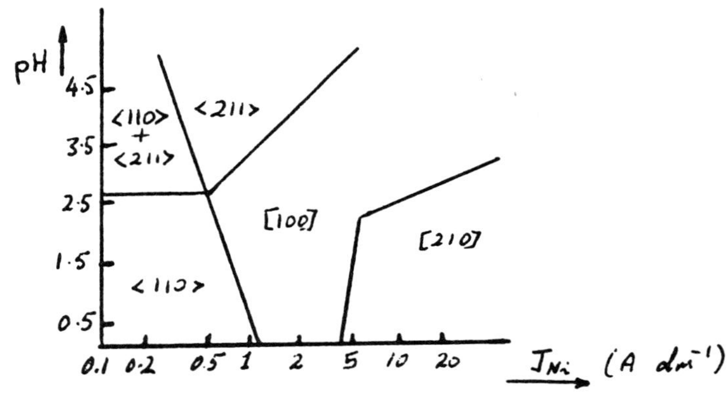

FIGURE 8 - Stability diagram for the various textures obtained in thick Ni deposits from an organic - free Watts bath at 50°C

COBALT OXIDE SPECTRAL SELECTIVE COATINGS

Authors : C.A. BOOSE, M. van der LEIJ

Contract number : ESA-C-057-NL

Duration : 3 years 1 July 1980 - 1 July 1983

Total budget : H.Fl. 781.520,- CEE contribution 100.000 ECU

Head of project : Ir. C.A. Boose

Contractor : Central Organization for Applied Scientific Research TNO

Address : Juliana van Stolberglaan 148
P.O. Box 297
2501 BD The Hague

Summary

The purpose of this project is to develop a durable selective coating based on cobaltoxide, from laboratory scale to full practice. The coating technique involved is based on simple electroplating methods, resulting in low costs spectral selective surfaces. The work is based on work of the Delft University of Technology concerning cobalt-iron oxide spectral-selective coatings (1).

Apart from upscaling, our work consists of experiments to improve the thermal stability as well as the corrosion resistance of the spectral selective system. Therefore, the effect of alloying the cobalt coating with molybdenum and chromium is investigated, using pulse plate techniques (2).

The durability of the coatings has been investigated by performing cyclic heat treatments in air as well as in vacuum at 200°C up to 500°C. The corrosion resistance of the systems has been investigated by means of the neutral saltspray test and the cyclic humidity test; a modified cyclic thermal / humidity test is described (Appendix A).

The main results can be summarized as follows:

- CoFe oxide coatings with a Cr intermediate layer are stable in air up to 300°C, the corrosion resistance being fair. After 400°C cyclic heating in air the absorptance decreases due to the change of the structure and the emittance increases, but they seem to stabilize at values of 0.82/ 0.14
- CoMo and CoCr coatings are suitable in air up to 200°C with superior corrosion resistance. Higher temperature cycles result in rapid degradation and poor anti-corrosive properties.

1. INTRODUCTION

The objective of this project is to produce a durable solar collector, based on the use of spectral selective surfaces of the cobalt (alloy) oxide type, using bright nickel coated stainless steel 304 as a substrate. The coatings are produced by electroplating techniques and should result in $\alpha_\perp/\varepsilon_\perp$ values of 0.9/0.1 or better. In this report the work on cobalt-iron, cobalt-molybdenum and cobalt-chromium is described. Also the first results of the sandwich plating technique are reported. This technique may overcome the problem of oxidation of the undercoat resulting in structure degradation.

2. EXPERIMENTAL PART

2.1 Phase 1 - Cobalt oxide and cobalt-iron oxide coatings

CoFe oxide coatings perform much better than pure Co oxide coatings (2). As already described in the preceding report (2), rather diluted CoFe "Watts" type electroplating baths produce coatings that are readily oxidized in a propane-oxygen burner (a few minutes at 600°C flame temperature). The resulting values for α/ε are very good reproducible and are around 0.95/0.08. Furnace oxidation (2 hrs at 400°C in air) produces values of 0.85/0.07 on the same CoFe coating.

Practical applications

After some preliminary tests, a prototype copper tube collector system of the Faculté Polytechnique in Mons (Belgium) has been coated according to the following procedure:

- alcaline degreasing, cathodic, 2 A/dm^2, 60°C, 2 minutes
- rinse
- acid dip in HCl 1/1 (activating)
- rinse
- bright nickel electroplating, 4 A/dm^2, 60°C, 30 minutes
- rinse
- CoFe electroplating bath with bath composition:

$CoSO_4$ 7 aq	:	45 g/l
H_3BO_3	:	4.5 g/l
$Fe_2(SO_4)_3 \cdot nH_2O$	:	0.9 g/l
H_2O_2 (30%)	:	1 ml/l
pH	:	2.2
temperature	:	55°C
current density	:	4 A/dm^2
time	:	6 minutes
air agitation		

After drying the cobalt-iron coating was oxidized by means of a propane-oxygen burner (flame temperature around 550°C). The whole oxidation procedure took about 10 minutes. The collector prototype is to be tested in Mons during the next project period.

2.2 Phase 2 - Other Cobalt (alloy) oxide coating

In this report period work on CoMo-oxide coatings has been continued and work on CoCr-oxide coatings has started.

2.2.1 CoMo oxide coatings

The most promising bath composition from earlier experiments (2) have been investigated both with respect to thermal stability and corrosion resistance. The following bath composition was used:

$CoSO_4$ 7 aq	:	75 g/l
NaCl	:	30 g/l
Na-gluconate	:	160 g/l
MoO_3	:	8 g/l
H_3BO_3	:	40 g/l
pH	:	3
temperature	:	55°C
current density	:	4 A/dm^2
time	:	4-8 minutes
air agitation		

With this bath composition CoMo coated test panels were made, using the standard bright nickel plated stainless steel base material. Fig. 1 shows the panels after electrochemical oxidation in m.n. benzene sulphonic acid sodium salt: 15 g/l, 10 Volts, 50°C, 1 minute. The first horizontal line had a plating time of 4 minutes, the second line of 8 minutes in the CoMo plating bath, the α/ε values being 0.92/0.10 and 0.93/0.15, respectively. Fig. 2 shows the same panels after various corrosion tests: The left vertical row: 165 hrs saltspray test, α/ε values changing from 0.92/0.15 (before) to 0.88/0.15 after. The second vertical row: cyclic heat treatment in air, 6 hrs 200°C / 6 hrs 20°C for two weeks; after that 165 hrs saltspray test, α/ε values changing from 0.92/0.15 (before) to 0.80/0.15 after. The vertical row at the right was submitted to the new combined cyclic heating/ humidity test (2 hrs cycle for 4 weeks), whereby the temperature ranges from 60 to 130°C and the humidity from practical zero to 100% (see also Appendix A). Under these conditions the α/ε values showed no significant change (0.90/0.15).

From these results the following conclusions can be drawn:

- Electrochemically oxidized CoMo oxide coatings show excellent stability in the combined cyclic heat / humidity test, fair stability in saltspray testing and lower performance as far as the absorption values are concerned after heat cycling at 200°C, followed by saltspray testing. The emission values remained on the same value in all tests (0.15).
- As reported earlier, no corrosion products could be detected. High temperature cyclic tests in air showed rapid desintegration of the structure and increasing ε-values from 0.15 to 0.40 (2).

Practical plating of black CoMo oxide layers

In co-operation with EBS, a number of collector panels of standard size (roughly 1.80 x 90 cm) were plated and are to be tested in the next period. The plating sequence (time in minutes) has been as follows: anodic degreasing 2' - rinse ½' - nickel strike 6' - rinse ½' - bright nickel 5' - rinse ½' - CoMo 6' - rinse ½' - oxide building 2' - rinse 2'. The whole treatment, including loading and unloading, takes less than 30 minutes. When operating with two rectifiers, one man could do 4 panels within one hour; at a rate of ƒ 80,- per hour, each panel costs ƒ 20,- or about ƒ 10,- pro m^2 (labour). Including profits, equipment and chemical costs, in our opinion a price of ƒ 20,-/m^2 is very realistic.

2.2.2 CoCr oxide coatings

The first CoCr oxide coatings could be investigated in this report period. As a basis we used the trivalent chromium plating bath as supplied by Harshaw, named "Alecra 3000". This bath contains trivalent chromium (approx. 20 g/l) and operates at room temperature and pH 2.5-3 with current densities from 4-40 A/dm^2.

By simply adding cobaltsulphate 7 aq to this bath we were able to deposit good CoCr alloys with 5-10% chromium. The CoCr deposits were oxidized by several methods and test panels are shown in Figure 3, before thermal cycling and saltspray testing. The two horizontal lines show test panels with decreasing Cr-content from left to right. The upper line shows flame oxidized panels (α's around 0.85), while the second line shows panels oxidized by electrochemical methods (α's varying from 0.70-0.95). All test panels were heat-cycled in air at 400°C during two weeks (except bottom right) and subsequently saltspray tested for 165 hours.

The results are shown in Fig. 4. With increasing cobalt content, the corrosion products increase (left to right) except for the last panel (bottom right), that was not heat cycled. These results indicate that CoCr oxide coatings behave similarly as CoMo oxide coatings: rapid degradation of the corrosion resistant coating when treated in air at 400°C.

Cyclic thermal / humidity tests, as described in Appendix A, show better results for the corrosion resistance of CoCr oxide coatings. However, in this case the peak temperature is only 130°C. It is to be expected that degradation will be less when sandwich plating is applied instead of alloy plating (alternating Co/Cr/Co/Cr coatings). This aspect will be studied in the next project period.

2.3 Phase 3 - Improving the thermal stability by using interlayers as oxidation barrier

2.3.1 Cobalt-iron (III) oxide coatings from a very diluted cobalt-sulphate bath

Experiments with diluted cobalt sulphate solutions have been continued. A dilution of our standard concentrated bath has been reported in the previous proceedings of the EEC Contractors' Meeting (2). The recent bath used here, contained: 45 g/l $CoSO_4.6H_2O$, 4.5 g/l H_3BO_3, 0.9 g/l $Fe_2(SO_4)_3 . nH_2O$ and 1 ml/l H_2O_2 in water. The bath temperature was 55°C, the acidity 2.2, the current density 4 A/dm^2 and the plating time 4-6 min. The concentration now is only 10% of our previous concentrated cobalt sulphate bath.

Oxidation methods

The oxidation methods have been kept the same as before (2), namely:
- furnace oxidation in air at atmospheric pressure at 400°C for 2 hours,
- flame oxidation with a propane + oxygen (excess) burner with flame temperature of about 600°C, for 1-2 minutes.

Chromium oxide diffusion barrier interlayer

Before the cobalt plating took place, a very thin coating of chromium (III) from the commercial bath "Alecra" was first deposited. The plating time was 30 seconds and 60 seconds. If the chromium may also be oxidized, for example into Cr_2O_3, then this oxide layer is expected to act as a barrier layer (3) against further diffusion of metal components to or from the substrate (nickel).

Results

The optical results of flame oxidized cobalt (+ iron III) coatings with chromium interlayers, are the same as the coatings from our concentrated bath: $\alpha_\perp$ = 0.94-0.95 and $\varepsilon_\perp$ (100°C) = 0.06-0.08. The furnace oxidized samples showed a lower absorptance of 0.80-0.85 and the same emittance values of about 0.07. With respect to thermal and mechanical stability of the cobalt (+ iron) oxide coatings, a plating time for the cobalt deposition of 6 minutes seemed to be the optimum.

Humidity tests

Some of the flame oxidized cobalt (+ iron) samples with and without chromium interlayer were exposed to humidity tests, as described in Appendix A. No corrosion products could be observed after two weeks of exposure. Also no change of $\alpha_\perp/\varepsilon_\perp$-values could be found.

Heating tests in air

Flame oxidized cobalt samples with chromium interlayers were subjected to thermal ageing in air at 230°C (1 week continuously), at 300°C (1 week continuously) and 400°C (4 weeks cyclically with 6 h on - 6 h off). Contrary to the results from the concentrated bath as reported earlier (2), the stability of the samples from the very diluted bath is much better: up to 300°C no significant change of $\alpha_\perp/\varepsilon_\perp$ was found and no pulverization of the coating took place. At 400°C the adherence of the cobalt coatings became somewhat weaker and in one case flaking occurred. However, the damage was generally less severe than we had before (2).

Though the absorptance decreased rapidly from 0.94 to 0.82 after a few days of heating at 400°C, no change of $\alpha_\perp$ was further measured after 2, 3 and 4 weeks. As we experienced earlier (2), the decrease of $\alpha_\perp$ was caused by the change of the microstructure. The emittances increased from 0.06 to 0.08 after 1 week, to 0.10 after 2 weeks and to 0.14 after 3 weeks and remained constant after 4 weeks. Compared to earlier work (1) these $\varepsilon_\perp$-values are much lower. The role of chromium has not yet been investigated in this period and will be studied in the next period.

In conclusion we found that flame oxidized cobalt (+ iron III) coatings from a very diluted cobalt sulphate bath have shown better results than the ones from our concentrated bath, with respect to thermal stability in air up to 400°C and humidity resistance. This is probably due to the presence of the Cr interlayer. Also the material costs of the diluted bath are lower than de concentrated bath, which can be considered as an advantage.

3. FUTURE ACTIVITIES

- Finding methods to prevent microstructure degradation during thermal ageing at high temperatures, e.g. sandwich plating, spinel formation.
- Continuation of the experiments with diffusion barrier layers, together with alcaline Co alloy plating.
- Post treatment processes to increase the corrosion resistant properties.
- Thermal ageing in air under direct concentrated solar radiation as a joint experiment together with the University of Houston, U.S.A.

4. REFERENCES

(1) W. Kruidhof and M. van der Leij, Solar Energy Materials 2 (1979), p. 69-79.

(2) C.A. Boose and M. van der Leij, "Cobalt oxide selective coatings", Solar Energy Applications to Dwellings, Series A, Vol. 1, p. 57-65.

(3) B.O. Seraphin, Thin Solid Films 39 (1976), p. 87.

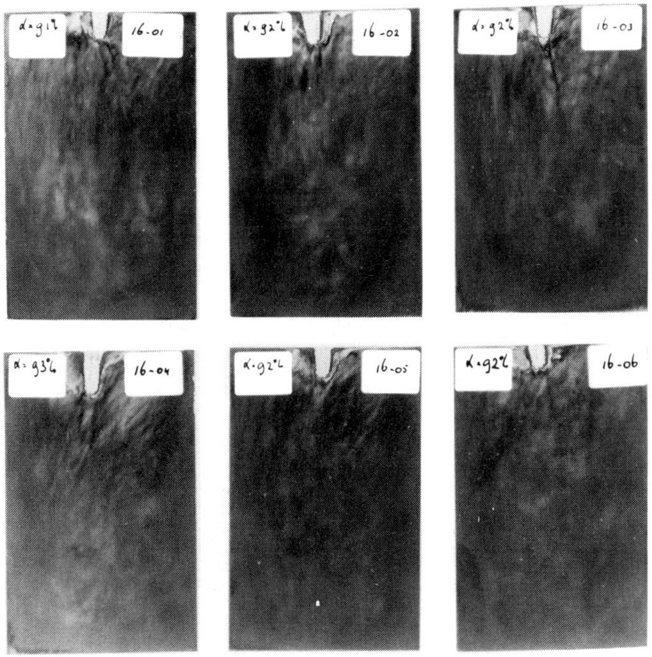

Fig. 1

Electrochemically oxidized **CoMo test** panels
1st horizontal line: 4 min CoMo plating time
2nd horizontal line: 8 min CoMo plating time

Fig. 2

Same test panels
1st vertical row: after 165 hrs saltspray testing
2nd vertical row: after 2 weeks heat cycling 200/20°C and 165 hrs saltspray
3rd vertical row: after 4 weeks cyclic heat/humidity testing

Fig. 3

CoCr test panels with decreasing Cr content from left to right
1st horizontal line: flame oxidized
2nd horizontal line: (electro)chemically oxidized

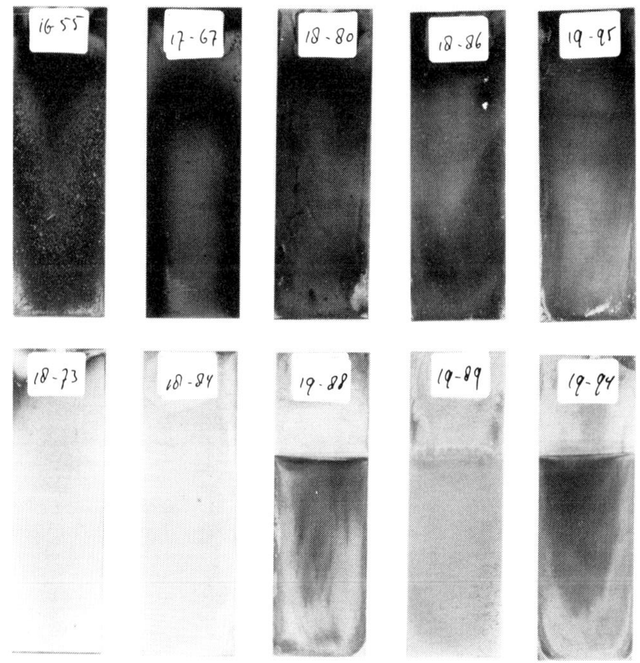

Fig. 4

Same panels, after two weeks heat cycling in air at $400^{\circ}C$ and subsequently saltspray tested for 165 hrs

Modified cyclical heating / humidity test method

Description of the test method for collector samples

In Fig. A-1 the test cabinet is shown schematically; the cabinet is about 3 m^3 and contains an air in- and outlet, an open watertank, a heating plate on which the samples are placed.

Cycling procedure

The switch cycles for the on/off-period of the air inlet and the heating are 90/30 minutes (switches X in Fig. A-1). During the off-period, the open watertank is heated (switch Y). This combination results in a dry period of 40 minutes (humidity $< 5\%$, temperature 90-130°C) and wet period of also 40 minutes (humidity 100%, temperature 60°C). As the whole cycle takes 2 hours, another 40 minutes are left for the intervals. The humidity and temperature run for the 2 hours cycle is given in Fig. A-2. The temperature overshoot shortly after each humidity rise is due to the switching-off of the air inlet at that moment.

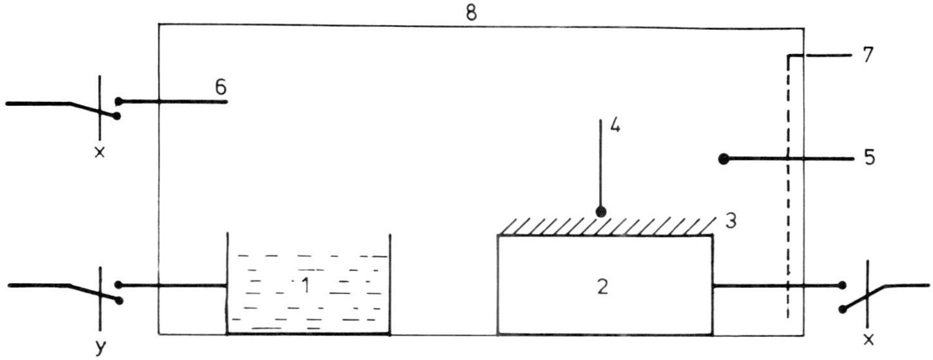

1 open watertank
2 heating plate
3 samples to be tested
4 temperature detection
5 humidity detection
6 air inlet (15 – 20°C)
7 air outlet
8 cabinet; approx. 3 m³

Fig. AI Schematic diagram of the heating/humidity test equipment

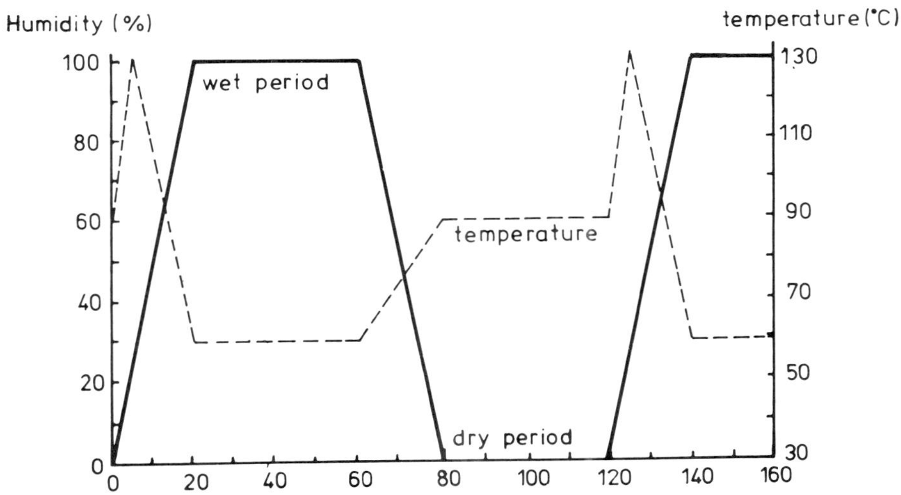

Fig. A II Temperature / Humidity cycle

STUDY AND CONSTRUCTION OF A SOLAR FOCUSING COLLECTOR USING A DEFORMABLE MIRROR, IN THE TEMPERATURE RANGE 100-200°C

Authors : J. BOUGARD, M. BOUIDIDA

Contract nr : ESA-C-054-B

Duration : 18 months, from 1.10.1981

Total budget : 3.906.000 BF CEC contribution : 50 %

Head of project : Prof. J. BOUGARD

Contractor : Faculté Polytechnique de Mons,
Service de Thermodynamique.

Address : boulevard Dolez, 31, B - 7000 MONS (Belgium).

Summary

A full scale collector in which the concentration of beam radiation is obtained by deformation of a thin glass mirror has been designed and built.
Optical measurements confirm that an energetic concentration ratio larger than 6 is obtained on a 80 mm width receiver for every incidence angle from -22.5°C to +22.5°C.
The sensitiveness of the concentration ratio against defocalisation and tracking error has been measured.
A thermal loop designed for the medium temperature range has been built.

1. Aim of the project

The aim of the project is to design, to build and to experiment a focusing collector, in the medium temperature range, from 100°C to 200°C, in which the focalisation of the sun rays is obtained by deformation of the mirror itself instead of, as usual, by a whole displacement of the mirror or the receiver.

2. Summary of previous results

Let us consider, as shown on figure 1, a cylindrical mirror, attached by its ends at two fixed supports A and B, permitting free rotation.

The focalisation of the beam radiation is obtained for every incidence angle if the shape of the mirror is ideally a parabol axially directed towards the sun.

This condition can be achieved by slightly deforming the mirror and displacing the receiver as shown on figure 1.

Moreover, experimental measurements performed on a test rig have shown that a good approximate deformation is obtained by pushing or pulling on the mirror in only two points, C and D.

For an incident angle variation from -22.5° to +22.5° , the energy concentration ratio is easily maintained larger than 6 with an interception efficiency of 85 %.
A seasonal sun tracking for such an east-west cylindrical focusing collector is thus feasible.

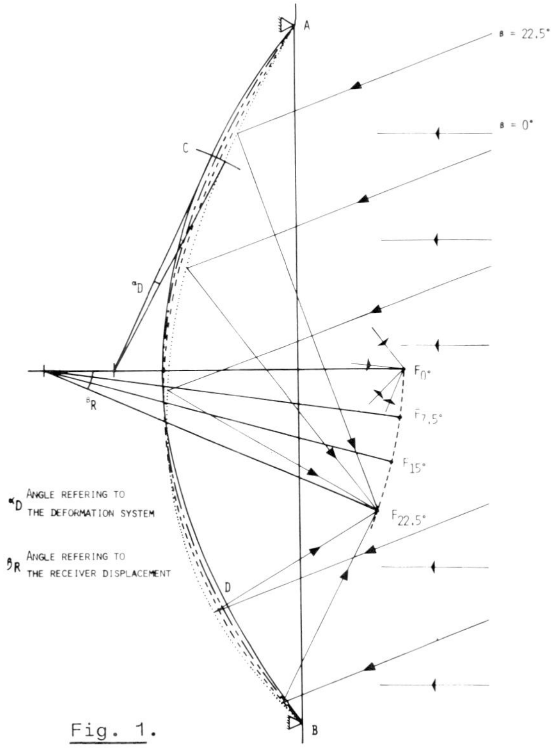

Fig. 1.

3. Prototype design

3.1. Deformation mechanism

The deformation of the mirror and the displacement of the receiver can be mechanically coupled and so achieved with a one degree of freedom mechanism activated by a single reversible motor.
The principle is shown on figure 2.

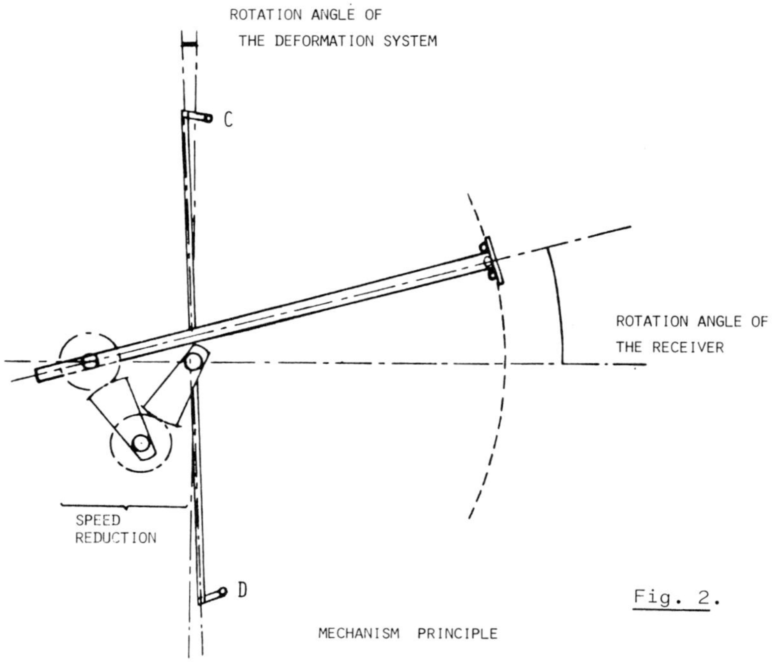

Fig. 2.

3.2. Mirror

The deformable mirror is a 0.8 mm thin glass mirror pasted on a galvanized iron sheet.

The specular reflectance is 0.9. The deformation forces are only 5 daN.

3.3. Receiver

The receiver (figure 3) consists of a folded copper tube soldered on a bronze plate so that the fluid flows firstly in the low irradiance region and after in the maximum irradiance part.

A black chrome selective coating was firstly electrochemically deposited. However, it has been impossible to obtain uniform coating, specially between the tubes.

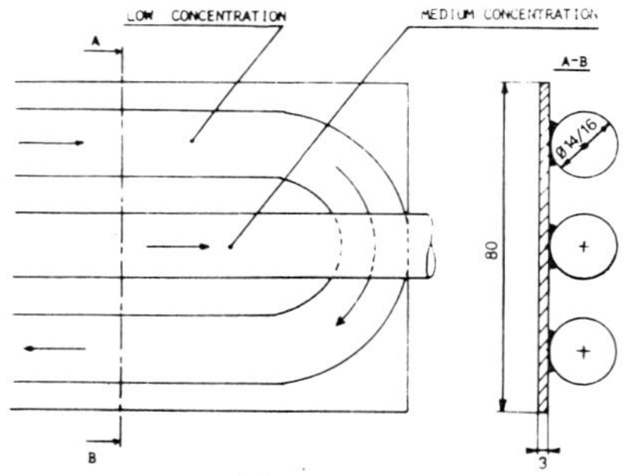

Fig. 3.

A cobalt oxyde selective coating, developped by TNO under CEC contract, was then studied. The optical properties, measured on a sample, were : $\alpha = 0.94$ $\varepsilon = 0.07$

The whole receiver is thus now being treated in TNO.

3.4. Prototype assembly

The photograph (figure 4) shows the different parts of the prototype : mirror, deformation rods, mechanism, casing, provisional receiver.

During sun tracking, the casing is fixed and a glass cover protects all the components against atmospheric injuries.

4. Prototype experiments

4.1. Concentration measurements

Let us define the average energetic concentration ratio :

$$\bar{C}_E = \frac{\int_{-\Delta R/2}^{+\Delta R/2} I(y)dy}{I_o \Delta R} \left(1 - \frac{\Delta R}{w_1}\right)$$

where ΔR = receiver width (80 mm)

w_1 = mirror width

I_o = incident irradiance

$I(y)$ = irradiance on the receiver at y ordinate

and the interception ratio η

$$\eta = \frac{\int_{-\Delta R/2}^{+\Delta R/2} I(y)dy}{\rho\, w_1\, I_o} \quad ; \ \rho = \text{mirror reflectance}$$

Figure 5 shows the variation of $\bar{C}_E$ and η against the receiver width for incidence angles of 0, 15 and 22.5°.
These results confirm the feasibility of obtaining a concentration ratio larger than 6 using a 80 mm width receiver.

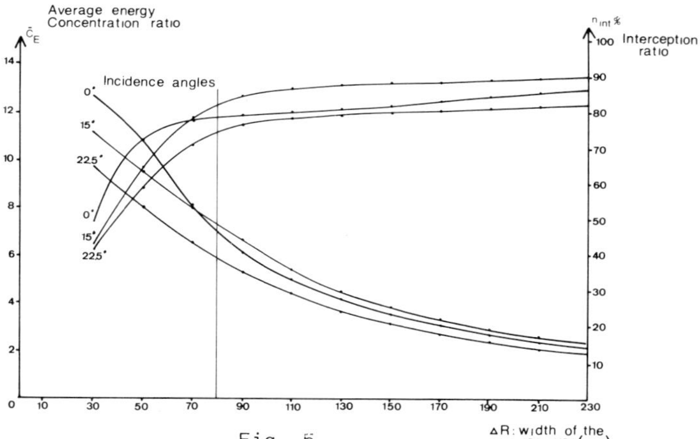

Fig. 5.

The sensitiveness of the concentration against a relative defocalisation of the receiver and a tracking error was also measured.
For instance, figure 6 shows the variation of the local concentration ratio on the receiver for a tracking deviation of ± 2° at an incidence angle of 22.5°.

These tests furnished the data for the design of the tracking system, the sensors being 9 photovoltaïc cells distributed along the height of the receiver.

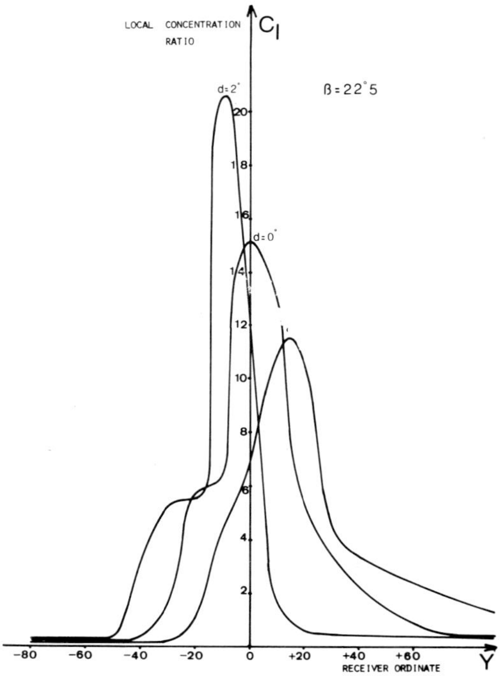

Fig. 6.

4.2. Medium temperature test rig

A thermal loop has been designed and built for testing the prototype in the range 100°C - 300°C (figure 7). The preliminary tests are beginning.

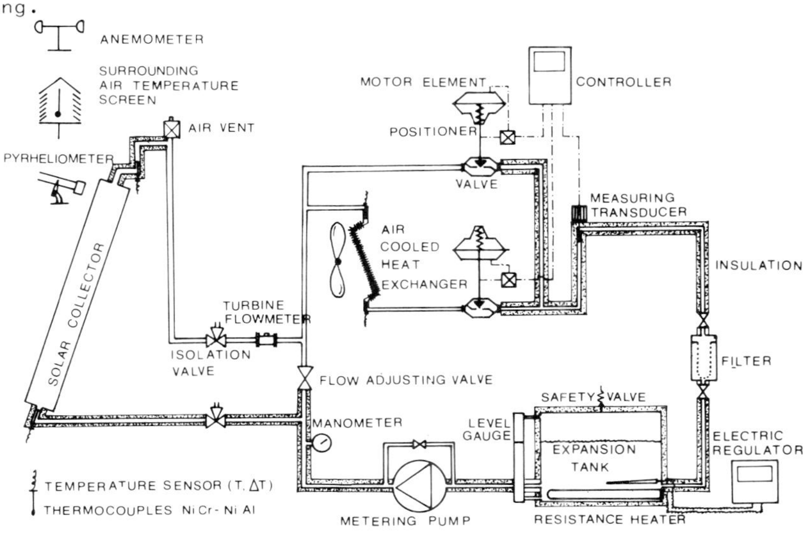

Fig. 7.

5. Future works

1. The fine distribution of the irradiance on the receiver will be measured and analysed using a photographic method.

2. The tracking system will be tested under solar simulation.

3. The thermal performance will be tested in outdoor conditions until 200°C.

4. Keeping the basic idea of obtaining concentration by deformation of the mirror, the possibility of maintaining fixed the receiver will be studied.
Figure 8 shows, for instance, the principle of a solution.
The corresponding deformation mechanism will be designed and tested.

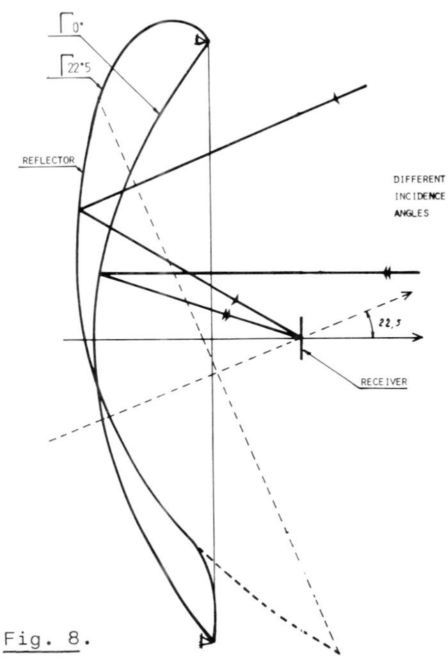

Fig. 8.

HIGH PERFORMANCE COLLECTOR FOR SMALL POWER SYSTEMS

Author : R. Hause

Contract number: ESA-C-011-80 D (B)

Duration : 18 months 1 July 1981 - 30 June 1982

Total budget : DM 160.000 CEC contribution: DM 80.000

Head of project: R. Hause, New Technologies Department

Contractor : Dornier System GmbH

Address : Dornier System GmbH
Postfach 1360

D-7990 Friedrichshafen 1

Summary

Technical plants, using solar insolation or basical energy, should be carefully designed and optimized in order to make best use of the intermittent and low density energy source. This project shall present the improvement, which can be obtained by using a high performance flat plate collector. For that purpose, the flat plate collector field with anodized absorbers of an existing solar absorption cooling plant will be exchanged by a selective coated one of the same type. The following operational improvements are to be expected.

- o Extension of the operation time of the cooling plant
- o Reduction of cold energy storage for the night cooling
- o Suitability of higher cooling water temperatures
- o Operation of lower solar insolation conditions.

Reliable data of improved operation behaviour however can be confirmed only after long time tests of the plant.

High Performance Collector for Small Power Systems

1. Introduction

The aim of the project is, to improve the properties of solar cooling plants by using high performance collectors.

The solar cooling test plant was erected at the National Research Centre in Cairo (Fig. 1).

Fig. 1: Solar Cooling Plant

It s main characteristics are:

Technical data:

Collector field

- Collector field area (effectiv) — 22.75 m^2
- Type of collectors — Dornier heatpipe flat plate collectors, anodized, double glazed

Cold store

- Volume of cold storage room — 12 m^2
- Nominal room temperature — 6 - 8 ^oC
- Weight of food (vegetables, fruits), to be cooled down daily — 300 kg

Absorption cooling machine

- Binary operating medium	ammonia/water
- Cooling energy output	3,0 kW
- Evaporator output	1,0 kW
- Storable cooling energy (night cooling requirement)	10,0 kWh
- Generator temperature	90 °C
- Evaporator temperature	2 - 3 °C
- Cooling water temperature	30 °C
- Daily cooling water consumption	10,5 m^3

The plant has shown good test results and has demonstrated all design data. Nevertheless, the operation and the future design of solar cooling plants can be improved considerably by using high performance collectors. In this case it was decided to use collectors of the same type, but with selective coated absorbers.

2. Utilization of selective coated collectors - Improving of efficiency

Fig. 2 shows a comparison of efficiencies between an anodized and a selective coated heatpipe flat plate collector at an insolation of 600 w/m^2.

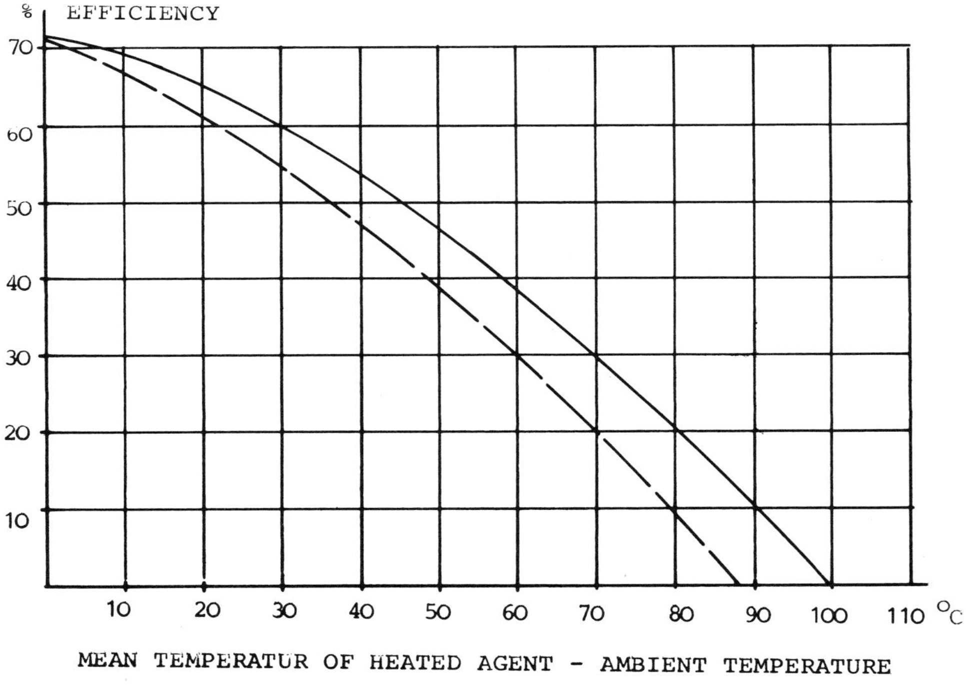

Fig 2: EFFIENCIES OF DOUBLE GLAZED DORNIER FLAT PLATE HEAT PIPE COLLECTORS AT 600 W/M^2 TOTAL INSOLATION

The improvement of the efficiency amounts to approximately 8 % (absolute) at a collector field temperature of 90 °C, which is also the generator temperature of the absorption machine, and a usual Cairo ambient temperatur of 30 °C. This addional collected energy can be used for different purposes.

2.1 Extension of the collector field operation time

The temperature characteristics of the generator respectively of the collector field during an operation day is plotted in Fig. 3

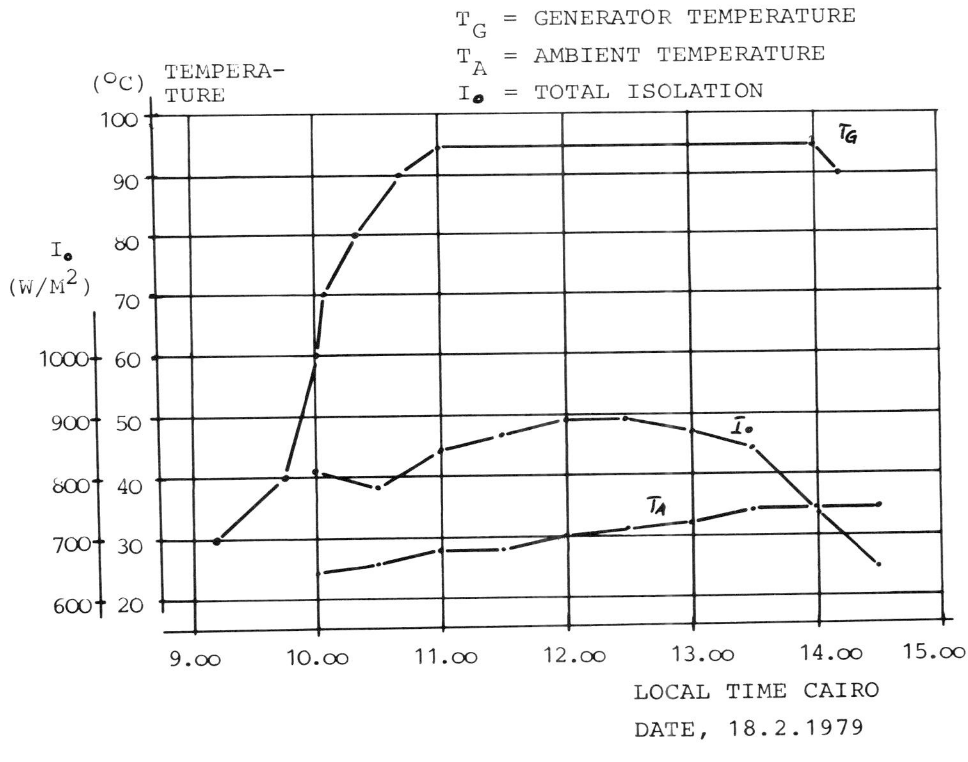

Fig. 3: MEASURED COLLECTOR FIELD (GENERATOR) TEMPERATURES OF THE SOLAR COOLING PLANT AT CAIRO

A selective coated collector field will start earlier in the morning and stop later in the afternoon with its nominal temperature. The extension of generator operation time can result in a reduction of the storage tanks for the night cooling requirements and a decrease of the manufacturing costs of the plant.

Additionally it has to be mentioned, that the operational conditions for the plant start up conditions of:

- Solar insolation = 700 w/m^2
- Ambient temperature = 25 ^oC

can be reduced by using selective coated collectors.

2.2 Increase of generator temperature

Fig. 2 shows, that at constant efficiencies (range 30 - 50 %) the collector field temperature increases approximately 10 ^oC (absolute) which corresponds to 20 - 30 % relative improvement.

This higher temperature can be used for different purposes.

- Higher temperature means higher output of cooling energy: If the generator temperature is kept at 90 ^oC, then the collector field area can be reduced. During the tests it can be realized practically by covering parts of the collector field.

- In Fig. 4 the generator temperature is plotted as a function of the evaporator respectively cooling room temperatur. Parameters are different cooling water temperatures. It is shown, that the increase of the generator temperature by 10 ^oC will allow a parallel increase of the cooling water temperature by 5 ^oC to approximately 35 ^oC of designed plant operating conditions. This result is essential for the application of these plant types in tropical areas, where in summer time the available cooling water temperature can reach more than 30 ^oC.

3. Realization of the project

The collector field is in production and will be integrated into the plant at Cairo in the second half of the year 1982. After function tests of the plant with the improved collector field, the long time tests are performed. These tests are expected to prove the designed data.

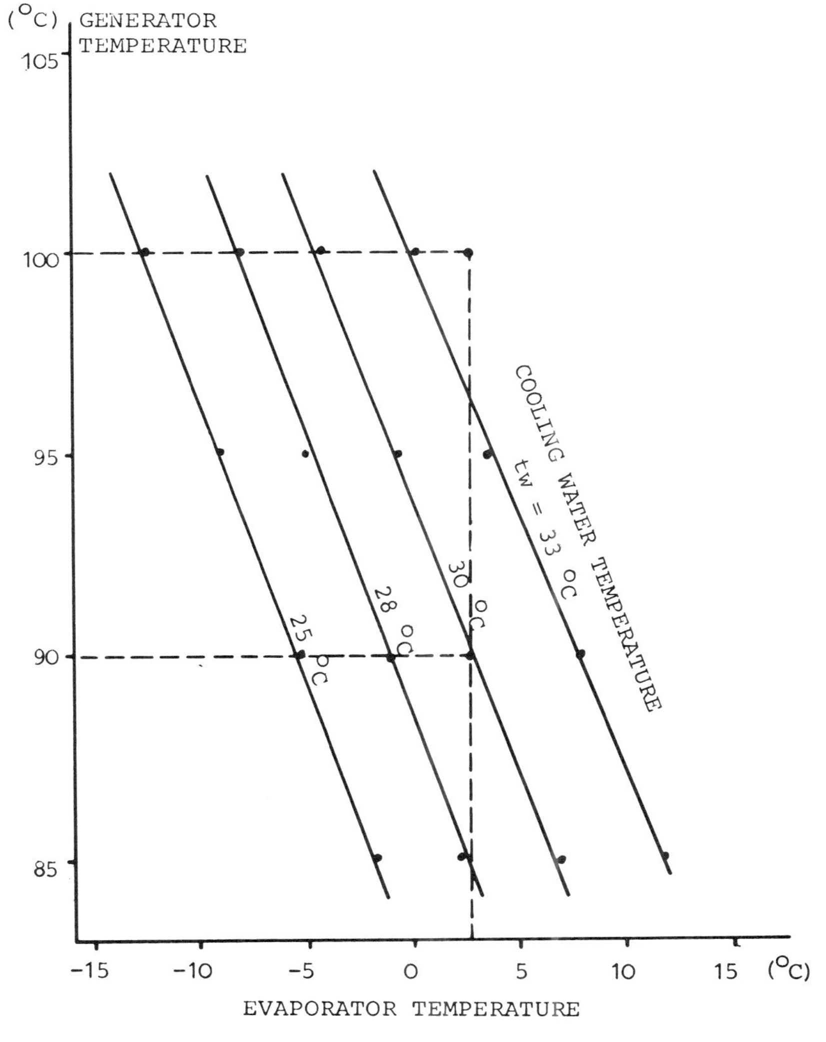

Fig. 4: CHARACTERISTICS OF AMMONIA/WATER ABSORPTION COOLING PROCESS

SECTION 4 - SOLAR COOLING MACHINES

Summary on solar cooling systems

Construction and acquisition of process data relative to a prototype of a solar absorption resorption refrigerator

Realisation and tests of a solar powered ice maker using the AC 35 - CH_3OH intermittent cycle

Self operating cooling system using solar energy - small power experimental plant

Development of an autonomous free piston refrigerating unit driven by Rankine cycle

SUMMARY ON SOLAR COOLING SYSTEMS

Author: C. den Ouden

1. Introduction

In the section 'Solar Cooling Machines' two research teams presented work with NH_3/H_2O machines, one solid absorption process for the production of ice was discussed and finally a Rankine-powered vapour compression cycle was lectured on.

2. Review of the presentations

The first presentation about 'The construction and acquisition of process data relative to a prototype of a solar absorption refrigerator' showed the complexity of a two-stages NH_3/H_2O absorption machine to allow lower collector temperatures to feed this machine (85 - 90 $^\circ$ C).
A 28 kW absorption plant was built, working at -32 $^\circ$ C in the evaporator, with an expected C.O.P. of 0.23. The problems with the starting up of this plant were mentioned. The test results will soon become available.

The second study 'Realisation and Tests of a Solar Powered Ice Maker using the AC 35-CH_3OH intermittant cycle' was presented by Dr. Meunier of CNRS, showing the use of an activated carbon (AC 35)-methanol pair for a solar powered ice-maker.
In the previous work the zeolite 13X - H_2O pair gave good results to obtain conservation temperatures of about 5 $^\circ$ C, but it does not provide low enough temperatures in the evaporator to obtain temperatures below 0 $^\circ$ C to make ice. A prototype unit was built to measure heat and mass balances in intermittant cycles. Based on measurements with this unit the dimensions of a future system to produce 25 kg of ice per day have been estimated. It seems that with 4 m^2 of collectors filled with ≈ 40 kg/m^2 of AC 35 this production of 25 kg per day can be achieved provided that there is sufficient cooling during the night to ensure a low adsorption temperature at the end of the night.

The third presentation, by Mr Velluet, dealt with NH_3/H_2O absorption machine to be used in developing countries for the production of ice. The self-operating cooling system is based on a device which drives the solution pump by expanding ≈ 20% of the NH_3 vapour coming from the generator , at 130 $^\circ$ C. The little scale of the produced prototype induced several difficulties for the proper running of the mechanical parts. At the present time they have changed their approach and are considering to build an efficient free-piston engine-pump group for operating the cooling cycle.

The final presentation about 'Development of an autonomous free piston refrigerating unit driven by a Rankine cycle', by Mr Vandendael, showed the last results obtained to develop a free piston refrigerating machine supplying a net output of 3 kW at 10 $^\circ$ C, with 25 $^\circ$ C ambient temperature, driven by solar heat at low temperature of ≈ 70 $^\circ$ C. Several tests were started with the first prototype machine with freon as working fluid, results are obtained about the functioning of the motor, some leakages have been observed and problems have been solved, pressure drops across the compressor valves have been found to be larger than calculated initially. In parallel with the experimental study a new computermodel giving a dynamic and thermodynamic description of the machine has been developed. The new computer programme is in good agreement with the experimental tests. A design for a second prototype will be made and built during the next months.

CONSTRUCTION AND ACQUISITION OF PROCESS DATA RELATIVE TO A PROTOTYPE OF A SOLAR ABSORPTION RESORPTION REFRIGERATOR

Author : M.ROSSI, F. UCCELLI

Contract number : ESA-C-122-I

Duration : 12 months - 1 July 1982 - 30 June 1983

Total budget : It.Lire 103.200.000 CEC contribution: 45%

Head of project : Aldo Dellacasa

Contractor : Società per Azioni Termomeccanica Italiana

Address : Società per Azioni
Termomeccanica Italiana
Via del Molo Nº 1
I - 19100 LA SPEZIA

Summary

The object of this research is the achievement of low temperatures for cooling purposes, i.e., storage of frozen products in a cold store (-32 C), by means of low enthalpy solar energy (90 C). This was made possible by adopting a multi-stage absorption refrigeration cycle.

The plant has been equipped with a cooling water system, a hot water circuit (solar simulator), a brine circuit (cold user simulator), and a safety valve exhaust line.

Due to the delay in the permission from the Authority, the test of the plant has not yet started.

Finally, the re-design of the plant for industrial purposes will start within October 1982. The aim is to simplify the plant and the operation on the basis of the experienced tests.

1. INTRODUCTION

The research TERMOMECCANICA ITALIANA S.p.A. (La Spezia) is carrying out, has the scope of planning and designing a refrigerating unit, of the absorption type, capable of delivering 28 kW at -32 C, exploiting heat energy at 85 C and using an ammonia-water mixture. By the thermodynamic cycle we selected, the highest coefficients of performance and the lowest consumption in mechanical power were obtained. In Fig.1 such a cycle has been sketched in a main cycle, A"B"C"D", and two secondary cycles, ABCD and A'B'C'D', the latter identifying a resorption system . Such a complex system was called for by the operating conditions.

The main data of selected cycle are summarized in Table 1 hereinbelow.

TABLE 1 - DATA OF SELECTED CYCLE

	A B C D	A'B'C'D'	A"B"C"D"
X_w (%)	0.22	0.40	0.33
X_s (%)	0.28	0.45	0.39
ps bar	3.3	1.1	7.0
pi bar	1.1	3.3	1.1
TC (°C)	65	33.8	70
TD (°C)	77.5	43	81
TB (°C)	33.8	4.3	13.9
TA (°C)	45	12.3	24.4
Tcond (°C)	--	--	13
Tevap (°C)	--	--	-32

where:

- X_w: weak solution concentration
- X_s: strong solution concentration
- p_s: pressure at the generator
- pi: pressure at the absorber
- TA, TB, TC, TD: temperature at point A,B,C,D, respectively
- Tcond: temperature at point condenser
- Tevap: temperature at point evaporator.

The total estimated performance is equivalent to 23 per cent, while power used for circulating the solution is 2.5 kW.

In Fig.2, the various parts of the unit as well as the piping system are represented in a diagram: generators are marked with "G", absorbers with "A", resorber with "R", evaporator with "E", heat exchangers with "S" and pumps with "P". A measuring and controlling system can also be found, providing for making easier the plant operation and carrying out all required measurements.

2. AUXILIARY CIRCUIT

The refrigerating unit has been connected to a number of auxiliary circuits, namely:

- cooling water circuit
- hot water circuit
- brine circuit
- safety valve blow-off system.

2.1. Cooling water circuit

Heat generated by dissolution of ammonia in the water, developing in the absorber "A" and in the resorber "R", was driven off by circulating, inside the units, cooling water from a cooling tower installed in the testing room at TERMOMECCANICA's workshops. The temperature of recirculated water from the cooling tower was kept at 32 C (design value), while two automatic valves provided to control the water flow rate through "A" and "R" according to the the plant requirements.

2.2. Hot water circuit (solar simulator)

The refrigerating unit designed for operating with a solar collector system, was supplied, during the experiments, with hot water from a diesel fuel boiler. This circuit, characterized by a simple construction, was carefully designed, since a successful operation of the refrigerating unit depends on it. In fact, any indesirable changes in the characteristics of the heat source would create a disturbance within the refrigerating cycles, thus altering the measurements in progress.

For this purpose, an automatic three-way valve was installed on the hot water inlet side to each generator (see diagram in Fig.3), normally actuated by a temperature controller connected to the weak solution outlet side. Such valve, which varies the hot water flow rate, and, accordingly, the heat gain to the generator, keeps the machine at the required rate according to the cooling load.

Another temperature controller, fitted to the boiler, provides for keeping the hot water at a constant set value. Moreover, also the flow rate of the water circulating inside each generator can be kept at a constant value (in this case, the temperature will undergo some changes). In this way, we can simulate the "transitory" periods required to set the solar collector system.

2.3. Brine circuit (cold user simulator)

The cold made available under operation by the absorption system needs to be transferred. To this purpose, brine (a 28% Ca Cl2 non freezing solution) is caused to circulate within the "E" evaporator, being cooled down at the expense of the evaporating ammonia. A like amount of heat is given back to the brine within an electric heater offering the following advantages:

- great reliability
- repeatibility
- easy switch off/on
- no control is required under operation
- heat load may be fractionated in several like steps.

The brine heater we adopted consists of eighteen elements, each of 1,700 W (30.6 kW in total). This device provides for keeping strictly constant the plant load, thus allowing for all basic setting to be made and performances to be defined. By subsequently switching on and off groups of heating elements, the variability conditions in a refrigerating room may be simulated and the reactions of the refrigerating unit in such conditions studied.

2.4. Safefty valve blow-off system

In compliance with the regulations of A.N.C.C. (Italian Boiler Inspection Authority), the Italian association in charge for issuing the safety specifications to be complied with both under construction and operation of pressurized gas plants, each piece of equipment is fitted with safety devices. Each pressure vessel is equipped, inter alia, with a safety valve to prevent any accident in the event of abnormal functioning of the system.

Former A.N.C.C. standards, which required evacuation of the excess gas directly to atmosphere, have been replaced by recent provisions requiring channeling of blow-offs and elimination of ammonia. So we were compelled to install a manifold collecting blow-offs from safety valves, ending in a distributor immersed in a tank filled with water, to permit absorption of gases without evacuating them to atmosphere.

3. STATUS OF THE WORK

Since some rules concerning safety of ammonia systems have recently changed, and due to slips ensuing from the fact that A.N.C.C., formerly under the State's control, has passed to a local control, we are still waiting for the permission to use the system.

At present, we have made all required modifications and implementations (such as, blow-off system, double fire fighting system, etc.) and are trying to locate the hot water boiler far out the refrigerating plant.

The final inspection by the Authorities responsible for giving the authorization to operate the refrigerating system is expected by the end of June 1982.

4. FUTURE PROGRAMME

The scope of this new Contract mainly includes re-designing of the refrigerating system in view of an easier utilization by non-specialized personnel. When realizing the pattern, element ensuring the greatest flexibility and the richest data recording have been selected, therefore, the prototype is extremely complicated and require qualified people for operation.

The criteria we'll follow when re-designing the plant, will be quite the contrary. First of all, we may rely upon a series of data we obtained by testing the plant under the most severe conditions. As far as the control systems are concerned, all main parameters to be checked during testing will be focused. In such a way, it will possible to select most reliable and simple automatic control devices. On the other hand, a more compact system will be realized, by using standard materials and vessels, which are cheaper than those used to construct the prototype.

Unfortunately, due to the obstacles we have just mentioned, the testing programme included in the previous Contrat, is expected not to start before July 1982; therefore, re-designing, expected in early July, will be deferred to next September-October, when the first prototype testing results are available.

In spite of this, we singled out a nuisance on the liquid film absorbers, which are causing us many troubles in dimensioning liquid distributors.

In the final design, such vessels will be replaced by absorbers of different type we are studying.

Taking into account the fact that performance of the testing on prototype and re-designing of the industrial machine may be carried out at a same rate, if of course a greater number of persons is involved, we believe the final date of this Contract, i.e., end of June 1983, may be kept.

FIG 1

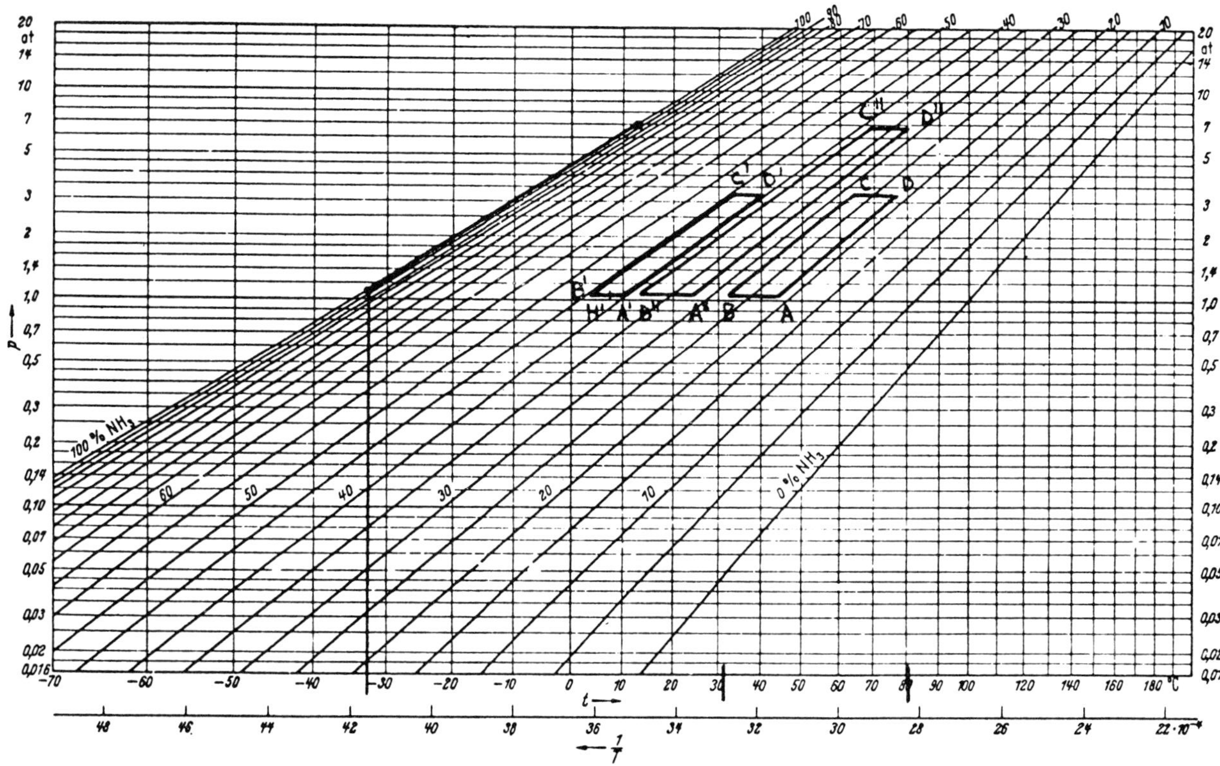

lg p, 1/T-Diagramm für Ammoniak-Wassergemische

Nach Bošnjaković und Wucherer,
entnommen: VDI 1006,
Kältemaschinen-Regeln: Absorptions-Kältemaschinen,
Berlin: VDI-Verlag G.m.b.H. 1943

Handbuch der Kältetechnik VII

Springer-Verlag, Berlin/Göttingen/Heidelberg

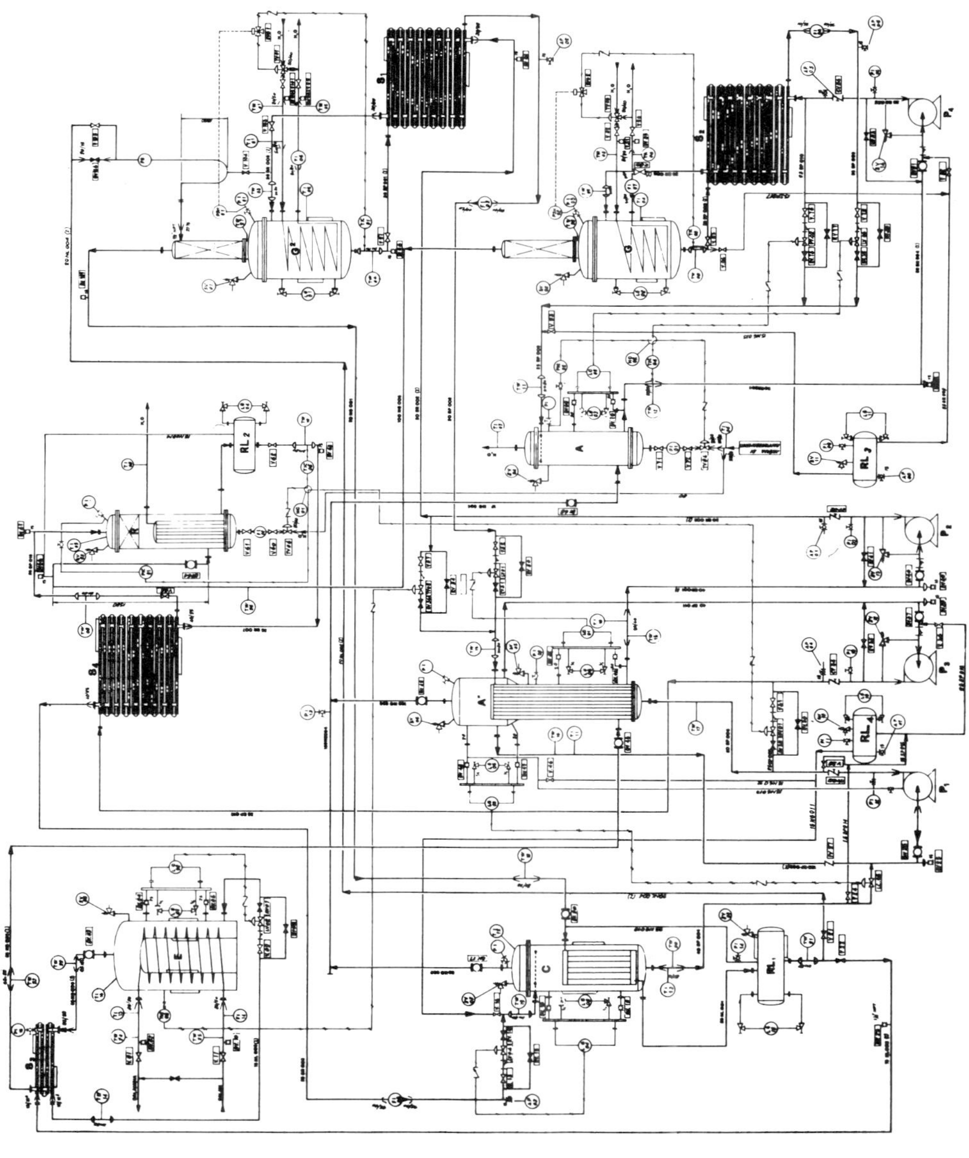

FIG

IMPIANTO FRIGORIFERO AD ASSORBIMENTO

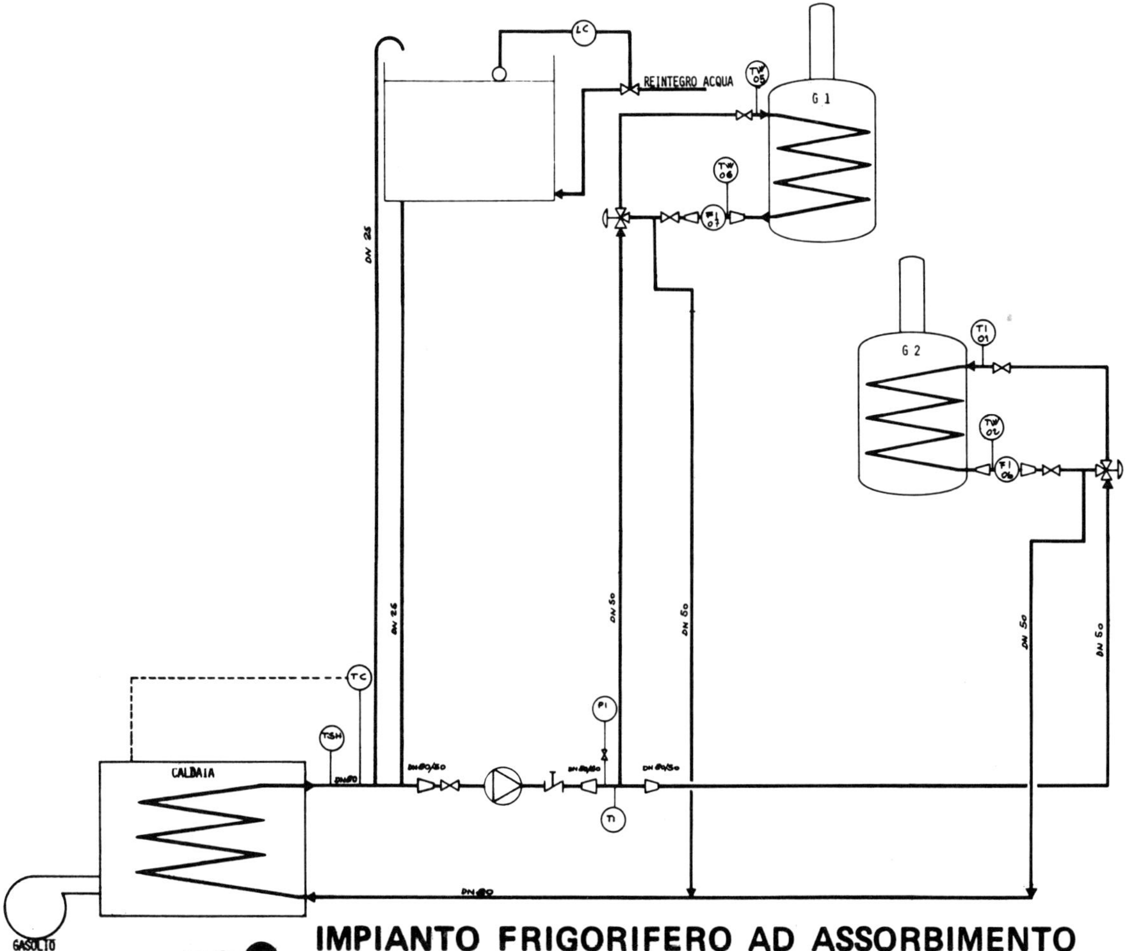

FIG 3 IMPIANTO FRIGORIFERO AD ASSORBIMENTO
SCHEMA ACQUA CALDA

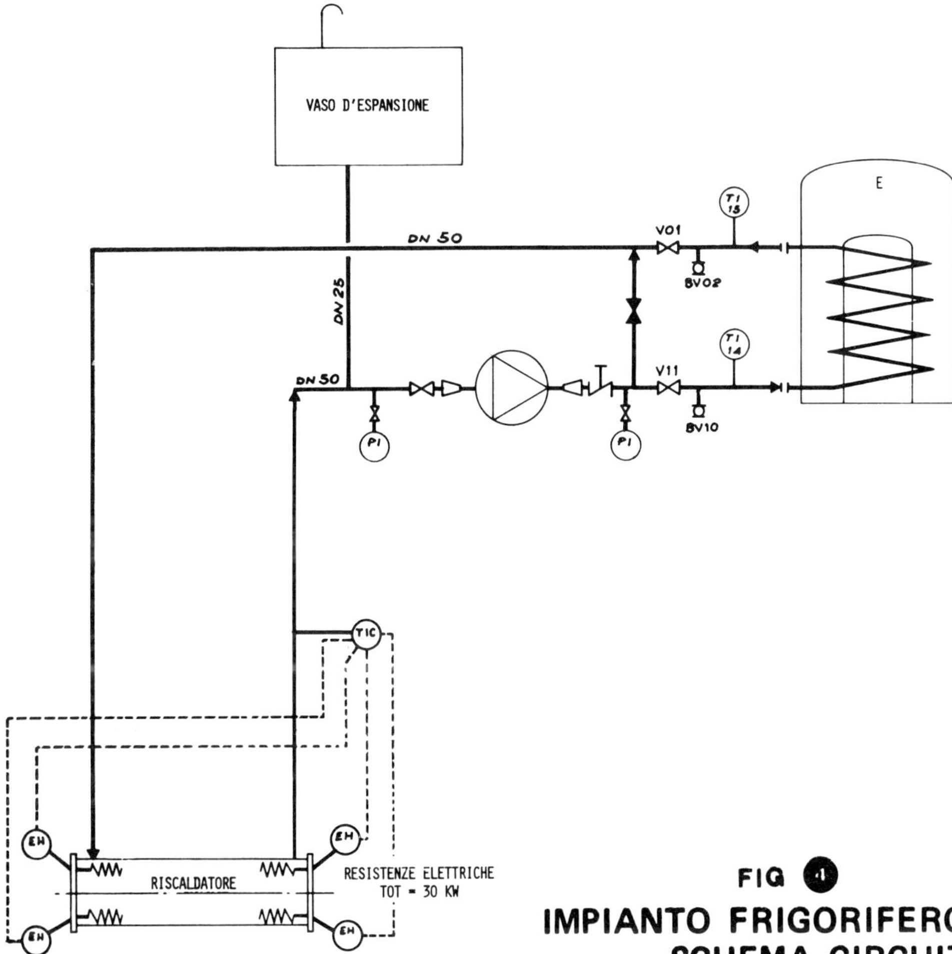

FIG 1
IMPIANTO FRIGORIFERO AD ASSORBIMENTO
SCHEMA CIRCUITO SALAMOIA

Fig. 5

Fig. 6

REALISATION AND TESTS OF A SOLAR POWERED ICE MAKER USING THE AC 35 - CH_3OH INTERMITTENT CYCLE

Authors P. GRENIER, F. MEUNIER, M. PONS

Contrat number ESA - C - 123 - F

Duration 24 months

Total budget 700 000 FF CEC Contribution 50 %

Head of project Dr F. MEUNIER

Contractor C N R S

Address Thermodynamique des Fluides
Bat. 502 Ter, Campus Orsay

91405 ORSAY

Summary

This paper presents the preliminary results obtained on the study of the activated carbon (AC 35) - methanol pair. The thermochemical parameters of the pair have been determined;with this parameters, the performances of the cycle have been calculated ; a comparison of this cycle with the zeolithe 13X - water cycle is then presented.
The unit for indoor measurements which has been built is then presented and the results of this measurements are discussed.
This thermodynamical study is the first step of a work which will include the realisation and the tests in situ of a solar powered ice maker of 25 kg of ice per day.

Work programme

Phase n° 1 : - Thermodynamical study
- Feasability study (indoor measurements)
status : achieved

Phase n° 2 : - Conception and realisation of the plans of a solar powered ice maker of 25 kg of ice per day.
status : underway

Phase n° 3 : - Fabrication of the machine and tests

Phase n° 4 : - In situ measurements in the south of France

Phase n° 5 : Analysis and conclusions.

INTRODUCTION

Thermodynamical cycles using solid adsorbents are good candidate for the purpose of solar cooling. So far, the zeolithe 13 X - H_2O pair has been tested and gives good results for the obtention of conservation temperatures of the order of 5°C. In the preceding contract two units using that cycle have been built (1) : (fig. 1)

- Individual refrigerator which has been tested two years
- A food storage unit of twelve cubic meters which is currently under tests.

Fig. 1 : The food storage unit of 12 m^3 and a small refrigerator using the Z 13 X - H_2O cycle

One of the limits of the Z.H_2O pair comes from the fact that it does not provide low enough temperatures in the evaporator so as to obtain temperature below 0°C and to make ice.

Another pair had been prospected : the Z 13 X - CH_3OH, but that pair did not give the expected results :

- its optimal temperature range is not optimized for our purpose
- a catalytic reaction occurs so that this pair cannot actually be used for cycles.

It was then important to prospect towards other pairs with the constraints :

- high temperatures compatible with flat plate collectors
- evaporating temperatures suited for ice making.

In this first intermediate report we present the work which has been achieved on the activated carbon (AC 35) - methanol pair.

1°) **The criteria for the selection of the pair**

On the fig. 2 we present a comparison between the activated carbon CH_3OH pair and other pairs (Zeolithe 13 X - H_2O, zeolithe 13X - CH_3OH). This comparison is presented in terms of log θ versus $(\Delta T/T)^2$ on the next curves :

- θ is the filling factor
- ΔT is the depression temperature : temperature difference between the solid and the liquid in the evaporation phase.
- T is the temperature of the liquid.

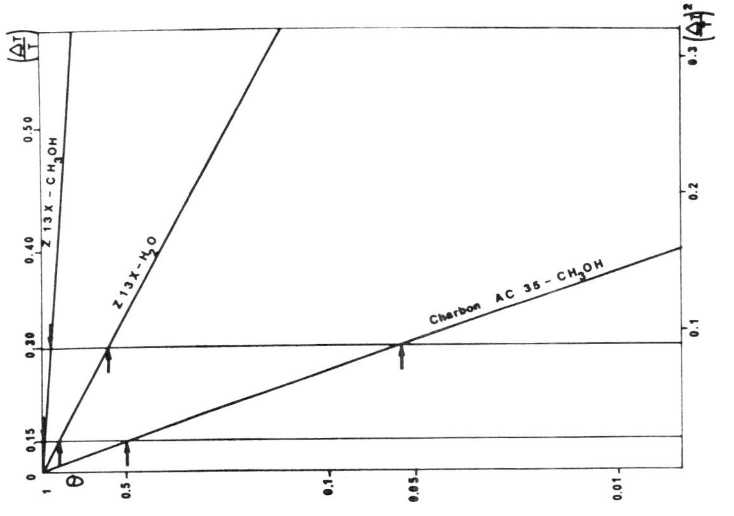

Fig. 2 : Log θ versus $\left(\frac{\Delta T}{T}\right)^2$ for various pairs.

It can be seen that in this representation we get linear lines. Taking typical values of the ratios $\Delta T/T_0$ in the evaporating phase and in the condensing phase this gives -as order of magnitude- the cycled mass of refrigerant fluid during a cycle. It is then the first criterium to conclude wether the pair is adapted or not to the desired cycles. From the curves in the figure 2, it is seen that for $\frac{\Delta T}{T} = \frac{40}{273} = 0.15$ in the evaporation phase and $\frac{\Delta T}{T} = \frac{90}{303} = 0.30$ in the condensation phase, the AC 35 - CH_3OH pair will cycle half of the total amount of adsorbed CH_3OH when the Z 13 X - H_2O pair will cycle about one third of adsorbed H_2O and the Z 13 X - CH_3OH pair will cycle less than 10 % of the adsorbed CH_3OH.
We see that on the basis of this criterium, the AC 35 - CH_3OH pair is better than the two others. But this criterium is not exclusive and the answer comes from the COP.

2°) **Physico chemical data**

Isotherms have been measured in the laboratories of the CECA[2]. This results are well represented using the following formula (Dubinin representation).

This leads to the following equations for the filling factor and for the heat of adsorption ΔH :

$$\theta = \exp - D \left(T \log \frac{P_s}{P}\right)^n$$

$$\Delta H = L + \frac{RT}{M} \text{Log} \frac{P_s}{P} + \frac{RT}{nDM} \left(T \log \frac{P_s}{P}\right)^{1-n}$$

- D and n are phenomenological parameters determined from the experimental measurements of the CECA : $D = 3.222 \times 10^{-7}$
  $n = 2.195$

- $\theta = \frac{m(T)}{m_o(T)} = \frac{W}{W_o}$

  . m(T) is the mass of CH_3OH adsorbed at the temperature T and the pressure p ; the saturation pressure for the liquid being $P_s(T)$.

  . $m_o(T)$ is the mass of CH_3OH which would be adsorbed at saturation at the temperature T.

  . W is the volume of adsorbed CH_3OH at the temperature T and the pressure p.

  . W_o is the volume of the pores : 0.407 l/kg.

- $\alpha = - \frac{1}{\rho} \frac{d\rho}{dT}$ is the dilatation of liquid.

- M is the molar mass.

A representation of the isosters is shown in the next figure :

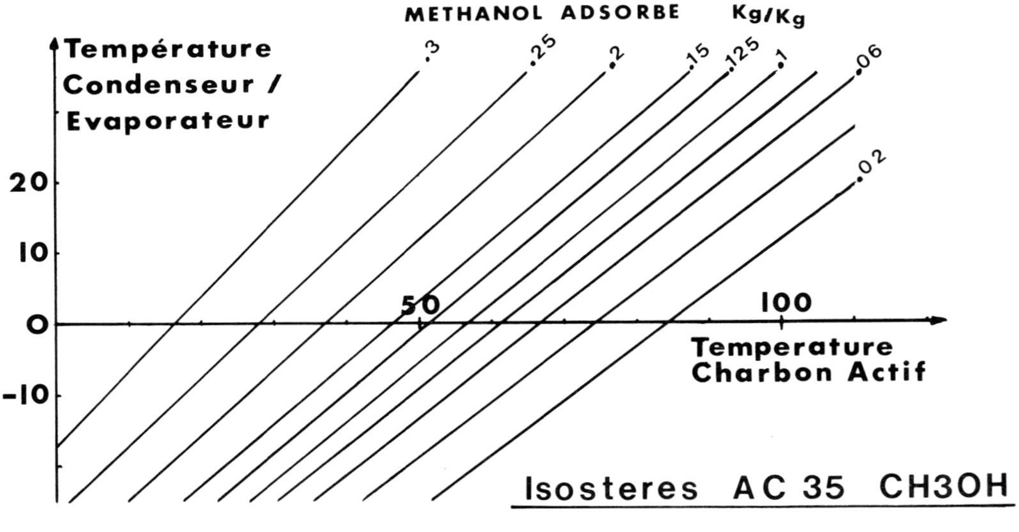

Fig. 3 : Equilibrium chart for the AC 35 - CH_3OH pair

A **table of the** values of ΔH is given versus the temperature and the mass adsorbed

m \ T	20°C	40°C	60°C	80°C	100°C
250	10.22	10.09	9.93	9.75	9.62
200	10.37	10.22	10.00	9.75	9.50
150	10.59	10.40	10.18	9.90	9.62
100	10.68	10.68	10.43	10.15	9.84
50	11.25	11.06	10.81	10.53	10.18
CH_3OH	8.73	8.53	8.27	7.95	7.57

TABLE I : Values of ΔH (kcal/mole) versus the temperature and the mass adsorbed

3°)**Theoretical predictions for the COP**

Once we get the thermodynamical and thermo-physical data of the pair, it is possible to calculate the thermal COP as well as the solar one through a numerical simulation programm using a computer (3).

Some results of these calculations are presented in the next figure where we plot the thermal COP for three various temperature conditions.

We can see on this figure :

- that the COP goes through a maximum versus the regenerating temperature. The temperature corresponding to the maximum lies in the range of 80 to 100°C.

- that the value of this maximum decreases when the depression temperature increases.

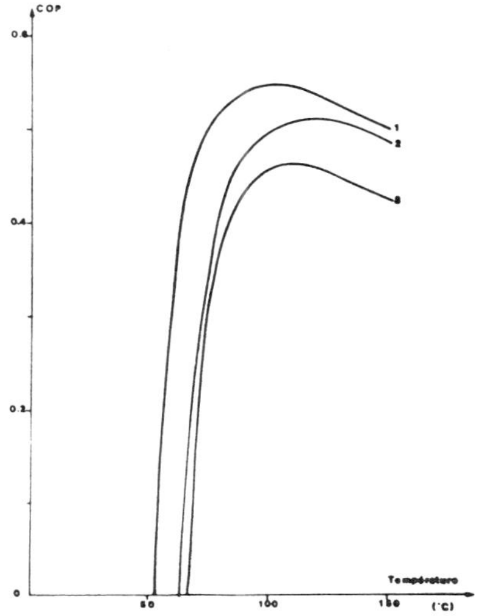

Fig. 4

Calculated curves of the thermal COP versus the regenerating temperature for three different conditions curve 1 - T_{ev} = - 10°C T_{cond} = 20°C T_{ads} = 20°C

curve 2 - T_{ev} = - 10°C T_{cond} = 30°C T_{ads} = 20°C

curve 3 - T_{ev} = - 20°C T_{cond} = 20°C T_{ads} = 20°C

4°) **Comparison of the theoretical COP values of the AC 35- CH_3OH and Z 13 X - H_2O pairs**

As we want to compare the AC 35 - CH_3O and Z 13 X - H_2O pairs for different values of the temperature of evaporation, the only way to make that comparison is to make it in terms of the ratios of the COP to that of the ideal Carnot cycle. We present such a comparison in the fig. 5 for the two pairs versus the depression temperature.

We calculated the ratio COP/COP carnot for different depression temperature, the temperature of evaporation being always the same (5°C for the Z 13 X - H_2O pair and - 5°C for the AC 35 -CH_3OH pair) ; we then took the maximum of the COP/COP carnot ratio and we plotted this maximum versus the depression temperature for each pair. This plot is presented on the fig. 5.

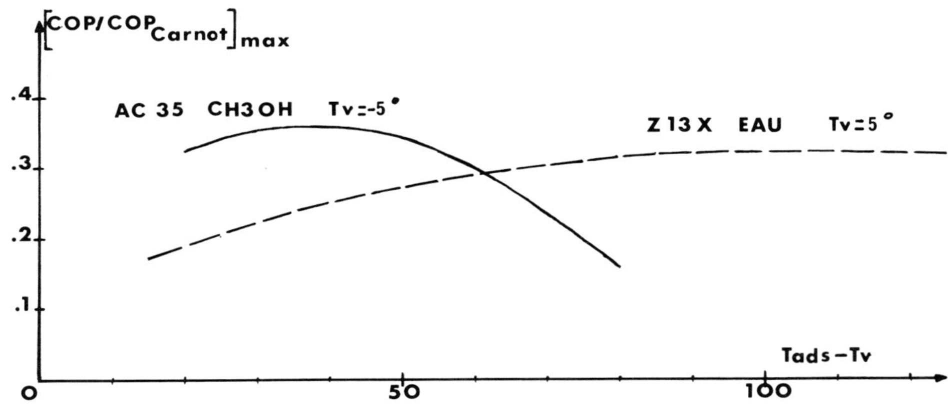

Fig. 5 : Variation of the maximum of the COP/COP carnot ratio versus the depression temperature for the Z 13 X - H_2O and the AC 35 - CH_3OH pairs.

As can be seen, the maximum is reached for a depression temperature of 40°C for the AC 35 - CH_3OH pair when the maximum is reached only for very high depression temperature ($\sim$ 100°C) for the Z 13 X - H_2O pair.

This stresses clearly that the AC 35 - CH_3OH will be a good pair only if the adsorption temperature during the night is not too high when this is not a problem for the Z 13 X - H_2O pair.

5°) **Realisation of a unit for feasability tests in laboratory**

A unit has been built to measure heat and mass balances in intermittent cycles (fig. 6) :

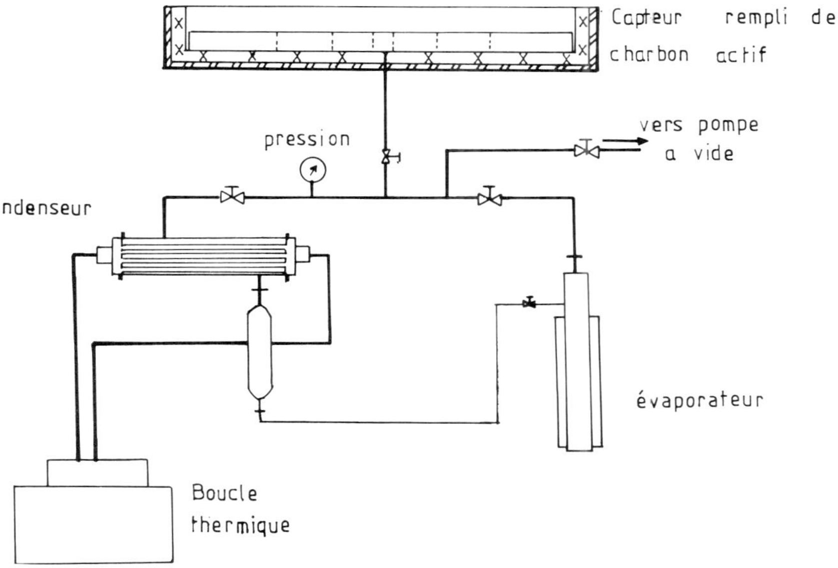

Fig. 6 : Schematic of the unit

This unit is made of :

- a 0.5 m^2 solar collector made of a copper box (4cm thick) filled wwith 9 kg of activated carbon AC 35.
- an evaporator built to make ice
- a condensor whose temperature is controlled with a regulated thermal loop
- a heat source made of eight lamps.

We measure :
- the temperatures
- the pressures
- the mass of methanol desorbed.

6°) **Results of the measurements**

Typical results are presented on the fig. 7.

With this experimental data it is possible to perform the heat balance and to determine the experimental COP of the machine (4).

The results of the measured COP are presented on the fig. 8 for three different temperature conditions.

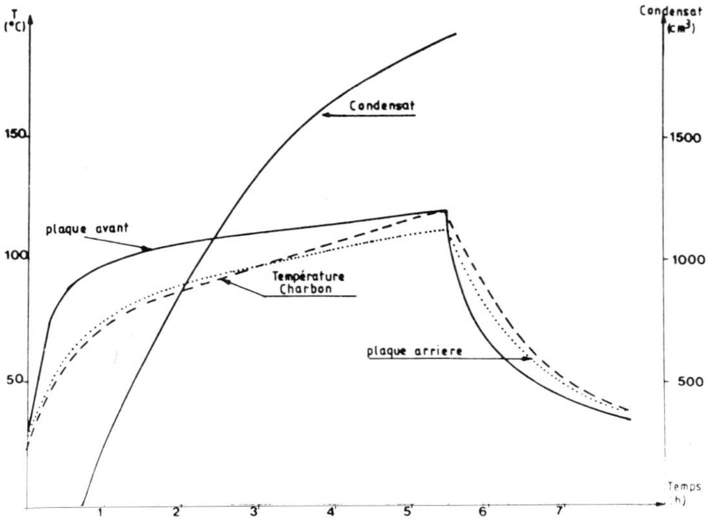

Fig. 7 : Experimental results of a desorption cycle for T_{ev} = - 10°C, T_{ads} = 20°C, T_{cond} = 30°C.

On the fig. 9, we present a comparison between the theoretical predictions and the measurements.

As we can see from this two figures :

- the COP decreases strongly when the depression temperature increases
- a reasonable agreement between measurements and calculations is obtained if we take for the calculations a temperature of condensation 7°C higher than the experimental one. <u>This ΔT = 7°C is characteristic of the irreversibilities in the machine during the cycle</u>.

7°) **Conclusion**

The AC 35 - CH_3OH pair is well adapted for solar powered (using flot plate collectors) ice making (T_{ev} - 10°C, T_{reg} ∿ 100°C) <u>provided</u> cooling during the night (through natural convection) of the collector filled with AC 35 is good so as to have the adsorption temperature at the end of the night (the only one which is important from the thermodynamical point of view) as low as possible.

The collectors which are under realisation will be 9 cm thick and filled with ∿ 40 kg/m^2 of AC 35.

The surface of the collectors will be 4 m^2 with the objective to have a nett production of 25 kg of ice per day (losses of the order of 10 kg per day, the gross production being ∿35 kg per day).

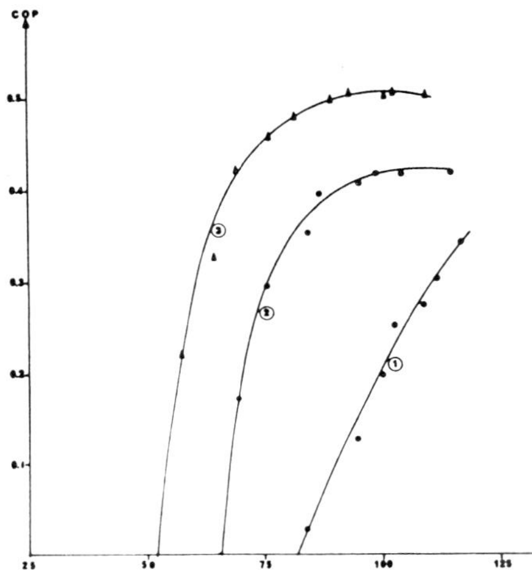

Fig. 8 : Measured COP for :

1 - T_{ev} = -24, T_{cond} = 30, T_{ads}=19.6

2 - T_{ev} = -10, T_{cond} = 30, T_{ads}=20

3 - T_{ev} = - 4, T_{cond} = 20, T_{ads}=20.5

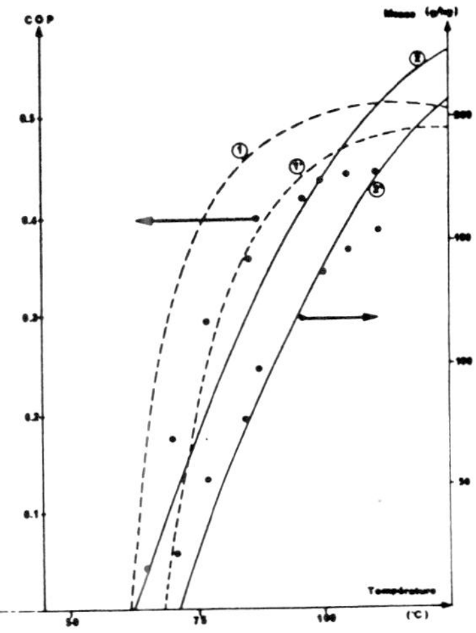

Fig. 9 : Experimental points for T_{ev} =-10, T_{cond} = 30, T_{ads}=20

. measured COP

. adsorbed mass of CH_3OH

Calculated curves

1 - COP T_{cond} = 30

1' - COP T_{cond} = 37

2 - Mass T_{cond} = 30

2' - Mass T_{cond} = 37°C

REFERENCES

1 - J.J. GUILLEMINOT, F. MEUNIER, G. MARSIN, D. ROYER. Proceedings of the EC Contractors Meeting Athens Series A, Vol. 1, p. 93. D Reidel Publishing Company.

2 - J. VANDERMEERSCH. CECA private communication.

3 - J.J. GUILLEMINOT, F. MEUNIER, B. MISCHLER. Rev. Phys. Appl., 15, 441 (1980.

4 - J.J. GUILLEMINOT, F. MEUNIER. Rev. Gen. Therm., 239, 825 (1981).

SELF OPERATING COOLING SYSTEM USING SOLAR ENERGY. SMALL POWER EXPERIMENTAL PLANT

Author : P. VELLUET

Contract number : ESA/C/050/F

Duration : 24 months 1.7.80 - 30.6.82

Total budget : 1 243 700 FF CEC Contribution : 34 %

Head of the project : P. VELLUET

Contractor : ARMINES

Address : 60, bld St Michel 75272 PARIS CEDEX 06

Summary

This paper presents an abstract of our work in the field of solar refrigeration. This study deals with a solar cooling system using a H_2O-NH_3 absorption cycle and especially designed to solve conservation problems in remote and developing countries.

To this end the following caracteristics were taken as initial data :

. The system may be as simple as possible in order to reduce cost and risks of troubles. As an example, no heat storage device is located between the solar collectors and the absorption machine boiler

. Secondly, no mechanical energy may be required by the system. The solution pump is driven by an engine which is fed by vapour of the frigorific mixture itself.

The main steps of the study are :

. Numeric simulation of the whole cooling system (solar collectors, frigorific absorption machine, cooled room)

. Complete optimization of the absorption cycle, with taking into account the parameters of the needs (temperature of the frigorific effect, external medium temperature)

. Seizing and design of a little scale prototype for a daily production of 20 kg of ice.

. Experimental study of the prototype running.

1 - THE STUDIED SYSTEMS

1.1 - Description of the system

The system is based on a H_2O-NH_3 frigorific absorption cycle which allows high efficiencies even for temperatures below 0°C. This cycle is adapted especially for solar cooling in remote countries of tropics. So we made it self-operating by including inside the frigorific loop an expanding engine in order to run the solution pump (see fig. 1), and this engine uses the pressure difference between the boiler and the absorber (patent). Another important feature of this system is the direct heating of the strong water-ammonia solution by solar energy. The boiler is located inside the solar collectors themselves. This is very interesting because no thermal medium loop and no auxiliary circulation pump are required.

1.2 - Interest of the system

Such a system presents a high reliability because there are few moving parts, which is important for using in remote countries.

In addition it is self-operating for a large range of cooling effect temperatures and it gives a high frigorific efficiency. As an exemple with 45°C and -8°C temperatures respectively inside the condenser and the evaporator, the frigorific efficiency is greater than 30 %. Let us note again that this value increases when the condenser temperature is smaller, or when the evaporator one is higher.

2 - THE SIMULATION AND THE OPTIMIZATION OF THE ABSORPTION CYCLE

2.1 - Numerical model

This self-operating frigorific system has formed the subject of a computer modelling. The input data are on one hand the thermodynamic properties of water-ammonia mixture, and on the other hand the values which permit the determination of the absorption cycle including the expansion through the engine. Among this values one can find the evaporation temperature or the minimum temperature difference between the two fluids inside heat-exchanger.

The model gives as results all the properties of the mixtures (concentration, enthalpy, temperature ...) at about 20 thermodynamic points of the loops. Figure 2 presents a diagram of the cycle drawn from the results of the numerical model.

Flows and exchanged powers are also computed and the last result is the frigorific efficiency of the cycle, i.e the ratio ε_f = evaporator power/boiler power.

2.2 - Using of the model

Firstly this model permits the optimization of the cycle. An example of this kind of results was presented in a earlier paper (1), in which the frigorific efficiency (ε_f) is represented versus the boiler temperature. With the evaporation and condensation temperatures given above, the optimum boiler temperature is 130°C.

Secondly, we can study the effect of each thermodynamic parameter. Figure 3 shows the effect of the minimum temperature difference (d) between the strong mixture and the weak one inside the solutions heat exchanger. The curves give the variations of the frigorific efficiency (ε_f), the power of the heat exchanger (P) and also its area (S). So an economical compromise between the frigorific efficiency and the area of this heat-exchanger appears clearly because S and ε_f both increase when d decreases : the price of the machine depends on these two parameters, the cost of the solar collectors depends on their area and then on ε_f, and the solution heat-exchanger cost is an increasing function of its area.

For the prototype we are building, we choose 4°C as the value of the parameter d, which gives a ε_f value near the maximum and a reasonable area of the heat-exchanger (0,53 m^2).

3 - THE PROTOTYPE ACHIEVEMENT

3.1 - Main features

By way of an example we choose to produce ice, what requires evaporation temperatures below 0°C, with a high ambiant medium temperature (condensator and absorber at 45°C).

In order to reduce the study costs, the experimental plant has a very little frigorific power. The forecasted daily production of ice is 20 kg, and the required solar collectors area is about 6 square meters. The collectors are focusing ones and they have E-W axis. The first tests of the system will be made with an electrical heating device.

3.2 - Components of the prototype

We firstly manufactured a gear engine-pump group (2). But the large presure difference (15 bars) and the small nominal power of the machine (18 W) set difficult leakage problems. At the present time, we are trying to build an efficient free-piston engine-pump group in spite of the difficult design of minute steam-valves for the motor.

As other components we have :

. the boiler, which is divided in 6 parts, and each of them is located inside a solar collector absorber
. The refiner, a packed column with Rashig rings,
. the condenser and the absorber, which are water cooled and made by a coiled tube inside a calender
. the evaporator, which is flat plate and plunged in water for the frigorific effect determination
. and the solution heat-exchanger which is a "tube-in-tube" type one.

Figure 4 shows the prototype during manufacturing.

3.3 - Instrumentation

Its chief object is the definition of the actual thermodynamic cycle obtained in this machine. But the instrumentation may also be used for detecting the no-proper runnings of the prototype and finding their causes. Then we can also compare the experimental results and the theorical ones, in order to valide the simulation modelling.

The parameters to be mesured are about 30 temperatures, 5 flow rates and 2 pressures.

4 - CONCLUSIONS AND FUTUR WORK

The theorical results of the study make this system very attractive for preservation in remote countries by solar cooling . But the little scale of the prototype induces particular difficulties for a proper running of the mechanical parts.

The next step of this study is the inclusion of an engine-pump group in the absorption machine and the test of the whole device.

5 - REFERENCES

(1) P. VELLUET - CEC Coordination Meeting of contractors - january 1981 (Brussels)

(2) P. VELLUET - CEC Coordination Meeting of contractors - november 1981 (Kavouri)

6 - FIGURES

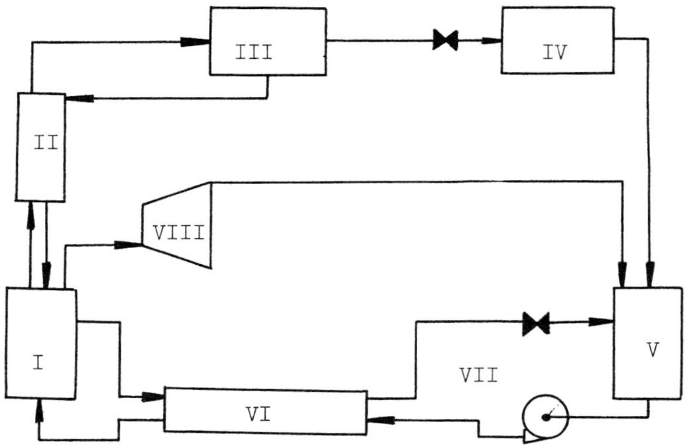

Fig. 1 - Scheme of the absorption machine :
(I) Boiler, (II) Refiner, (III) Condenser, (IV) Evaporator
(V) Absorber, (VI) Solution heat-exchanger, (VII) Solution pump
(VII) Expanding engine

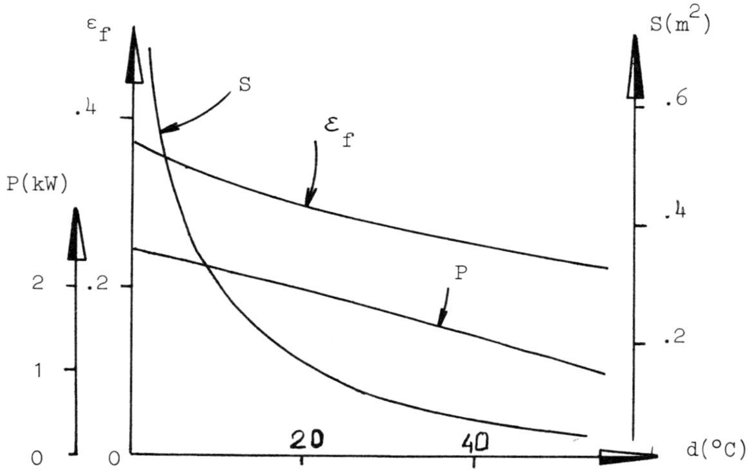

Fig. 3 - Effect of the smallest temperature difference inside the solution heat-exchanger (VI), on its power (P), its area (S) and the frigorific efficiency (ε_f)

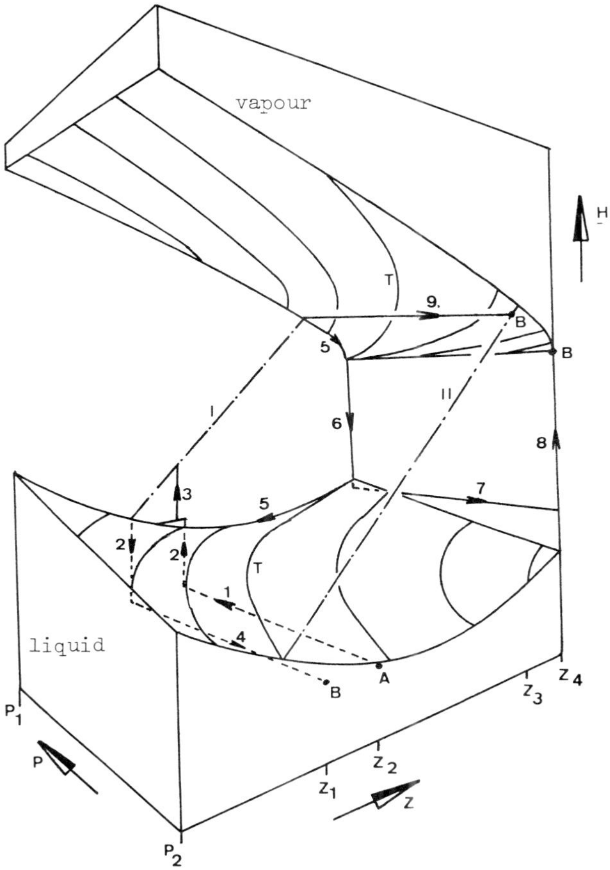

Fig. 2 - Absorption cycle with included engine drawn in (Z,P,H) space (composition, pressure, enthalpie) :

P_1 : condensation pressure
Z_1 : weak solution composition
Z_2 : strong solution composition

P_2 : Evaporation pressure
Z_3 : Engine vapour composition
Z_4 : Frigorific vapour composition

A : Absorber Exit
1 : Pumping of the strong solution
2 : Solutions heat-exchanger
3 : Boiling
I : Boiler exit equilibrium
4 : Expansion of the weak solution
5 : Rectification

6 : Condensation
7 : Expansion of ammonia
8 : Evaporation
9 : Expansion through the engine
B : Absorber Inputs

Fig. 4 - Self-operating absorption prototype during construction

1 - boilers
2 - solutions heat-exchanger
3 - evaporator

DEVELOPMENT OF AN AUTONOMOUS FREE PISTON REFRIGERATING UNIT DRIVEN BY RANKINE CYCLE

Authors : Y. VANDENDAEL & D. VOKAER

Contract nr : ESA-C-046 B

Duration : 25 months - 1.6.81 - 30.6.83

Total budget : FB 9980000 - CEC contribution: FB 4570000

Head of Project : A. JAUMOTTE - J. BOUGARD - D. VOKAER

Contractor : Université Libre de Bruxelles

Address : Institut de Mécanique Appliquée
Avenue F.D. Roosevelt 50
B - 1050 Brussels (Belgium)

Summary

A cooling effect can be achieved by using a so-called "Three Source machine" either of the absorption or the Rankine type. Whereas absorption machines require higher temperatures to work correctly, the ability of Rankine powered heat pumps to operate at lower boiler temperatures makes them particularly attractive for use in solar energy contexts.
It has been previously demonstrated that a single free piston machine can act as compressor and pump, for the purpose of achieving two (direct and inverse) Rankine cycle.
This paper is a summarized progress report of the research from November 1981 to May 1982. The last results obtained from the study of the first prototype of the free piston machine are discribed, and a new version of the modelling computer program is presented. Finally, the future work is proposed, according to the timetable of the proposal.

1. Aim of the Project

The research deals with a free piston refrigerating machine supplying a net output of 3 kW at 10°C, with 25°C ambient, driven by solar heat at low temperature : 70°C.

Previous contracts (528-78-1 ESB and ESA-C-017 B) led to the design of a prototype, which will be comprehensively studied in order to bring the necessary improvements and to refine the modelling programmes, with the future aim of building and testing a new prototype.

2. State of the work-summary

Informatic tools have been developed which can optimize the components of the thermodynamic cycle represented Fig. 1 a,b.

The major originality of the system lies in the way which has been chosen to design the mechanical components : expander, compressor and pump. All these functions are achieved at the same time by a patented free piston machine, whose principle is illustrated in Fig. 2.

At the last coordination meeting of contractors (Athens, November, 1981), the results of the following investigations were presented :

- study of the expansion process
- study of the pressure drops
- review of all machine configurations (as far as the distribution system is concerned)
- results of preliminary tests carried out with air and freon 114 as working media.

3. New developments

3.1. Experimental investigations

Comprehensive tests of the first prototype have been started with freon in the new configuration of the distribution system (fig.3), and the following results have been recorded :

(A) - The motor works satisfactorily; however its mass flow is more important than previously stated, owing to the "total admission" (which means that the high pressure driving fluid, is admitted along the whole stroke), as a consequence the condenser has to work over its maximum capacity, and its temperature increases.

(B) - Slight leaks have been observed across the compressor valves, another arrangement has been found, which avoids this problem.

(C) - The pressure drop across these valves has been found to be more than calculated initially, so that the mass flow rate delivered by the compressor is too small.

(D) - Nevertheless, the machine produces a cooling effect, but of course less than expected (see conclusions).

The experimental (P,V) diagrams of the motor and compressor are illustrated Fig. 4.

3.2. Theoretical developments

In parallel with the experimental study, a radical change has been brought to the modelling program of the machine.

(A) - The previous version of the program was based on a so called "total approach"; in other words :

- the thermodynamic cycles of both the motor and the compressor were introduced as data.
- all loses were taken into account by coefficients which modify the shape of these thermodynamic cycles.

(B) - The new approach is based on a "step by step" concept : each following effect (useful and losses) is discribed by a physical equation; the following basic equations are then solved step by step :

- the momentum equation
- the first law of thermodynamic applied on each chamber
- the equation of state of freon 114
- the heat exchange by convection between the cylinder and the gas
- the heat balance applied to the whole system
- response time of the electro-valves
- leakages trough piston rings
- friction .....

The computer model finally gives a full (dynamic and thermodynamic) description of the machine, as a function of time.

All the coefficients used in the equations have been experimentally determined by the study of the prototype now available.

Fig. 5 shows a comparison of the (P,V) cycles resulting from the new simulation program and the cycles actually recorded on the experimental rig.

Table 1 compares the experimental results and the computer simulation results.

4. Conclusions and future work

The experimental and theoretical investigations have been carried out according to the proposed timetable.

It should be stressed again that all these tests are still being done with the first machine ever built, whose dimensions have been shown to be bad. The main objective now is not to optimize the performances, but to point out the difficulties and to imagine better solutions for the future.

The new computer program is in very good agreement with the experimental tests.

Using the simulation we are now designing the second prototype which will be built during the next months.

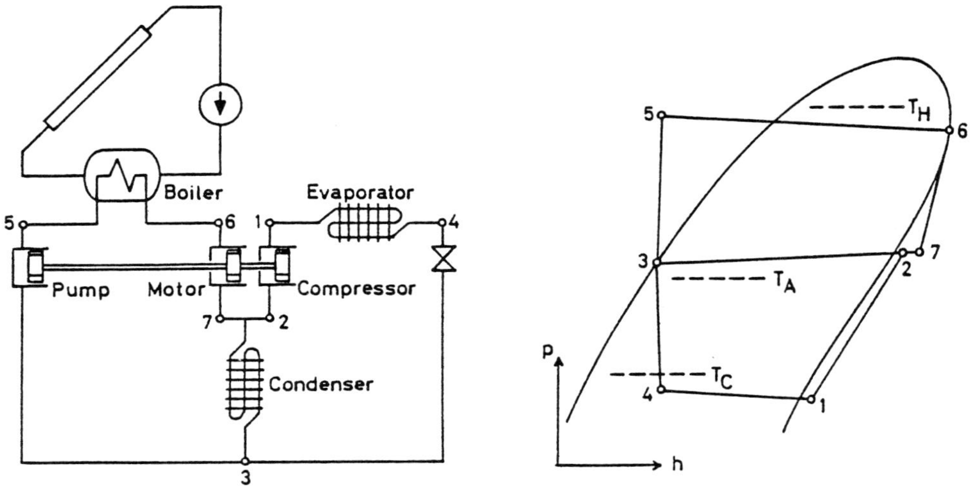

Fig.1.-a-Lay out of the installation.
-b-Thermodynamic cycle in (log p,h) coordinates.

Note : The two Rankine cycles have a common condenser both for simplification and economic purposes.

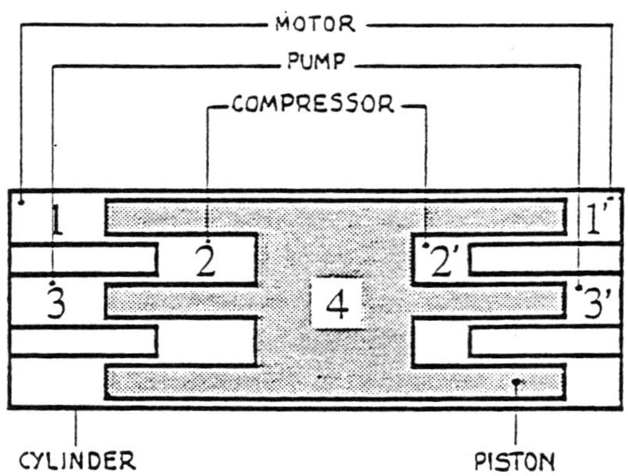

Fig.2 - Patented shape of the free piston, combining different functions.

Note : The force exerted on the piston by the driving fluid is directly transmitted to the fluids in the opposite chambers.

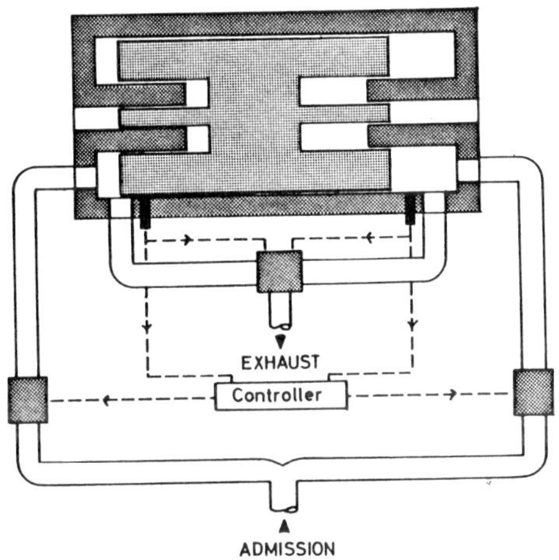

Fig.3 : New configuration of the distribution system

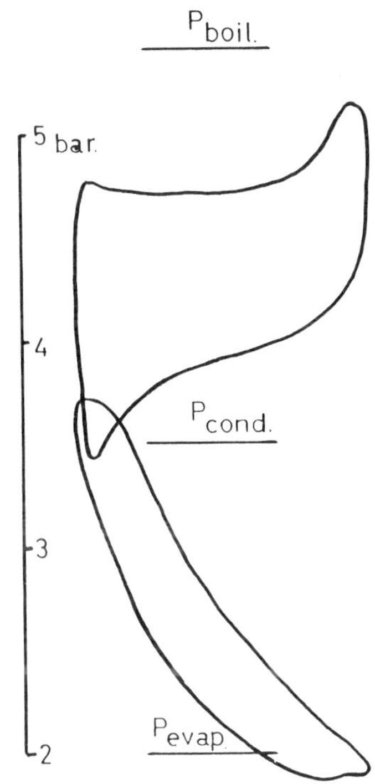

Fig.4 : Experimental (P,V) diagrams

Fig.5 : COMPARISON OF EXPERIMENTAL (----)

AND THEORETICAL (——) RESULTS.

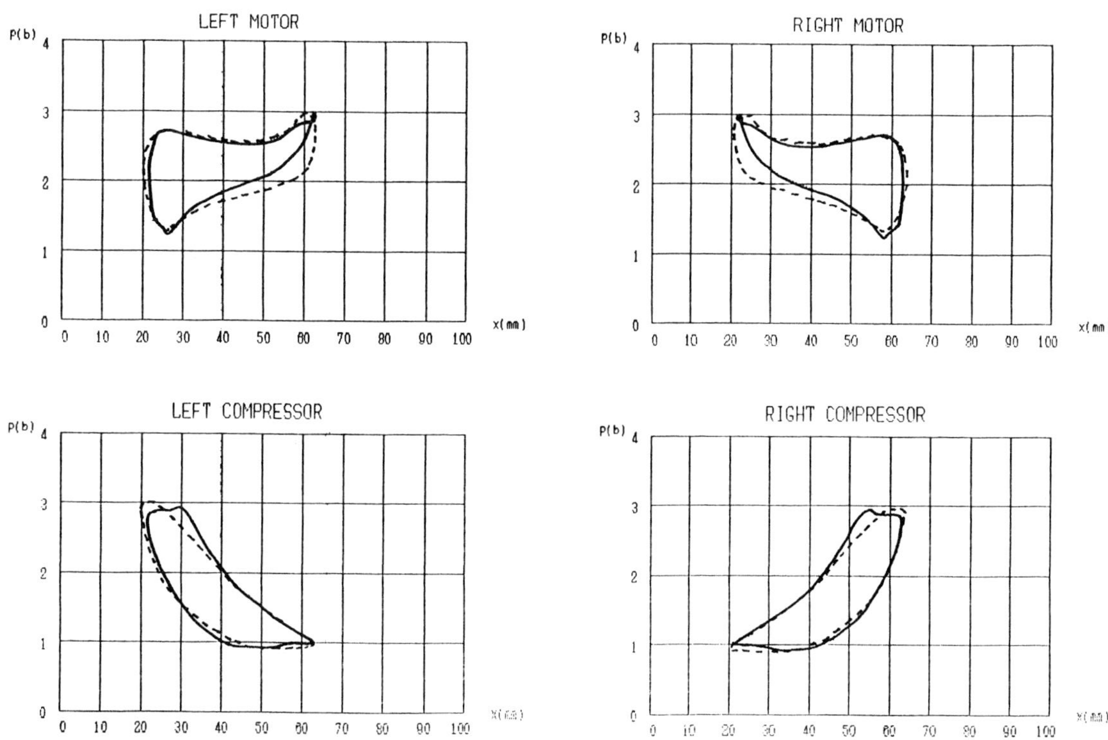

	THEORITICAL	EXPERIMENTAL
MECHANICAL CHARACTERISTICS.		
-STROKE (MM.)	42.	43.
-FREQUENCY (HZ.)	4.8	4.8
FLOW RATES.		
-MOTOR (GR./S.)	10.	10.
-COMPRESSOR (GR./S.)	.6	.5

TABLE 1. COMPARISON BETWEEN THEORITICAL (SIMULATED BY COMPUTER) AND EXPERIMENTAL RESULTS.

SECTION 5 - HEAT STORAGE

Summary on heat storage

A. Subsection : Stores using latent heat, heat of adsorption, etc.

The development and optimisation of cost effective thermal energy storage systems for solar space heating by means of a microprocessor controlled test facility

The development of a solar energy installation for space heating and cooling based on hygroscopic materials

Energy storage in chemical reaction - study of a chemical heat pump

Salt gradient solar pond for solar heat collection and long term storage

B. Subsection : Miscellaneous

Evaluation of thermal storage for solar heating systems

Physical model for the study of mass and energy transfers in the non-saturated layer of soil located above a solar energy storage

C. Subsection : Long-term heat storage

Long-term solar heat storage through underground water tanks for the heating of housing

The heliogeothermal doublet : a real scale test facility

Field test to investigate the performance of an undeep prototype seasonal heat storage system with a heat capacity for 100 solar houses using the soil as the storage medium

Seasonal heat storage in underground warm water stores (Construction and testing of a 500 m^3 store)

SUMMARY ON HEAT STORAGE

Authors: E. van Galen and R. Marshall

1. Introduction

In the section on heat storage there were less contributions compared to the last contractors meeting because several contracts which were reported of last time have been successfully concluded.
There were three presentations on storage concepts using latent heat, heat of adsorption etc. and for achieving interseasonal storage five concepts are in progress, including salt gradient solar pond, saturated soil storage, aquifer storage with heat pumps, burried tank water storage and a membrane lined pit water storage.
Progress was seen to have advanced from the design stage to the construction phase, in time for experimental results to be gathered over the forthcoming winter period.

2. Review of presentations

The first presentation of Dr. O'Callaghan of Cranfield Institute of Technology showed some interesting results obtained by characterising the behaviour of stratified water storage systems in a dimensionless manner based on a large number of tests performed in his microprocessor controlled test facility. Some other work on phase change stores was reported as well.

The next presentation by Mr Verdonschot of TNO discussed the progress of work on a short term solar storage system based on hygroscopic materials. The work developed towards a system design in which all year operation is guaranteed: storage of heat during wintertime and cooling during the summer period. Preliminary calculations show that the yearly useful solar contribution of such a system can be approximately 300 kWh/m^2 year for Dutch climatological conditions. Applications under more sunny and humid conditions might be feasible.

The research presented by Faculté Polytechnique de Mons - on a chemical heat pump working between 5 $^\circ$ C and 100 $^\circ$ C with $CaCl_2/NH_4Cl$ and $CaCl_2/MnCl_2$ made an important progress with the design of a heat exchanger, which meets the requirements of the storage materials. This storage system too is to be used for a combined heating and cooling system.

The last presentation within the section short term storages was presented by R. Marshall. He discussed a newly developed simplified design method for the evaluation of thermal storage systems. He showed some interesting conclusions with respect to system design. It should be worthwhile to consider the validation of his model within the European Modelling Group activities.

The first presentation related to seasonal or long term heat storage concepts from the University of Sussex discussed the progress of work on salt gradient solar ponds. Work on a laboratory scale model (0.3 m^3) was reported of and the development of a numerical model and a densitometer for measuring the salt density gradient. The next stage envisages a small scale 4.5 m diametre pond located outdoors before going on to the construction of an intermediate scale 150-200 m^3 solar pond.

On saturated soil storage work is proceeding in two areas. Firstly, presented by Prof. Jouanna of the University of Montpellier, a study of the thermal behaviour and mass transfer of the insulating non-saturated top soil, necessary for the computation of heat losses from above the saturated soil layer storage area. A carefully planned experimental set-up to study this effect was presented.

A second project - presented by Mr Wijsman of TNO - deals with a pilot seasonal heat store serving the needs of some 100 houses. A boiler simulates the collectors, and a cooler the building loads. Most encouraging was the fact that a way to reduce the cost of this project by some 50% had been found with only little loss in performance. The project has moved into the construction phase of a pilot plant.

On aquifer storage there was a presentation of Mr Iris from Armines. The project he discussed was a full scale project to couple inexpensive solar collectors with aquifer storage in the Paris area, using heat pumps to upgrade the energy to useful temperature levels. The project has well advanced into the construction and instrumentation phase.

On buried tank water storage there was a presentation of Prof. Marinelli. He proposes three approaches: above ground water storage, partially buried storage and buried tank water storage. Both the computer codes for each case and the economics for each case have been derived. Design methods are under development. Although the buried storage leads to a poorer financial situation because of the need for excavation, initial studies indicate that all three methods can be attractive.

In the last presentation, of Mr Hansen from the Technical University of Denmark, a pilot plant of 500 m^3, consisting of a membrane lined water pit storage device was discussed. This device is under construction after a full investigation of the soil conditions and considerations of the detailed design aspects.

A boiler, again, will be used to simulate the collectors.

THE DEVELOPMENT AND OPTIMISATION OF COST EFFECTIVE THERMAL ENERGY STORAGE SYSTEMS FOR SOLAR SPACE HEATING BY MEANS OF A MICROPROCESSOR CONTROLLED TEST FACILITY

Authors : M.A. BELL, R.R.COHEN, B.J.W.MANLEY, P.W. O'CALLAGHAN, R.J. WOOD

Contract number : ESA/S/039/UK

Duration : 36 months 15 July 1980 - 14 July 1983

Total budget : £189,500 CEC contribution £94,750

Head of project : Dr P.W. O'Callaghan

Contractor : Cranfield Institute of Technology

Address : Cranfield, Bedfordshire, England

Summary

This paper summarises the progress made with the development of three thermal energy storage technologies: stratified water, encapsulated phase change material and saturated salt solutions. Results are presented of experiments on a full scale, 2.8 m^3, water store and on smaller scale versions of the other two stores. Computer models of each store have been developed and their validity is discussed. Initial results from a solar heating system optimisation program are presented.

1. INTRODUCTION

Three thermal energy storage (TES) technologies are being developed within this project, these are:-

1. Stratified water - base case for economic and thermal comparisons
2. Packed beds of encapsulated phase change materials
3. Saturated salt solution storage systems.

After extensive proving at a laboratory scale, the stores are being evaluated at full scale on the microprocessor controlled test facilitty as though they were part of a solar space heating installation. At the same time computer models of the stores are being validated so that they can be incorporated into annual simulations of complete solar heating installations which are used to determine optimised systems.

2. STRATIFIED WATER STORE

2.1. Experimental Results

The $2.8m^3$ cylindrical store was fitted with a distributor at top and bottom to minimise vertical mixing of incoming water. Step input tests on the store according to the ASHRAE procedure have been completed for a wide range of temperatures and flowrates and the results are shown in figure 1. The peak performance occurs when the conflicting requirements of a low flowrate to prevent vertical mixing caused by the penetration of the inlet jet and of a high flowrate to minimise vertical diffusion during the charge or discharge process are best satisfied. The performance of the store is essentially the same for charge and discharge under the same flow conditions.

In a solar heating system the process of discharging the store is likely to be similar to a step input discharge in so much as the water returning from the heating distribution system will be colder than the bulk temperature of the store and can therefore be encouraged to enter the bottom of the store. In the charge process however the water will leave the solar collector at a variable temperature and must be encouraged to find its own temperature level in a stratified store - for this purpose an inlet manifold consisting of a vertical tube containing many simple one way valves is being tested on a small scale store. Results from this work will be available shortly. Meanwhile the store is being tested under simulated solar cycle conditions for a prolonged period representing a predominantly discharge time of year. When the test is completed the performance of the store will be compared with predictions from a computer model.

2.2 Computer Modelling of the Water Store

The dominant constraint for most solar system simulations is the cost of computation. Consequently only the most rudimentary models for the storage system have been considered. The most appropriate is the segmented tank model. The computed response of the store to a step input is shown in figure 2 as a function of the number of fully mixed zones used by the model. These results have energy balance errors amounting to less than 1%. In figure 1 the response using a 1 zone (fully mixed) and a 10 zone model are shown in terms of thermal effectiveness in comparison with the experimental results. The model can be refined by incorporating thermal diffusion between the zones. The reduction in thermal effectiveness that this causes is shown in figure 1 for the 10 zone model. The effect is undetectable for the flow conditions which are likely during a charge process and very small for discharge conditions. However, the effect of diffusion will be important during storage dwell periods so that it is only during these times that the considerable increase in computation time that is required can be justified. The model can be further refined in order to try to cope with the variable inlet temperatures which will occur during the charge process. To illustrate this the response of the store has been computed for a process in which the step change in temperature is reduced by 50% after 50% of 1 time constant during a step input test. A comparison between a fully mixed and a stratified response to this temperature inversion is shown in figure 3. The true response will lie somewhere between these depending upon the performance of the bouyant feed system involved, if any. It should be noted that any bouyant feed system under variable inlet temperature conditions is unlikely to perform better than a distributor under step input conditions. This factor will weigh in favour of reducing the collector flow rate so that one is operating at the higher Archimedes number end of the range for the charge process.

In conclusion it appears that the 10 zone model without diffusion might be satisfactory for the operational range of interest and the next stage of the work will be to validate the model against the experimental results from running the store under simulated solar cycle conditions using a bouyant feed system for the charge process.

3. ENCAPSULATED PCM STORE

3.1 Experimental Results

Extensive testing is in progress of a small scale (10 MJ) store using food cans as the encapsulant and sodium thiosulphate pentahydrate as the PCM. Two can shapes are being evaluated for heat transfer performance. The larger one has a volume of 224ml, an aspect ratio of 0.84 and gives 59 m^2 of surface area per m^3 of store, the other has a volume of 127 ml, an aspect ratio of 1.26 and gives 65 m^2/m^3. The problem of protecting the can against internal and external corrosion has been taken up by a UK can manufacturer who now believe they have a cost effective solution.Long term cycling of the PCM in the store has proved that phase stabilisation is necessary to prevent segregation of the dihydrate. Synthetic hydrophilic polymers have been tested for this purpose and initial results, shown in Figure 4, show no degradation after 10 cycles. Both possible nucleating agents for the PCM, i.e. anhydrous sodium sulphate and activated charcoal have proved successful.

The heat transfer performance of the store has been evaluated for various flow conditions for the inlet water and the results are shown in Figure 5. Each step input test was conducted with a temperature swing approximately equidistant about the phase change temperature. The tests were continued until steady state so that it was possible to determine the experimental TSC. The variation in storage capacity relative to the theoretical maximum is shown in figure 4 as a function of the number of cycles. The relative storage capacity fluctuates, with a mean value of 91% for charge and 83% for discharge but there is no downward trend. The data from a representative experiment is given in figure 6. ASHRAE performance coefficients ranging from 90-95% on charge and 85-90% on discharge were achieved for bed Reynolds numbers between 20 and 60, which is the likely operational range for a solar heating system. The effectiveness reduces with increasing Reynolds number; this is because the Number of Transfer Units (NTU) is approximately proportional to $Re^{-0.3}$. The packed bed Reynolds number is defined as

$$Re = \rho D_c \bar{u} / \mu$$

where ρ = fluid density
D_c = diameter of a sphere having the same surface area as a can
$\bar{u}$ = plug flow velocity through the bed
μ = fluid viscosity

3.2 Computer Modelling of the Packed Bed Store

The problem in modelling packed beds is in determining the heat transfer coefficient between the transfer fluid and the PCM. Responsive models optimise the coefficient in order to produce a best fit with experimental data. Predictive models calculate the coefficient by assuming it is correlated to the bed Reynolds number. Both approaches are being used. In each case the bed is assumed to consist of a discrete number of independent layers. The computed response of a packed bed is shown in figure 7 as a function of NTU and Stefan number, which is the ratio of latent heat to sensible heat for the bed and which for most practical systems lies in the range 2 to 4. A comparison between the output of a responsive model and an experimental result is shown in figure 8. A deviation of up to 3°C occurs due to supercooling and attempts are now being made to account for this in the model. More experimental data is required before the predictive model can be used with confidence.

3.3 Future Work

After the model of the store has been validated with respect to the small scale experimental store, it will be incorporated into the solar heating system optimisation program (see section 5). Using this program the optimum size for this type of store will be determined. The store will then be constructed and tested under simulated solar cycle conditions. Initial computations suggest that the store capacity should be about 300 MJ. The design of such a store is given in table I.

volume of can (ml)	127	224
wt of PCM/can (Kg)	.22	.41
no.cans/layer	235	121
no. layers	18	19
total no. cans	4230	2299
total wt. PCM (kg)	930	936
bed voidage	31%	35%
bed volume (m^3)	.78	.83
energy density (MJ/m^3, kWh/m^3)	385,107	361,100
surface area density (m^2/m^3)	65	59

TABLE I Design parameters of 300 MJ encapsulated PCM store using two different cans (Temperature swing = 35°C)

4. THE SATURATED SALT SOLUTION THERMAL ENERGY STORE (SSSTES)

4.1 Introduction

Tests using the experimental prototype SSSTES have continued aimed at solving two remaining practical problems in respect of its operation viz.
- design of a satisfactory low-speed positive displacement pump to enable the store to operate in a stratified mode
- start up problems in respect of pumping

4.2 Positive Displacement Pump

A positive displacement pump was constructed using a PTFE cylinder, stainless steel piston and large synthetic rubber-lined valves. The construction is shown in figure 9. The pump's performance was first tested using water. It appeared to be satisfactory and was therefore installed into the experimental store. During its test using the SSSTES its performance was satisfactory until the temperature of the solution fell to just below the upper saturation temperature, from which point onwards the pump's performance deteriorated until it virtually ceased to function.

On removing the pump it was found that crystals had formed and grown in the pump preventing one of the valves from closing.

On the basis of these results a stratified discharge mode has been rejected as a viable operational strategy for a SSSTES.

4.3 Fully Mixed Mode

Fully mixed operation is achieved by maintaining a comparatively high circulation rate of the solution or slurry during both charge and discharge using the centrifugal pump previously used during charge only. Fully mixed discharge has been successfully demonstrated, although two minor problems occurred:-

1. The heat exchanger became blocked. This was thought to have

occurred because of the very cold inlet water temperature (11°C) used to cool the solution and is not expected to occur when return water tempeatures in the range 20-25°C are used.

2. Seizure of the centrifugal pump owing to crystal formation after shut down. The location and configuration of the pump meant that the crystals which had formed in it were not melted on the initiation of the charge cycle.

The performance of the SSSTES in the fully mixed mode compared with a fully mixed water TES is shown in figure 10. An accurate computer model of the fully mixed operation has been developed and the computed response is also shown in figure 10. The model will be incorporated into the solar heating system optimisation in order to determine the optimum SSSTES size.

4.4 Future Work

The experimental SSSTES is currently being altered to relocate the centrifugal pump so that it will be heated on the initiation of the charge cycle. The new configuration is shown in figure 11.

Provided this change eliminates the problem a viable system should result and design of a 300 MJ prototype will proceed immediately.

Some calculations have been done which indicate that the scraper and pump could be driven at the same speed. This would considerably simplify the system and reduce its cost. This option will be further examined.

5. OPTIMISATION OF SOLAR HEATING SYSTEMS

A multivariable cost optimisation program has been developed operating on the following independent variables: store volume, collector area, store insulation volume, collector flow rate and flow temperature to the distribution system. The program includes an annual simulation with hourly input data. The hourly building load is calculated by a separate program based upon the response factor technique. The objective function is the difference between the cost of a solar heating system with auxiliary and the cost of a conventional system over a 20 year period. In figure 12 sensitivity surfaces of solar fraction and annual cost are shown for a system with a collector area of 20 m^2 and a distribution temperature of 30°C, using a fully mixed water store. The minimum cost excess of the solar system is £250/annum. The optimum storage volume is 2.7 m^3. The optimum collector flow rate is much higher than conventional practice but no account has been taken of the parasitic pumping costs. These will be included in future optimisations. The next stage of the work is to run the program using the computer models for the three storage technologies being developed.

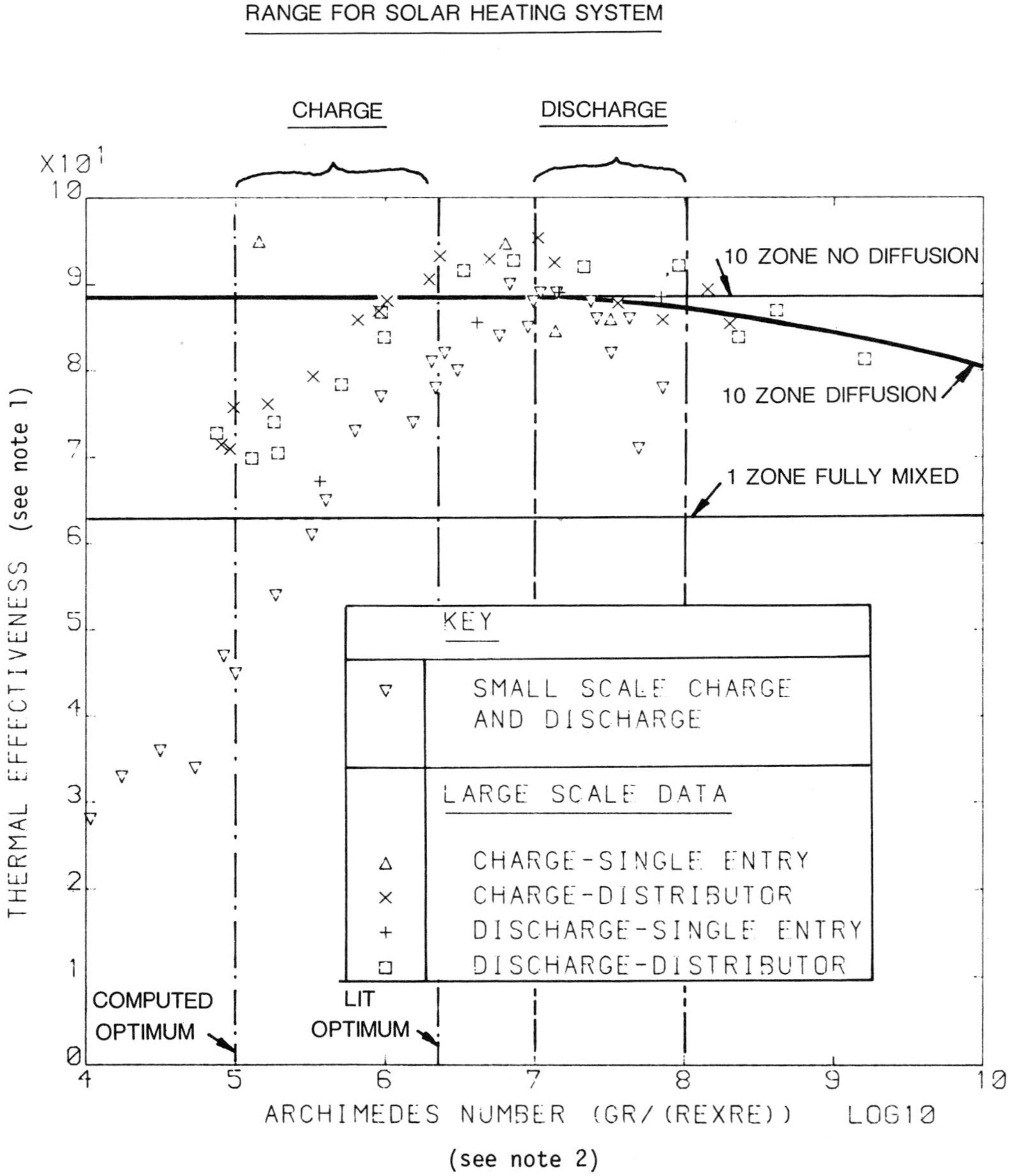

Note 1: This is the amount of energy charged to or discharged from the store in one time constant as a percentage of the theoretical storage capacity.

Note 2: A flow parameter representing a ratio of bouyancy to inertial forces.

FIG. 1 - Effectiveness vs Archimedes number (hot water storage)

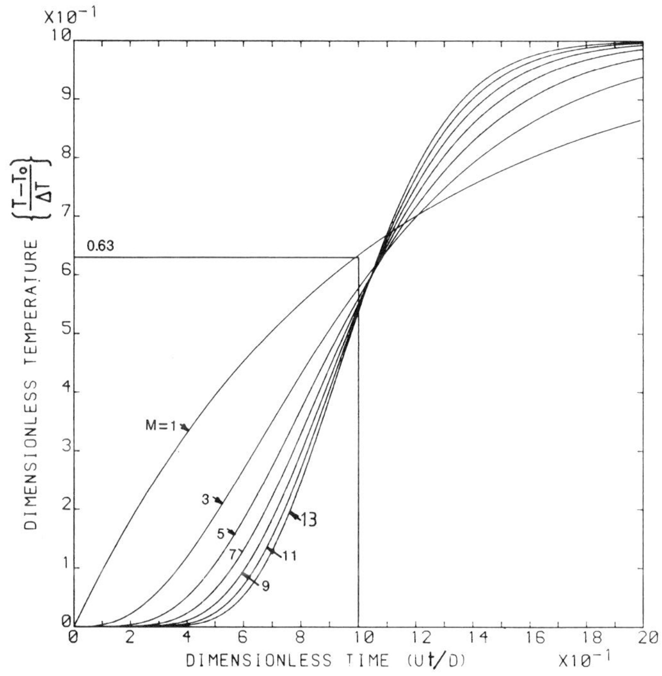

FIG. 2 - Step response of segmented tank model

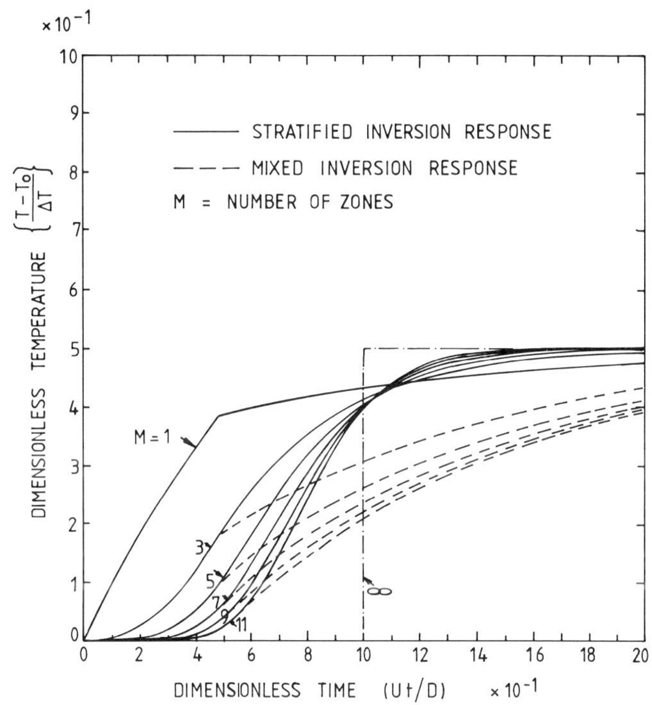

FIG. 3 - Computed fully mixed and stratified response to a temperature inversion

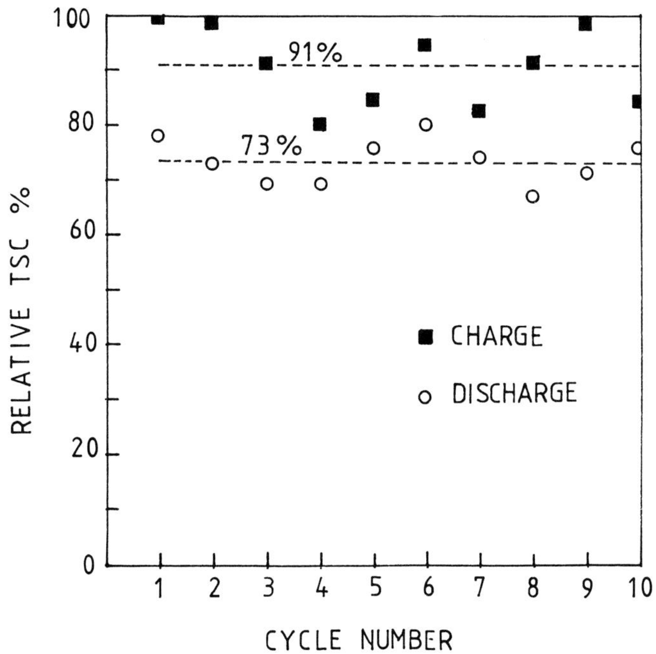

FIG. 4 - Relative TSC vs cycle number for packed bed

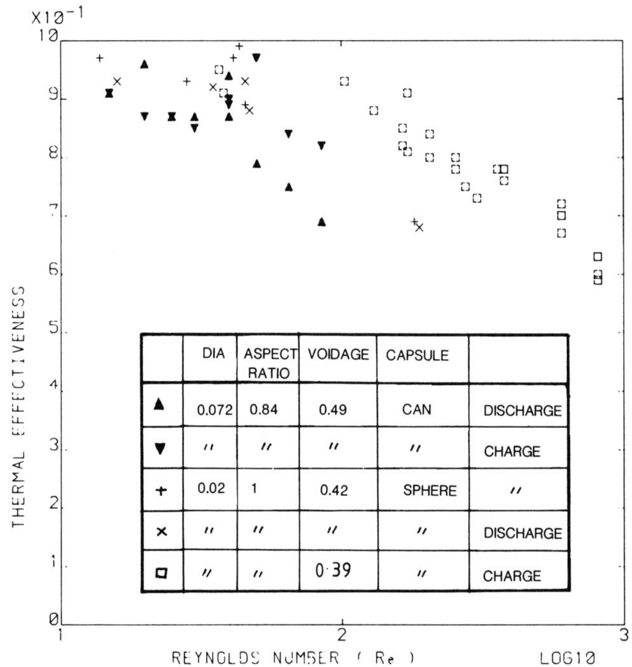

FIG. 5 - Effectiveness vs Reynolds number

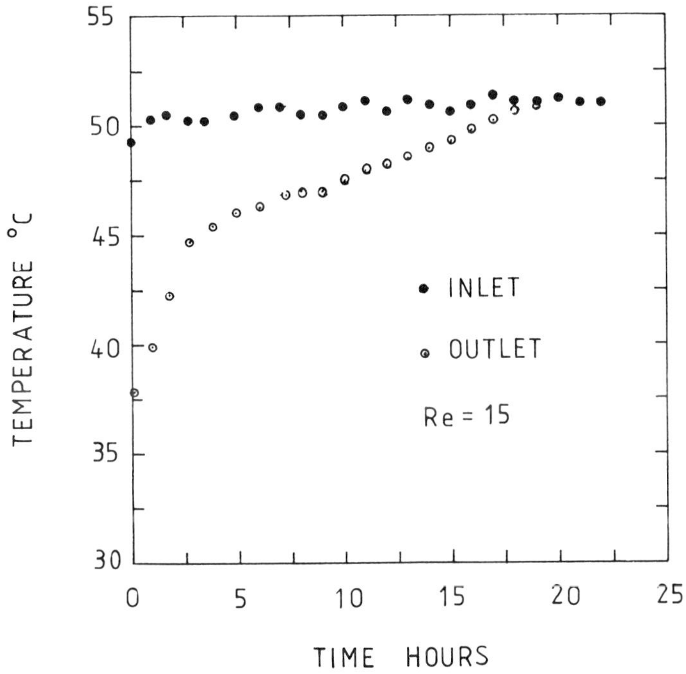
55
50
45
40
35
30
0
5
10
15
20
25
TEMPERATURE °C
TIME HOURS
INLET
OUTLET
Re = 15

FIG. 6 a

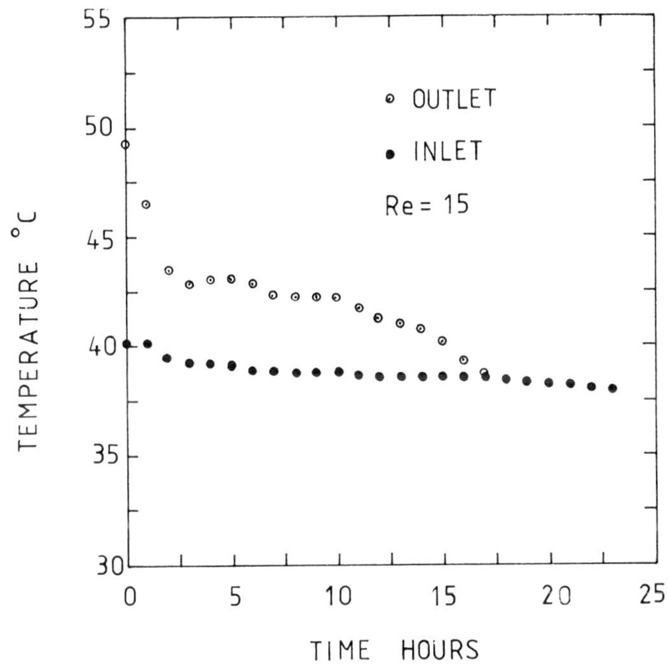
55
50
45
40
35
30
0
5
10
15
20
25
TEMPERATURE °C
TIME HOURS
OUTLET
INLET
Re = 15

FIG. 6 b

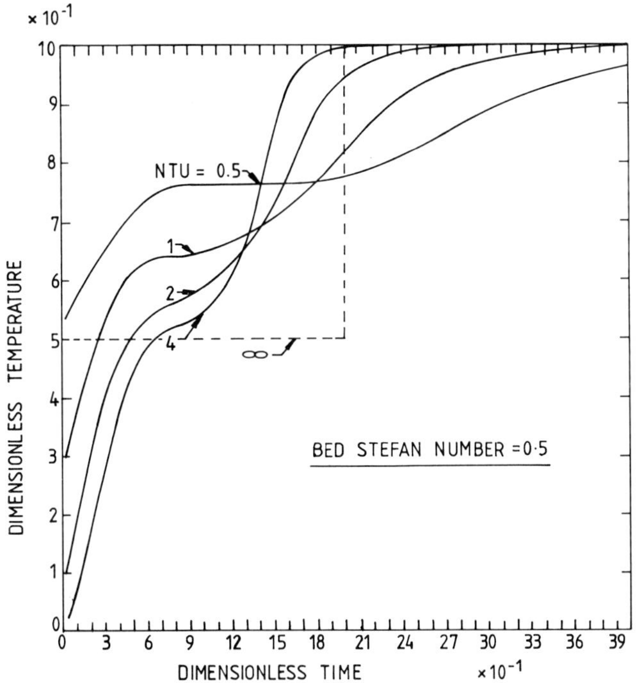
× 10-1
DIMENSIONLESS TEMPERATURE
NTU = 0.5
1
2
4
∞
BED STEFAN NUMBER = 0·5
DIMENSIONLESS TIME
× 10-1

FIG. 7a

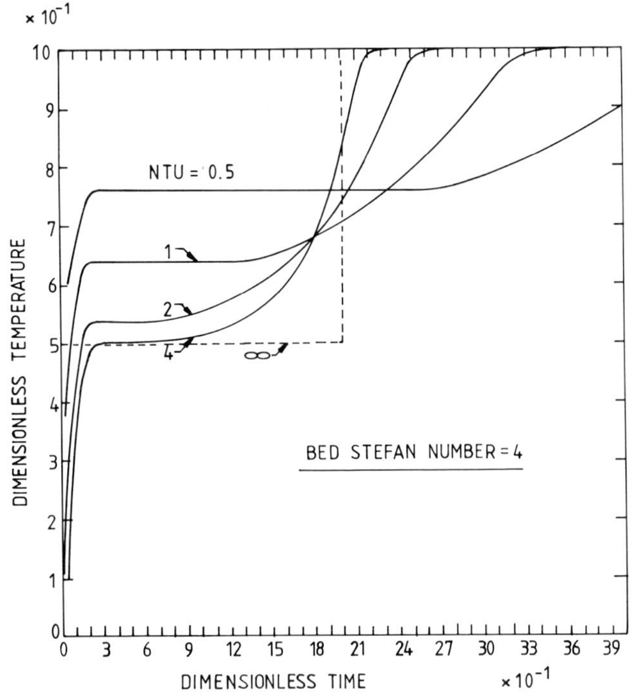
× 10-1
DIMENSIONLESS TEMPERATURE
NTU = 0.5
1
2
4
∞
BED STEFAN NUMBER = 4
DIMENSIONLESS TIME
× 10-1

FIG. 7 b

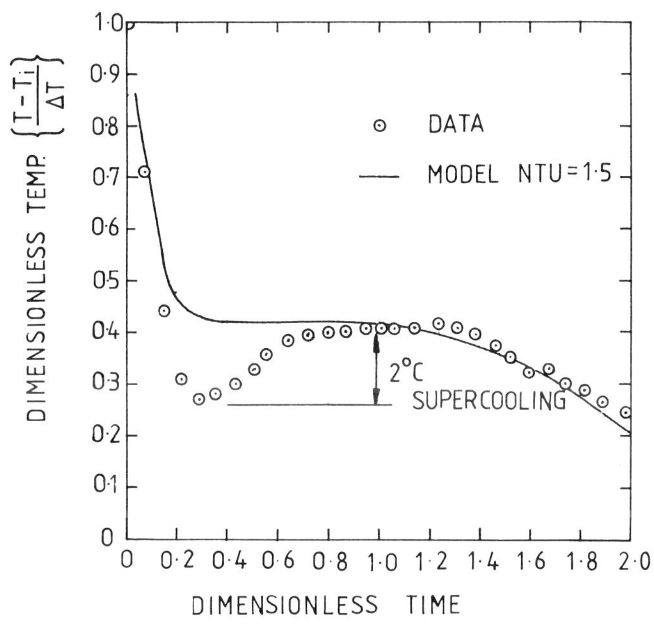

FIG. 8 - Comparison of experimental and computed results for a discharge experiment

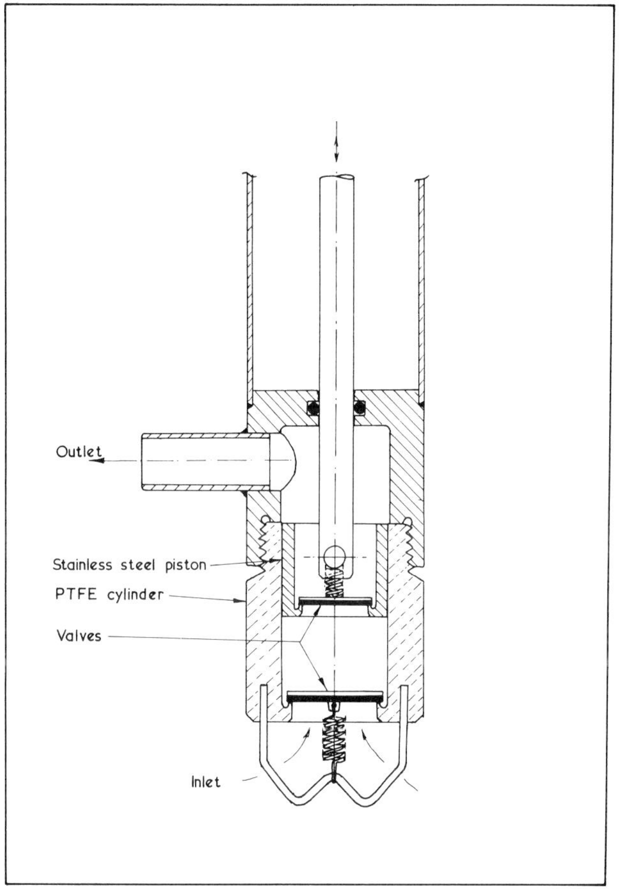

FIG. 9 - Positive displacement piston pump - test module

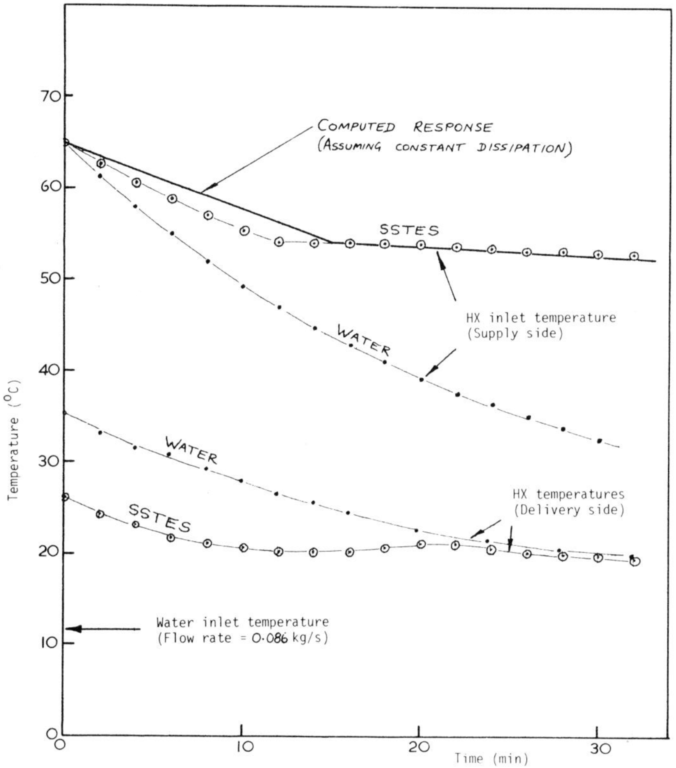

Fig. 10 - Performance of the SSTES in the fully-mixed mode compared with a fully mixed water TES

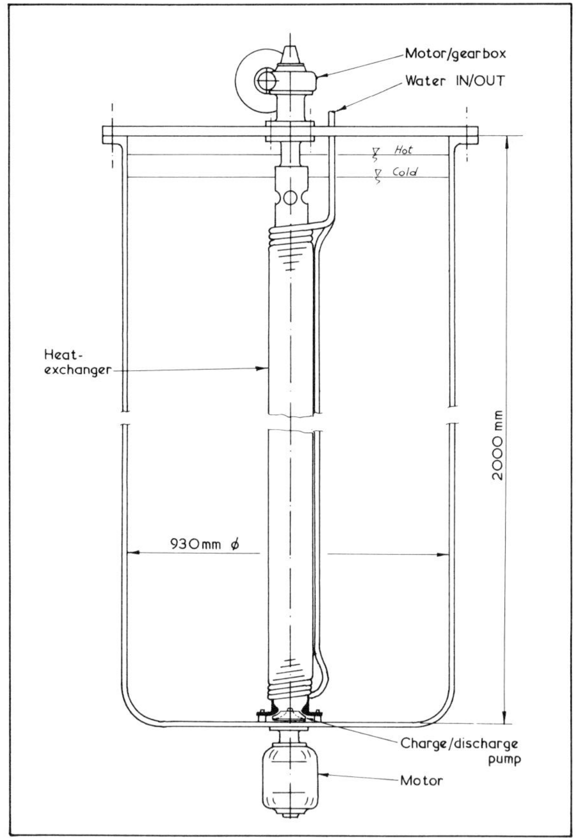

Fig. 11 - Proposed design of a SSTES discharging in a fully-mixed mode

SOLAR FRACTION (%)

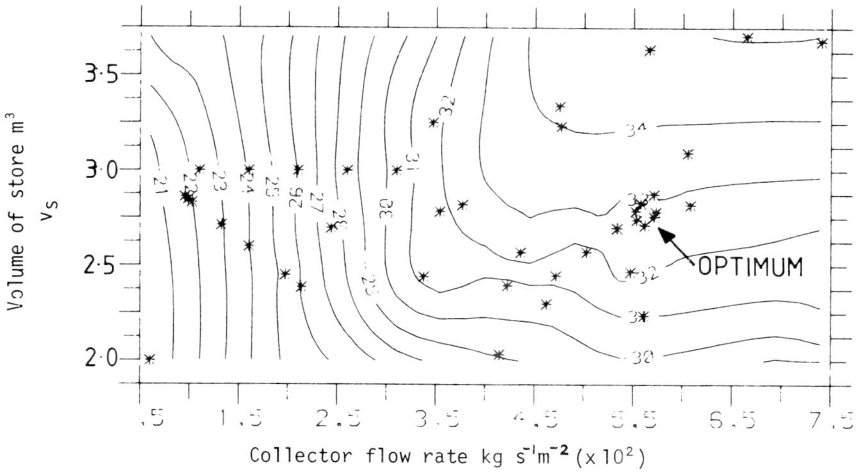

Annual cost difference between conventional and solar system (£/year)

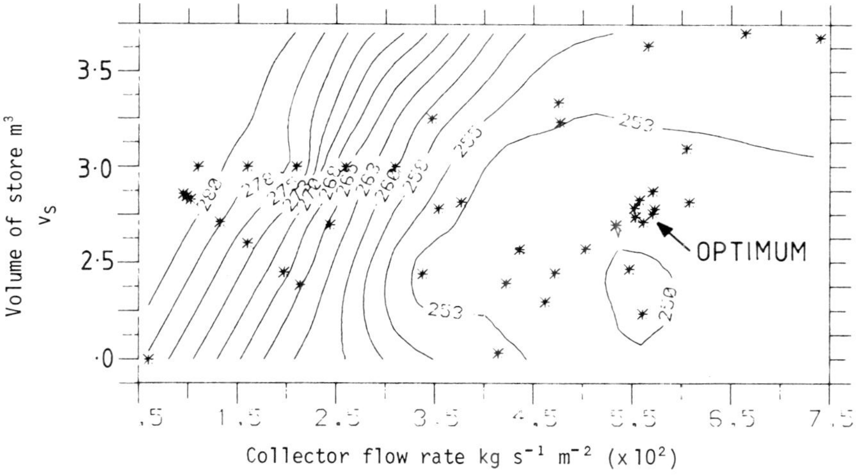

Parameters
- space heating only 7335 kWh
- Distribution temperature 30°C
- ηo = 75% } single glazed
- $Uo = 5\,W\,m^{-2}\,k^{-1}$ } single glazed
- Collector cost = £120/m²
- Store cost = 240 $vs^{-0.35}$ (£/m³)
- Fuel cost = £0.023/kWh (equiv. oil)
- interest rate = 10%
- fuel inflation = 3% real terms

Figure 12 Sensitivity analysis assuming 20 m² collector and a fully mixed store.

THE DEVELOPMENT OF A SOLAR ENERGY INSTALLATION FOR SPACE HEATING AND COOLING BASED ON HYGROSCOPIC MATERIALS.

Author	: J.K.M. Verdonschot
Contract number	: ESA-S-125-NL (B)
Duration	: 21 months, 1 October 1981 - 30 June 1983
Total budget	: Dfl. 333.000,-- CEC-contribution: 50%
Head of the project	: Ir. C. den Ouden
Contractor	: Institute of Applied Physics TNO-TH
Address	: P.O. Box 155, 2600 AD DELFT, The Netherlands

Summary

In a previous study [1] we have investigated a heat storage system based on the adsorption of water in solid adsorbents. The general conclusion from that study was that the application of such storage systems is hardly feasible due to the relative unimportant function of storage in air-based solar heating systems which results in small improvements over a rock-bed or a water storage device. Therefore the application of adsorption heat storage systems cannot be justified in solar heating installations where the system only operates during winter-time. The energy output of a solar adsorption system can be enlarged by using it also during summer-time which then results in a continuous load during the whole year. In many climates both heating and cooling are required depending on the season. A well designed dessicant system has the capability of delivering heat or cold depending on the operation mode which makes the application for combined heating and cooling interesting.

For the Dutch circumstances such an energy demand only occurs in office buildings. In order to make a first approximation concerning the energy output of the solar dessicant system we have carried out some calculations for a Dutch office building. The result of that study is very promising because it appears that the yearly usefull solar energy can be around 300 kWh/m^2 (1000 MJ/m^2) collector for the combined heating and cooling system. This means that 30-40% of that thermal energy demand can be delivered by solar energy. Also a first cost benefit analysis gives promising results which might be even better in more sunny climatic regions. In the near future the solar adsorption system performance will be studied in more detail.

1.1 Introduction

The research work on the application of hygroscopic materials in solar energy has started in 1978 in our Institute. After having determined the physical properties of adsorbents a study was carried out on the development of an adsorbent heat storage device based on the principle of water adsorption and desorption. In that study several computer models were developed in order to calculate the adsorbent behaviour under dynamical circumstances. These models were validated by means of laboratory experiments. An important part of the research work were the system studies on air-based solar heating installations in order to compare the adsorbent storage device with other storages like water and rock-bed systems. That study has been reported in [1] and the results can be formulated as follows:

1. It appears that the application of sophisticated heat storage systems in air based solar heating systems might only be interesting for houses with a very small heat load (< 20 GJ/year). For heat loads above 50 GJ/year the solar energy gained by air-collectors is mostly used directly (by-passing the storage) for house heating. In that case the function of the storage system is almost negligible. There is ofcourse an influence of the daily heat load pattern. When this load pattern is such that most of the energy demand is in the evening, it appears that the adsorption system gives only a small increase in heat output compared to water and rock-bed systems. This is mainly due to the much smaller heat transfer rate of the adsorbent.

2. When an adsorbent storage system for house heating is compared with a water storage device at a yearly heat load of 20 GJ, it appears that the differences are small. This means that an adsorbent storage will be too expensive because the system is more complicated and demands a larger amount of fan power.

3. In a house heating system the solar energy delivered to the water heater is an important part of the total solar energy constribution. Therefore the storage system is mostly used both for space heating and domestic hot water. In an adsorbent storage system this can not be integrated which makes the total system much more complicated.

As a result one can conclude that the adsorbent storage system has several practical disadvantages.

1.2 Application of hygroscopic materials for cooling

A great deal of interest has been generated in the area of solar heating in residential and industrial buildings. The subject of solar cooling is relatively untouched. There are carried out some investigations of solar energy application on a modified conventional absorption machine.

The application of hygroscopic materials for cooling has a number of advantages compared to other cooling methods especially when solar energy is applied. The reasons for investigating such a system design are;

- The required energy is a form of thermal energy of a temperature level which makes solar system applications interesting. Hygroscopic materials can be regenerated at temperatures around 50-60°C.
- The solar collector can be used for heating in winter and cooling in the summer period.
- The cooling system is a mechanical simple design which uses air as the working fluid.
- In the Dutch climate the necessity for cooling is only applicable to office buildings. Most of the office buildings have rather large window area which causes overheating in the summer period. In fact the indoor

climate in the office building is controlled during the whole year. On the other hand there are a number of factors which make the application of adsorbents for cooling more interesting than the application for heat storage only.

- The cooling load of a building is most of the time in phase with the solar irradiance on the collectors which means that for cooling purposes the storage mode can be eliminated. This means that for an adsorbent cooling system a relative small amount of hygroscopic material is needed because a rotary dessicant system can be applied.
- The pump energy for driving the adsorption-desorption process is much smaller compared to the heat storage application because less material is applied.
- Such a dessicant system can be easily integrated in common air-conditioning installations.

Because of the above mentioned results we concluded that the application of adsorption materials for heat storage only has hardly any advantages. Therefore the aim of our research work on hygroscopic materials has changed from the heat storage application to the application for cooling and air-conditioning. For the Dutch circumstances there are mainly application possibilities in office buildings. These buildings mostly have a cooling /air-conditioning demand in the summer and a heating demand in the winter period. For the solar energy installation this would imply that in summer cooling and in winter heating can be provided which would ofcourse increase the usefull energy output.

The cooling installation driven by solar thermal energy is a so-called open-cycle system. This 'open-cycle' implies that the cooling process does not have the same beginning and end-conditions of the air when this process is drawn on a psychometric chart. The most important installation components are;

- evaporative cooler
- air-to-air heat exchanger
- dessicant system

These components can be combined in various ways. Two principally different open cycle systems proposed by Nelson [2] are given in figure 1.

In the ventilation system the supply air for cooling the conditioned room is in fact ambient air. This air is first dehumidified and afterwards sensibly cooled in the regenerator and evaporatively cooled below the room-air temperature in the humidifier. The exhaust from the conditioned room is used to cool the supply air and regenerate the dehumidifier. Therefore this air is first evaporatively cooled and then heated in the heat exchanger and the energy source. After the dessicant system is regenerated this air is exhausted to ambient.

In the recirculation system the same physical processes take place. Only in this system the room air is first dehumidified, then sensibly and evaporatively cooled and reintroduced into the conditioned room. In the regenerating stream, ambient air is first humidified and afterwards heated in the heat exchanger and by the heat source. After the dessicant system is regenerated this air is exhausted to ambient again.

The systems as shown in figure 1 are designed for cooling only. In our case we want to apply solar air-collectors as the heat source whereas these collectors will also serve as a heat source for space heating in the winter period. Therefore in figure 2 is given a scheme of a combined solar heating and cooling system as it might look like in a final design. In that system,

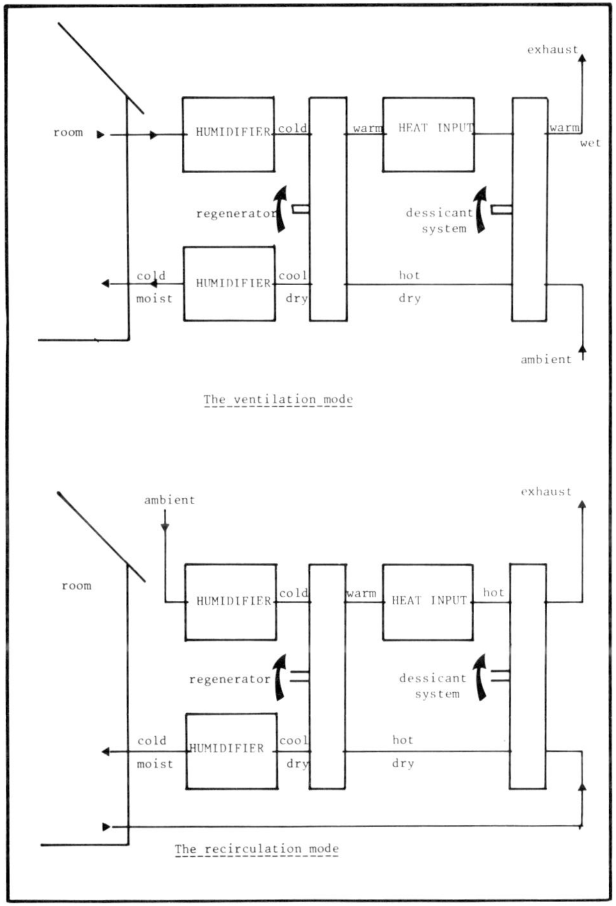

Fig. 1 - Open cycle dessicant cooling systems

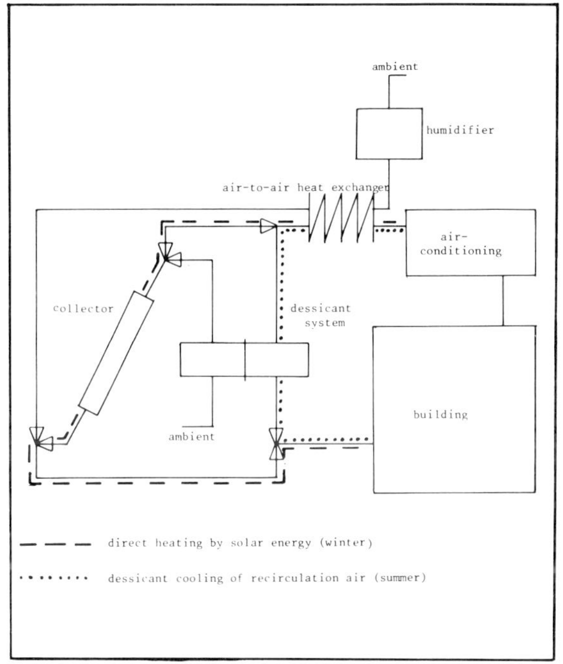

Fig. 2 - Scheme of a combined solar heating and cooling installation

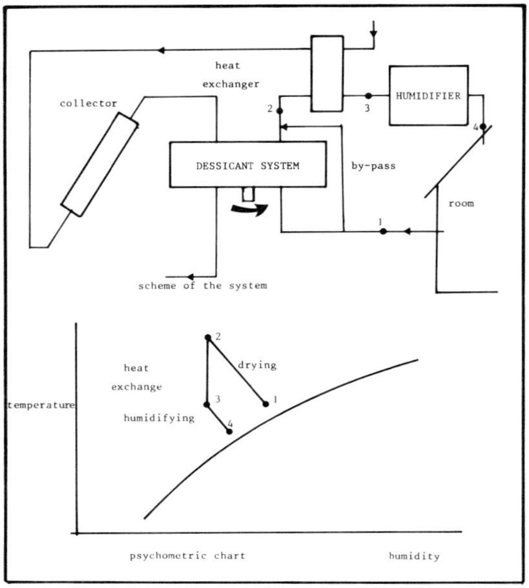

Fig. 3 - Principle of the solar cooling installation

which is based on the recirculation system of figure 1 for cooling, only the direct use mode for space heating by the solar collectors is applied. When the adsorption system should also serve as a heat storage device the total system design will be somewhat more complicated. It should be mentioned that this possibility of heat storage will be studied with respect to the total system design and cost in relation with the usefull energy output.

It should also be clear that the installation as shown in figure 2 is just one out of many possibilities. However the basic components will be the same. In the next half year a detailed study will be carried out to determine the most promising system design. Until now we have carried out some preliminary calculations for such kind of a system.

1.3 Performance of solar cooling systems based on hygroscopic materials

As this is just a preliminary study which is carried out to mark the possibilities for cooling, we did our calculations for just one type of installation design. A scheme of that installation is given in figure 3 together with the air states in the recirculation stream. In fact this is a simplified form of the recirculation mode from figure 1. Ambient air is passed through a heat exchanger and a collector array and afterwards used to dry the adsorption material. On the other side, relative wet room-air is dried by the adsorption material and this hot dry air is cooled in the heat exchanger. This air will then pass a humidifier where it is evaporatively cooled below the room-air temperature. From the calculations it appears that recirculation air is heated around 10°C during the drying process. Afterwards this air of around 32°C is cooled down to a temperature of around 13°C which is the room supply air temperature for cooling. The calculations are carried out for a representative Dutch office building. Such an office building consists of a great number so-called room-modules. Heat transpory between the room-modules within one office building is

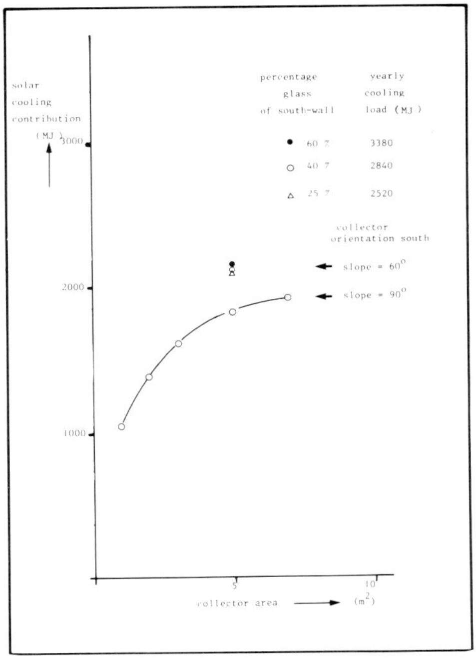

Fig. 4 - Usefull solar energy as a function of the collector area for one module

neglected. This means that heat transport only occurs at the front-side. This front-side consists of a window and a concrete wall.

The cooling load of the building is strongly determined by the internal heat production and the incoming solar radiation. The important parameters for calculating the cooling load are;

- orientation of the window
- operation hours of the lightning
- number of persons within a module
- operation of the blinds
- ventilation rate of the office building
- outside meteorological conditions
- thermostat-setting and relative humidity of the air

For the dessicant system we have applied the already verified model for the adsorption heat storage system which is based on the following assumptions;

- the dessicant is assumed to be packed as a porous homogenious medium
- the physical properties are constant
- the air-condition in the bed is only varying in the flow direction
- there is an infinite masstransfer coefficient between the air and the hygroscopic material.

The humidifying and dehumidifying process as it occurs in the rotating dessicant system is described by two adsorption systems and intermittently changing the air stream for each packed bed. In our calculations we have applied this method because the existing computer model would require very small time-steps which would largely increase the computer time in order to calculate the performance of a rotating dessicant system. This is due to the small amount of adsorption material in such a system. Due to this simplification we introduce a thermal and mass inertia in the dessicant system which does not occur in the practical situation. In the next half year of the research program we intend to study this problem in more detail and problably will improve the calculation model with respect to the rotating dessicant system.

The evaporative cooler path is calculated from an adiabatic change of the air water properties. The psychometric changes follow the constant enthalpy curve.

The results of calculations on the above described system are given in figure 4. In that figure is given the usefull energy output for cooling as a function of the collector area per module.
The applied collector has the following specifications

- single glazing, spectral selective coating, maximum efficiency $(\alpha\tau) = 0.78$ heat loss factor $U_1 = 3.8 + 0.02\ \Delta T$.

The lowest line gives the usefull cooling energy for a room module with 40% glass. The total cooling load is in that case 790 kWh (2800 MJ)/year The collector is in the vertical position and has an orientation towards south. This means that the collector is placed agains the south-wall. In that case a solar collector array of 3 m^2 delivers about 56% of the energy for cooling.

Changing the glass area from 25 to 60 percent of the south facing wall results in an increase of the cooling load from 700 - 900 kWh (2500-3400 MJ) per year. The usefull solar energy delivered is rather independent of this change in cooling load as can be seen in the same figure. A collector array of 5 m^2, south facing and a slope of 60° delivers 570 kWh (2050 MJ) per year which is a solar contribution of 60 to 80%.

From those calculations we can conclude that tha yearly usefull solar energy for cooling under Dutch circumstances will be 150 - 180 kWh (540 - 650 MJ) per m^2 collector. Taking into account the heat load of an office building we can expect a yearly solar contribution for direct heating of 100 - 150 kWh per m^2 collector. This is a very promising result because the solar collector can be used both in summer and winter which results in a yearly usefull energy output of 250 - 330 kWh (900 - 1200 MJ) per m^2 collector.

1.4 Application possibilities for combined heating and cooling using adsorption materials

If we assume that the ultimate combined heating and cooling system will look like figure 2 then this does mean that the common air-conditioning installation is extended with the following components;

- solar collectors
- dessicant system
- air-to-air heat exchanger
- evt. an extra humidifier

As this installation design is rather simple, extra regulations devices will be limited. The above mentioned components are in fact existing industrial products which are comparable with a common air-conditioning system in relation to maintenance, reliability and life-time. For this system we assume that no heat storage is applied. The necessity for heat storage will be studied in more detail.

In order to get an idea concerning the economical feasibility we calculated the extra investment necessary for the extra components as mentioned above. This investment is for a common Dutch office building around Dfl. 1.800,-- per room-module. The yearly solar contribution per room-module will be in the order of 600 - 700 kWh (2.1 - 2.5 GJ) which is about 20% of the heat demand and 40% of the cooling load. It should be mentioned that the investment cost are primarily determined by the collector cost (≃ Dfl. 1.000,-- per module). Compared to the total investment cost for a heating and cooling installation of a middle-class office building (volume = 25.000 m^3) the extra investment is about 30% of that cost.

It should be noted that these numbers apply to the Dutch circumstances. Probably such a solar system will be even more suitable in Southern Europe, where the cooling load and solar irradiance are both higher. In that case such systems could be applied in dwellings instead of the conventional air-conditioning installation which would largely extend the application possibilities.

1.5 Conclusions

- The application possibilities for solar cooling based in hygroscopic materials seem to be very promising, especially when a combined heating and cooling installation is applied.
- For Dutch circumstances these systems can be applied in office buildings while in more sunny climates the application possibilities can be much larger.

Literature

1 Ir. J.K.M. Verdonschot
Thermal heat storage in hygroscopic materials for air-based solar heating systems.
Institute of Applied Physics TNO-TH, 21 October 1981, Delft
Report nr. 803.220.

2 J.S. Nelson, W.A. Beckman, J.W. Mitchell and D.J. Close
Simulations of the performance of open cycle dessicant systems using solar energy.
Solar Energy vol. 21, pp. 273 - 278, 1978.

ENERGY STORAGE IN CHEMICAL REACTION

STUDY OF A CHEMICAL HEAT PUMP

Authors	: R. JADOT, Y. DEBLESER
Contract nr	: ESA-S-036-B
Duration	: 30 months Contribution EC : 50 %
Total budget	: 7.250.000 BF
Head of the project	: Prof. J. BOUGARD
Contractor	: Faculté Polytechnique de Mons
Address	: rue de Houdain, 9, B - 7000 MONS (Belgium).

Summary

The aim of this research is the study and realization of a prototype of a chemical heat pump based upon the differential absorption of ammonia gas between two inorganic salts. The absorption or desorption are used to heat or to cool a dwelling.
The couples of salts able to be component of a chemical heat pump, working between 5°C and 100°C, are $CaCl_2/NH_4Cl$ and $CaCl_2/MnCl_2$.
In this paper, are shown the test results to determine heat exchange coefficients which are necessary to design heat-exchangers.
Some usual heat exchangers have been tested but a new particular model of exchanger was designed to meet requirements of the powdery salt.

1. Introduction

Solar energy can be stored in some chemical reversible reactions between 20°C and 100°C. The most interesting of them, from a volumetric capacity point of view, are those between condensed and gaseous phases. Such a reaction may be written :

$$A_{s\ or\ l} + B_g \rightleftarrows AB_{s\ or\ l}$$

A chemical heat pump is based on two coupled reactions of this sort. Condensed phases are stored in separated tanks in contact with sources; the gaseous phase assumes heat and mass transfer between them by reacting differently in each tank (decomposition in one, synthesis in the other) according to the temperatures of the sources. With a good choice of reaction couples, it is possible to transfer heat from a cold source to a hot one. That is the working principle, represented on figure 1, of a chemical heat pump.

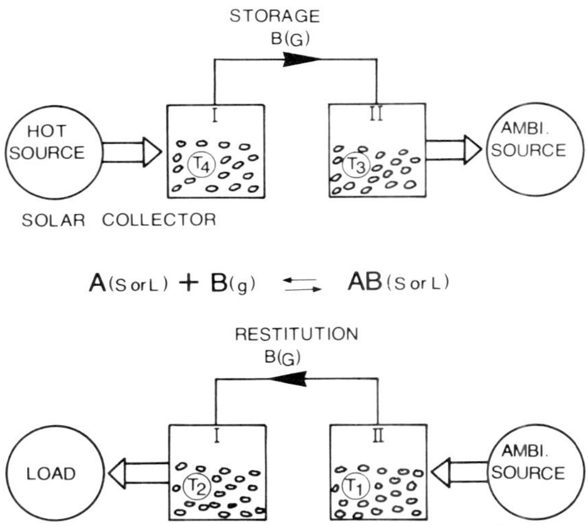

FIG.1

For the reactions chosen in this research, the energy transfer is assumed by gaseous ammonia. From the figure 1, if we have for the following temperature sources :

$$t_1 = 5°C \qquad t_3 = 20°C$$
$$t_2 = 40°C \qquad t_4 = 80°C$$

Then the working conditions of a chemical heat pump on equilibrium pressures of ammoniated salts are :

$$p_{II}\ (5°C) > p_I\ (40°C)$$
$$p_I\ (80°C) > p_{II}\ (20°C)$$

These requirements are filled for couples of inorganic salts which are chosen : $CaCl_2/NH_4Cl$ or $CaCl_2/MnCl_2$.

The preceding paper explained problems to design a prototype. These technological problems come from 4 principal causes :

1) important volumetrical expansion of $CaCl_2$ during absorption ($\Delta V/_V \simeq 4$)
2) heat exchange with mass transfer
3) heat recuperation through an insulating medium
4) corrosion

Middle scale prototypes have been made to study those phenomenas. The aim of that is to make signigicant measurements of heat exchange coefficients. According to these considerations, a maximum height of the reacting solid medium, for instance, $CaCl_2$ bed, has been evaluated to 2 or 3 cm.

2. Description and design of the prototype

2.1. First, some measures have been made on classical heat exchangers and they gave us realistic data for heat exchange coefficients.
For that, temperature curves have been plotted during absorption.
Figure 2 shows curves obtained with a plane exchanger.

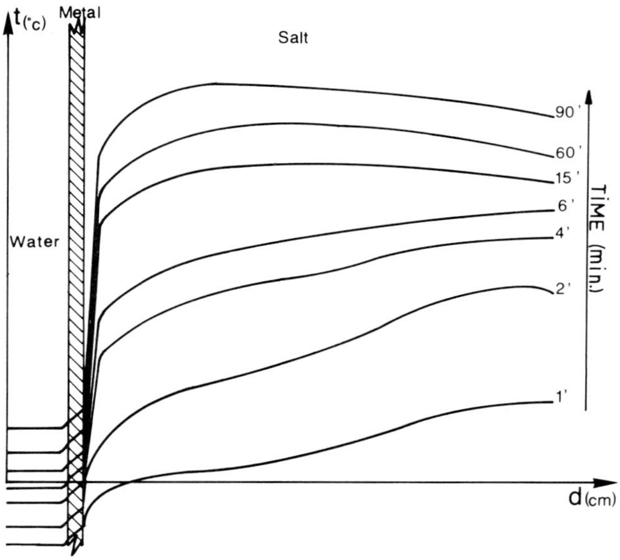

FIG.2

In working conditions, the diagram in figure 3 allows us to calculate a global coefficient of transmission Kgl (in $W/m^2{}^\circ C$), and to know the available power per mass unit of salt.

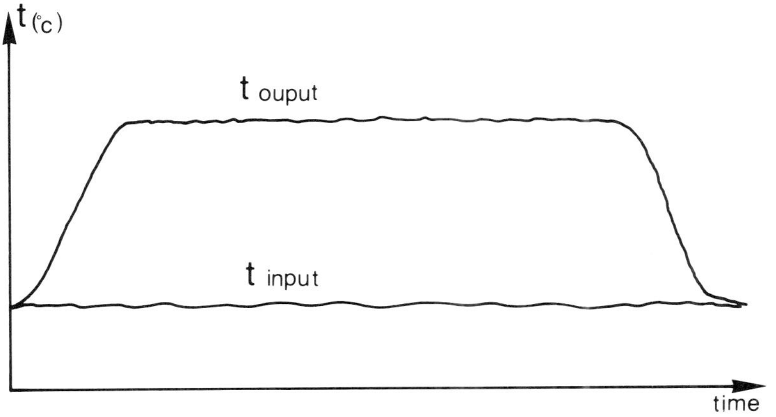

FIG.3

For that, the transmitted heat flow Ø is determined by measures of water temperatures in exchanger and by the measure of water flow.
Global coefficient of transmission is given by the logarithmic delta t method :

$$K_{gl} = \frac{Ø}{S \,.\, \Delta t_{log}} \text{ in } W/m^2.°C$$

where

$Ø$ = heat flow in W

S = exchange area in m^2

$$\Delta t_{log} = \frac{(t_{out} - t_{in})_{water}}{\ln \dfrac{t_{reaction} - t_{in.W}}{t_{reaction} - t_{out.W}}}$$

For the most efficacious exchanger, measures of global exchange coefficient range between 30 and 40 $W/m^2°C$.
For each type of exchanger, a global heating power per mass unit of salt can be calculated. Nevertheless, that feature depends on the type of exchanger, on its surface area, on the degree of accomplishment of chemical reaction, ... Regulation of such a system to produce heat is very easy because a changement of gaseous flow is enough to modify rapidly the level of temperatures in the reactor.
The response time is relatively short, lower than 3 minutes.
In practice, it is sufficient to turn a tap to modify water temperature at the end of the exchanger, or to hold an uniform temperature. To regenerate the cycle, hot water is injected in the exchanger to take ammonia out of salt. During desorption, temperature stays at a constant level (figure 4) and the end of this level indicates the end of desorption.

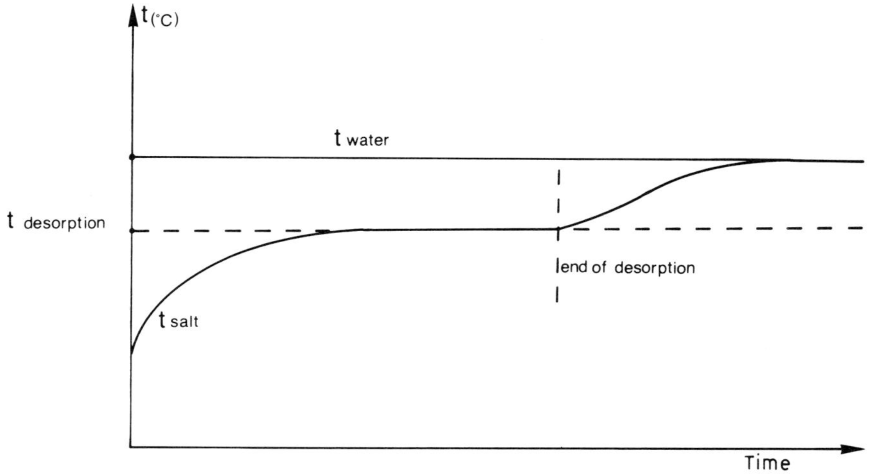

FIG.4

2.2. Technological point of view.

The aim of this part is to find an efficacious exchanger in the case of our application. After having tested classical exchangers, it appeared that no one of those devices was right for this purpose for 3 reasons :

1) During absorption, salt increases 4 times its volume. Preventing that expansion causes a compaction of salt and a stop of the chemical reaction. Exchangers with very close radiating plates are destroyed by this compaction; increasing the interval between radiating plates should prevent this phenomenon but also it should decrease the surface area of exchange.

2) Heat generation is created by a mass transfer. Reactive gas must reach each part of the salt easily; in practice that forbids beds higher than 3 cm.

3) Heat extraction through a powdery medium, thus insulating, is limited by the global coefficient of transmission Kgl. So, exchange surface must be important to have a good efficiency.

In short, a good heat exchanger in this case needs to be as follows :

1. Exchanger must have a large surface in contact with salt, without using radiating plates to avoid compaction.

2. Salt must be distributed in their beds (2 or 3 cm high). That means using multistaged exchangers to avoid vertical self-compaction.

3. Each exchanger stage must present a large exchange surface without partitioning.

At this research stage, a special type of exchanger has been made to satisfy all the preceding criterious. Figure 5 shows a stage of this exchanger.

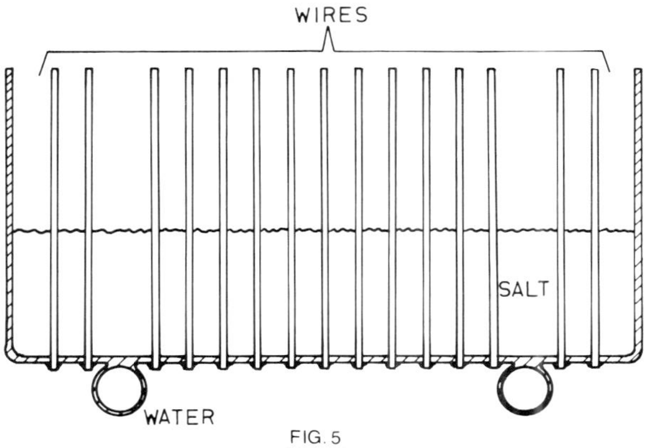

FIG. 5

Each stage is made of a system of vertical wires which are fixed on a plane support. Wires allow vertical expansion and present a large exchange surface. Advantage of such a system is to offer a large surface per mass unit of salt without compartmentation, like shown in the following comparative board (figure 6).

	plane exchanger	radiating plates exchanger	wires exchanger
S (m^2/m^3 of salt)	5	15 to 20	35

FIG.6

In each case, bed thickness doesn't exceed 2 or 3 cm. Another advantage of the last system is that wires fill only 3 % of the total available volume. Supplementary restraint is that each part of exchanger must be removable to load or to unload salt; that is why a configuration with axial symetry will be chosen.

3. Experimental results

1. The first result is confirmation of the good calorific capacity of $CaCl_2$ salt. One kilo of $CaCl_2$ can store about 2.200 kJ on an average. If no compaction occurs, the capacity of storage is not changed after some absorption-desorption cycles except after the first cycle. That comes from the first desorption which is not complete because temperature level (80°C) given by solar panels is not enough to desorb quite ammonia ($6NH_3$ instead of 8). The following figure 7 shows the evolution of heat recuperation in function of cycle number.

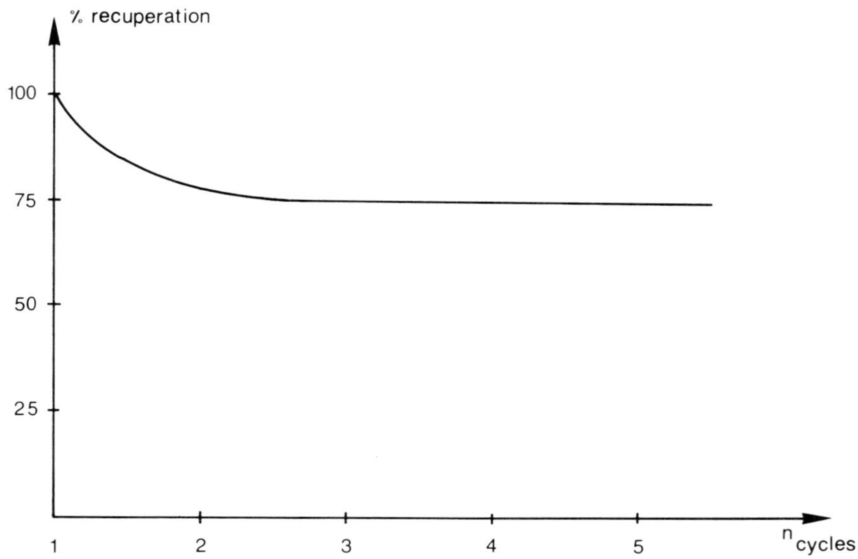

FIG. 7

2. Systematic measures of temperatures and of water flow on each type of exchanger allow us to find values of transmission coefficient ranging between 35 and 45 W/m^2°C. This limit given by use of salt is an important data to design heat exchangers. Here is a comparative board figure 8) which gives : (1) heat flows generated in working conditions by different types of exchangers containing similar weight of salt (± 1 kg),
(2) global coefficients of heat transmission kgl,
(3) specific areas of heat exchange S.

	plane exchanger	radiating plates exchanger	wires exchanger
Ø (w)	**40**	**200**	**300**
Kgl (w/m^2.°c)	**10**	**20**	**40**
S (m^2/m^3 of salt)	**5**	**15** TO **20**	**35**

FIG. 8

4. Conclusion

Part of heat transmission problems has been resolved by using a new kind of exchanger. Wires, used like radiating plates, ensure a good heat transmission through the salt without any compaction.

At the present time, we are working on the design and realization of such a prototype which will contain 200 kilos of salt and which will be able to generate a heat flow of 20 kW in working conditions.

SALT GRADIENT SOLAR POND FOR SOLAR HEAT COLLECTION AND LONG TERM STORAGE

Authors : V PHILLIPS, P J UNSWORTH, N A AL-SALEH

Contract Number : ESA/S/038/UK

Duration : from SEPTEMBER 1980 to AUGUST 1983

Total Budget : £39,810 EEC Contribution : 50%

Head of Project : Dr P J Unsworth, School of Mathematical and Physical Sciences

Contractor : University of Sussex

Address : Physics Building
University of Sussex
Brighton BN1 9QH
England

Summary

Work is described concerning the instrumentation, thermal modelling and laboratory tests on a salt gradient solar pond to be used for heat collection and storage. A densitometer capable of measuring the salinity to approximately one part in 10^4 is described. It balances the upward buoyancy force on a small disk like buoy, by a downward magnetic force, from a solenoid acting on a permanent magnet attached to the buoy. The position of the buoy is servo controlled and the current in the solenoid gives a direct read out for the density.

SALT GRADIENT SOLAR POND PROJECT

1 INTRODUCTION

Prior to building a 150-200m^2 salt gradient solar pond for solar heat collection and storage, we have been developing instrumentation, and developing a numerical model to assist in optimising the design shape and depths of the insulating (non-convective) and storage (convective) zones of the pond.

So far we have tested a laboratory solar pond in which the storage layer reached 60°C, developed the numerical model for solar ponds and their surroundings, and designed and tested an instrument to measure the salinity gradient. Data will be collected from a 4.5m diameter pond this summer, then a 150-200m^2 pond will be build in the following year.

2 INSTRUMENTATION

There are two major parameters to measure in pond work: The salt density profile within the pond, and the temperatures in and around the pond. Forces originating from surface wave motion and local thermal convection tend to erode the salinity gradient. It is therefore vital to measure these effects in the pond in order that evasive action be taken before they destroy the salinity gradient.

3 DENSITY MEASUREMENT

A densitometer fig (1) has been designed and built and is undergoing extensive tests to determine its capabilities. It was designed to measure density changes to one part in 10^4.

This is a direct reading instrument in which the upward buoyancy force on a small plastic buoy is balanced by a downward magnetic force exerted on a small permanent magnet attached to the buoy by an external solenoid. The current required to stabilise the vertical position of the buoy provides a direct measure of the buoyancy force and hence the density. The position of the buoy is servo controlled by a bridge balanced LED/photoresistor arrangement which controls the solenoidal current.

The equation of motion governing the buoy is

$M\ddot{z} + \alpha\dot{z} + kz = (M - V_L\rho_L)g - |\mu|\frac{\partial B}{\partial z}$ where M is the mass of the buoy, α the damping constant (mostly viscous), k the weak spring constant due to small wires attached to the buoy, V_L the volume of liquid displaced by the buoy, ρ_L the density of the liquid, μ the magnetic moment of the permanent magnet and $\frac{\partial B}{\partial z}$ the axial field gradient.

In the steady state, the equation reduces to:

$g(M - V_L\rho_L) = |\mu|\frac{\partial B}{\partial z}$. The field gradient for a short solenoid is given by $\frac{\partial B}{\partial z} = \frac{\beta\mu_o NI}{2}$ where β is a factor dependent on coil geometry and the distance between the coil and the magnet and is given by

$$\beta = \frac{1}{[z^2 + a^2]^{\frac{1}{2}}} - \frac{1}{[(L - z)^2 + a^2]^{\frac{1}{2}}} + \frac{(L - z)^2}{[(L - z)^2 + a^2]^{\frac{3}{2}}} - \frac{z^2}{[z^2 + a^2]^{\frac{3}{2}}}$$

where L is the length of coil and a the radius.

For a resolution of one part in 10^4, a conveniently controllable maximum current of 1A, the magnetic coil and buoy parameters must then be related by:

$\frac{10^{-4}|\mu|\mu_o\beta}{2g} = \frac{V_L}{N_L}$ where N_L is the number of turns of wire per unit length of the solenoid.

As convection starts in layers approximately equal to 1cm thick, to see these instabilities arising, the buoy has to sample horizontal distances of less than 1cm, which limits buoy thickness.

In order that the magnetic force be independent of position, the servoing position should be where $\frac{\partial^2 B}{\partial z^2} = 0$. The axial field gradient $\frac{\partial B}{\partial z}$ for a coil of varying length and radius was computed, and a coil of length 2cm and radius 1cm was chosen, which gave a uniform $\frac{\partial B}{\partial z}$ near the end of the coil, enabling the physical size of the buoy to be kept within the 1cm height restriction. At this position, the magnet is also self-centering due to the high radial held gradient $\frac{dB}{d\rho}$. Change in position of the buoy is sensed by the change in resistance of a photoresistor, illuminated by a light emitting diode (LED) mounted on the buoy. To minimise the effects of temperature coefficient, changes in opacity of the water, and LED ageing the output of the LED/photoresistor combination is compared in a bridge circuit with the output from a similar pair with fixed separation. The bridge is balanced when the distance between the first LED/photoresistor pair is equal to the predetermined fixed distance of the second, see Fig (1).

4 PERFORMANCE

The instrument was calibrated in a constant density enviroment. Solenoid current and density were shown to be linear to one part in 10^4. The second measurement made was to observe the system noise, see fig (2). A programable voltmeter connected to a microcomputer was used to take second time averaged values of voltage across the coil for an hour. The signal to noise ratio has a value of the reciprocal of 5 parts in 10^4. By averaging over periods of 5 minutes, this effect can be reduced to within 1 part in 10^4.

The third measurement was to investigate the transient response of the system. (See Fig (3)). This was observed by introducing an electronic step function to the system, and storing values of voltage across the coil on a digital storage oscilloscope. The system is just under critically damped which is good for stability. The settling time (t_s) is 1.6 seconds and the overall damping ocefficient α/M calculated from the results of this experiment is 0.67 sec^{-1}.

The fourth test was on the temperature dependence. Over the range 0–40°C there was no measurable effect on current. When other more suitable sealants are found the temperature will be raised to the maximum pond temperature of 100°C.

The densitometer can also be used to monitor vertical pond water movement due to local convection or wind induced oscillations, since it responds to both buoyancy forces and viscous forces. A fifth test was therefore made to determine its sensitivity. When measuring vertical components of water velocity, a sensitivity of 10^{-2} mm/sec was recorded. An oscilloscope trace for surface wave motion is shown in fig (4).

5 PROBLEMS

The major problems have been associated with the materials used in the construction of the instrument. Due to the corrosive hot environment one is restricted to using stainless steel and plastics. The plastics that have a low thermal expansion have poor adhesive properties. Careful consideration had to be given to maintaining good electrical insulation.

The buoy made from polypropelene was painted black to keep ambient light reflection to a minimum. A more suitable but more expensive solution would be to use molybdenum polypropelene.

The accuracy of the instrument is affected to a certain degree by the spring constant (k) for the wires. Yet again a more suitable solution would incur great expense. Thinner non sleeved wire with a double coating of special varnish could be used in order that electrical insulation be preserved.

6 TEMPERATURE MEASUREMENT

A thermocouple multiplexer has been built and switches up to 32 copper constantan thermocouples. These have been used in the interior pond to determine the temperature profile. Problems were first encountered with protecting the thermocouple junctions from the hot salt water enviroment. But this has now been overcome by isolating the thermocouple from the salt water with a closed ended stainless steel tube.

7 TANK FILLING

A tank of dimensions (91 x 57 x 57 cms) was filled (1). In order to establish the salinity gradient, the tank was filled with high density salt solution (17%), to the height of the convection layer plus half the insulation layer. A diffuser was positioned horizontally at the convectionlayer/insulation layer interface, fresh water was injected whilst the diffuser was brought up to the rising surface in steps of 4cms for every surface shift of 2cms. The final surface had zero salt concentration, and the density profile measured by weighing samples from each layer is shown in fig (5).

Before filling some experimentation was performed on flow rates of fresh water into salt water. The maximum rate for a semi-circular diffuser of radius 20cm was found to be 550 cc/min, before turbulence below the diffuser occured.

Within 5 hours of filling a surface layer appeared which was 2cm thick. This layer increased in size to 10cm after a month of pond heating.

8 TANK PERFORMANCE

The tank sides were insulated using 8cm of glass fibre and the bottom with 2.5cm of polystyrene. The tank was heated using two 1 kW tungsten halogen lamps, which gave an average insolation of 300 W/m^2 to the pond after filtering out most of the infra red component with an intermediate filter. This consisted of a layer of water 1cm thick. The final temperature profile is given in fig (6), and follows closely to the profile predicted by the one dimensional simulation model.

9 THE COMPUTER MODEL

A computer model has been designed as a tool to analyse the effect on thermal pond behaviour from changing physical parameters in and around the pond. Also it will be used to predict and produce a semi-empirical equation for heat losses from different shaped ponds. This program can then be used to determine an adequate site for a solar pond.

The parameters affecting pond behaviour are : the depths of the lower convection zone (LCZ), upper convection zone (UCZ), distance of the pond bottom to ground water level, and the thermal conductivity of the ground (K_g).

The model consists of a pond placed in the earth whose wall boundary is part of a hemisphere with a movable centre above the ground. A water table of varying height from the pond bottom gives a constant temperature boundary.

The interior of the pond is treated as a one dimensional problem with infinite conductivity in the horizontal direction. It is assumed that all light incident on the pond wall is absorbed. The pond sees the earth as a poor conductor whereas in computing the temperatures in the earth, (a 2 dimensional problem), the pond is seen as a heat sink or source.

The heat input to the pond is calculated over hourly intervals, assuming no heat flow across the boundary between pond and earth. (The pond surface/air boundary is assumed to be modelled by a heat transfer coefficient of 30 $Wm^{-2}k^{-1}$). The earth heat flow is then computed for the same time interval of an hour, treating the pond as a constant temperature boundary. The heat flow from the pond into the earth is calculated at each point on the boundary for each time step. The effect of this heat outflow from each layer is then calculated and applied as a temperature correction, taking into account the thermal losses into the ground (see fig 7).

Equations are modelled by implicit finite difference techniques. Calculating temperatures in the pond is a simple problem taking up little computing time. But calculating temperatures in the earth requires an alternating direction method for which spherical geometry restricts the size of the time step, initially to around one hour, for the method to be stable[2]. More work is to be done on defining the stability of this method.

CONCLUSION

The work has concentrated on testing building and improving a new method of measuring salt concentration, building and observing the effect of sunlight on a laboratory solar pond, and developing a numerical program to model thermal behaviour.

Testing will be carried out this summer on a 4.5 diameter pond.

REFERENCES

1. Zangrando, F. "A simple method to establish salt gradient solar ponds", Solar Energy, Vol. 25, pp 467-470

2. Peaceman, D.W., and Rachford, H.H., JR, J. Soc. Indust. Appl. Math. Vol. 3, No. 1, March 1955

3. Smith, G.D., Numerical Solutions of partial differential equations finite difference methods, Oxford Press.

The Densitometer

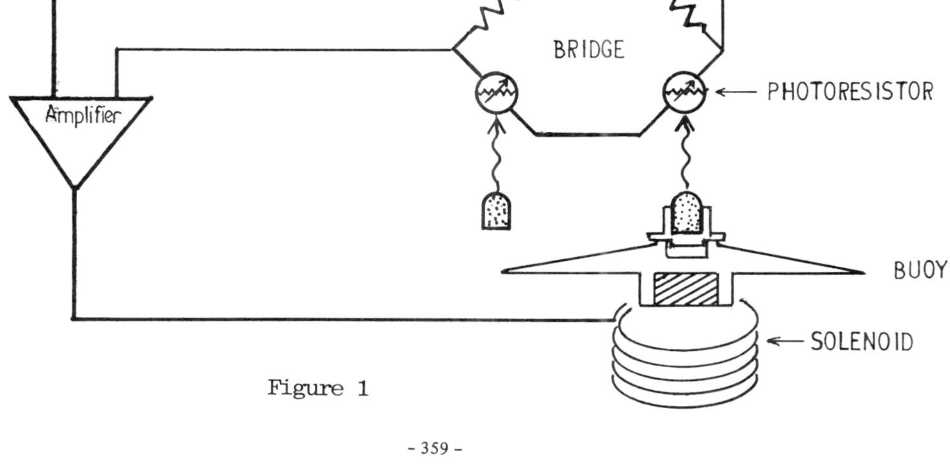

Figure 1

System Noise

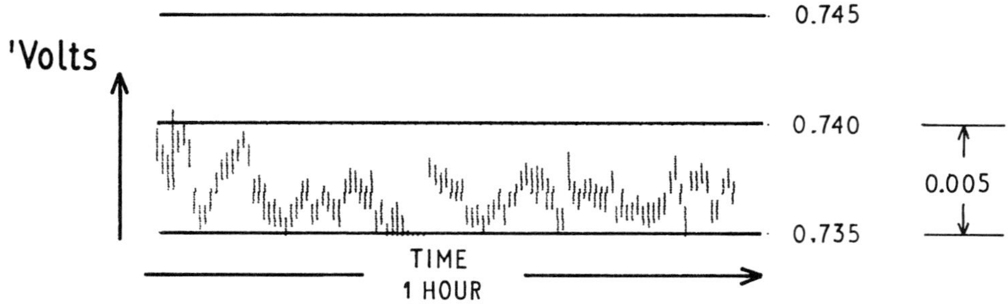

Fig 2.

Transient Response

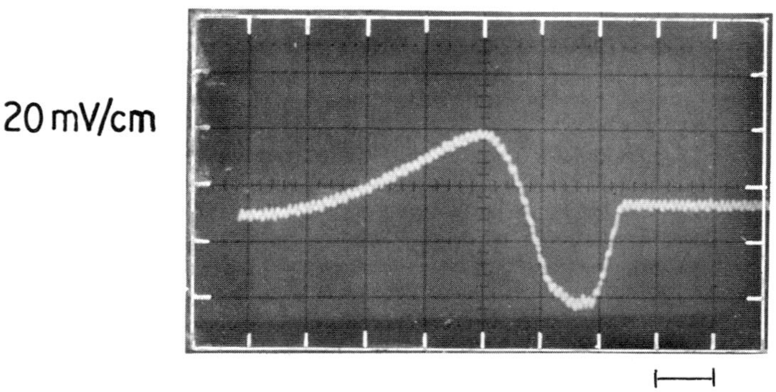

Fig 3. 0.2 sec/cm

Response to Surface Wave Motion

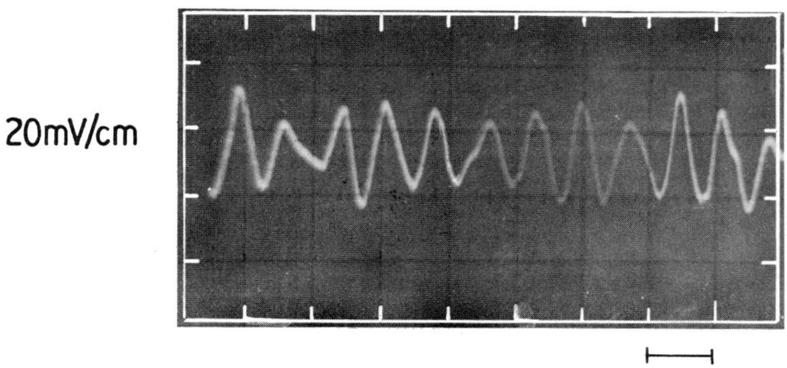

Fig 4. 0.5 sec/cm

DENSITY PROFILE

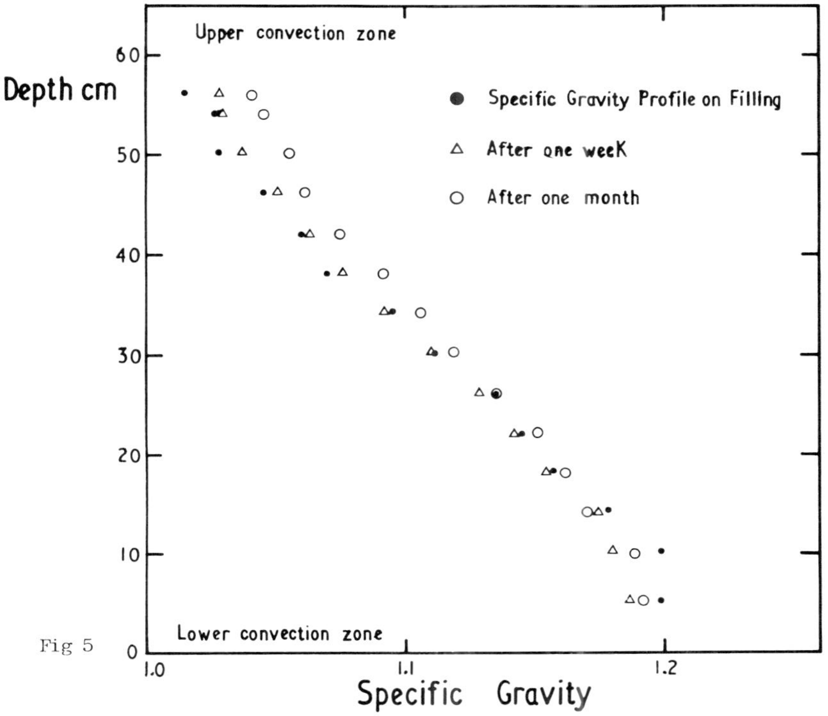

Fig 5

TEMPERATURE PROFILE

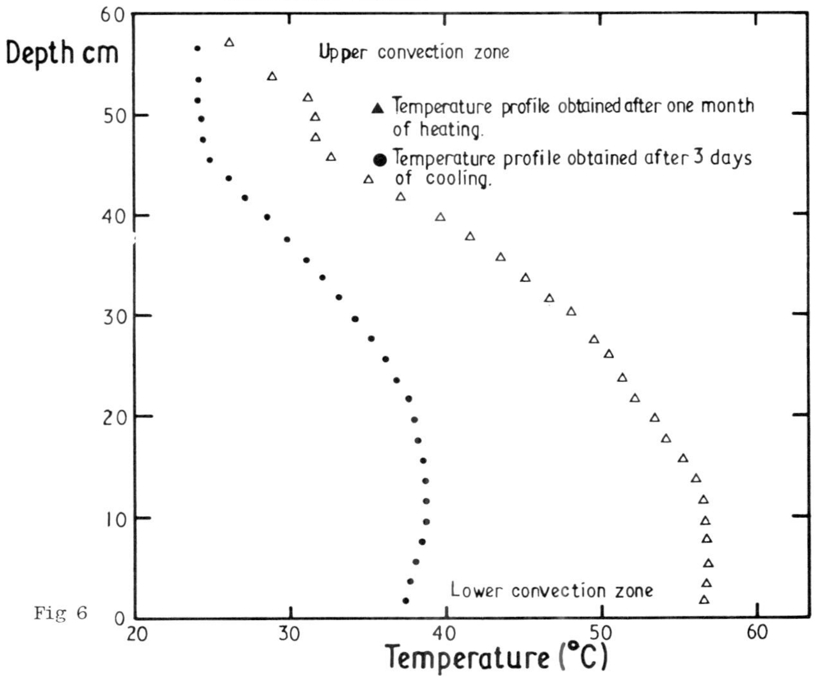

Fig 6

Pond/earth interface boundary representation of pond/earth interface by isothermal and adiabatic boundary conditions.

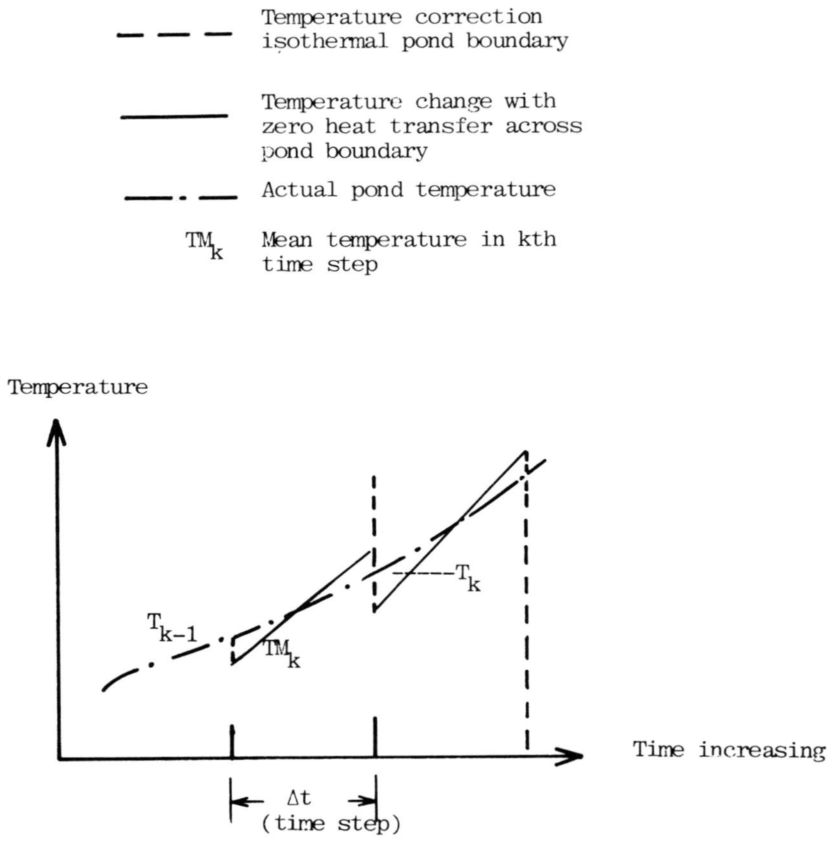

Figure 7

EVALUATION OF THERMAL STORAGE FOR SOLAR HEATING SYSTEMS

AUTHOR : R. H. MARSHALL

CONTRACT NR : ESA-S-121-UK

DURATION : From: January 1982 To: June 1983

TOTAL BUDGET : £49 900 CEC Contribution: 50%

HEAD OF PROJECT: Professor B. J. Brinkworth

ORGANIZATION : Solar Energy Unit
Dept Mechanical Engineering and Energy Studies
University College Cardiff

ADDRESS : Dept Mechanical Engineering and Energy Studies
University College Cardiff
Wales, U.K.

SUBCONTRACTORS : None

Summary

Under the previous contract work, ESA-S-041-UK(N) and E/5A/CON/1023, the role of various alternatives to fully mixed water storage has been investigated by means of theoretical computer simulation studies of stratified water storage, infinite NTU fully mixed and finite NTU stratified phase change storage. On the basis of thermal performance alone, no clear cut optimum storage technology could be found because the choice of a storage device is influenced by the choice of the collector quality and distribution system. The last sub-task of that work, therefore, was the derivation of a Simplified Design Method so that the choice can be made on a comparative cost basis.

The present contract has as its unifying goal the derivation and experimental validation of one (or more) storage parameters which can relate component test results to long term system performance. The present paper, therefore, indicates the progress to date in defining a storage parameter, the Rs parameter, which occurs as one of 6 fundamental non-dimensional parameters which form the basis of the Cardiff SEU Design Method. The derivation of the method is outlined, the influence of the heat distribution system is discussed, and the best correlations (to date) are given for the open loop, the closed loop and the combined systems respectively.

1.1 Introduction

Four sub-tasks within the present contract are to be undertaken:

a) upgrade the existing Cardiff prototype pcm store and test facility by

   i) replacing the previous plastic swimming pool panels which served (well, in fact) as the heat exchanger by a fin and tube heat exchanger with a larger heat transfer capability ($UA \sim 2000\ W\,K^{-1}$)

   ii) upgrade the loop to permit dynamic simulation of both a source loop and load loop

   iii) upgrade the loop by means of a microprocessor for control, data acquisition, data logging and evaluation

b) define appropriate test procedures based on a generalised storage model for component testing whose parameters relate to long term system performance

c) validate the model using both component tests and dynamic system tests

d) compare the component test results in dynamic simulations to indicate the sensitivity and hence areas for improving the performance of both the storage device and the system within which it operates

The last sub-task, in fact, is the starting point, as considerable experience using simulation has already been gathered from comparisons of fully mixed water storage against alternative storage devices operating in a wide variety of systems (1). The role and type of the storage device has been seen to be of secondary importance compared with the role of the collector quality (U/η_o), and the method of auxiliary injection (series or parallel) as reflected by a characteristic minimum useful temperature level, Tminu. Thus a simplified design method, valid for fully mixed water storage, has been derived at Cardiff which relates the solar fraction, E, to six non-dimensional parameters, Ma, Ke, $p(\tau)$, $q(\tau)$, Ls, Rs. Attention therefore is focussed on the last, the storage parameter, Rs. The simplified method is derived and the correlations to date for the open, closed, and combined systems are given. Future work will be aimed at verifying the correlations experimentally and extending the parameters, e.g. Rs, to include other storage devices.

2.1 A Proposed Non-Dimensional Performance Estimator

The last task remaining on the previous contract was to develop a generalised, location independent analytic form for the solar fraction as a function of the system parameters. Initially we envisaged making three major improvements to the $\bar{\phi}$-F-Chart method, see Duffie and Beckman (2), using the detailed system and various storage sub-system models in hour by hour simulations to generate the data points.

1) to express the solar fraction (hereafter referred to as E) as an analytic *explicit* function of the system variables as opposed to the *transcendental* expression necessary for $\bar{\phi}$-F-Chart

2) **to improve** the correlation of the prediction **method for poorer** climates typical of countries lying above 43° North. (Recall the F-CHART method did not predict as well the performance in the rainy states of Oregon, Maine and Washington State and that the correlations were derived for latitudes below 43° North)

3) to include alternative storage devices of arbitrary capacity within the correlation

It is evident that this sub-task represents in itself a major effort - too much, in fact, to be treated in full detail as just a sub-task within the previous contract. The ground work was undertaken by Jeff Kenna, a doctoral candidate, as part of his PhD thesis. A brief review of the approach is given in order to tie together the principles of energy collection, storage, and use.

The starting point for a proper design method is the fundamental system equation for the reference system. As preceding reports to the Commission have demonstrated, there is little, if any, improvement in the solar fraction for any alternative storage device when compared with a fully mixed water store coupled to a parallel auxiliary system (1). Therefore, the Design Method is based on the non-dimensional equation for fully mixed storage with the goal toward extending the results to include alternative storage devices.

We begin with the differential equation for a fully mixed storage device which ignores all pipe losses, component capacities, pump power inputs, etc., i.e.

$$MCp\frac{dT}{dt} = F'' \, Ac\, \eta_o \, \{G + \frac{U}{\eta_o}(Ta - T)\}^+ + p(t)\, \dot{m}Cp^*_\ell \, \{Tc - T\} + q(t)\, \overline{\dot{m}Cp\xi}\, \{Trm - T\} + UsAs\, \{Tas - T\} \quad \ldots . (1)$$

On the right hand side of eqn(1) the first term accounts for the collector input for positive values of the bracketed term with F'' implying a modified F'' to take into account a collector loop heat exchanger. The second term represents the dhw draw-off governed by the variable hourly load pattern p(t) referenced to a daily load draw off rate through a separate heat exchanger. Similarly the third term accounts for the sh draw-off subject to an on-off controller, q(t) = 0 or 1, which depends both on the sh load pattern and the availability of stored energy above a minimum useful temperature, Tminu. The last term accounts for the store loss to an ambient about the store, Tas, which may differ from the outdoor ambient, Ta, or the required room temperature, Trm.

The next step in the development is the most crucial because, from the beginning, we insisted on looking for a suitable non-dimensionalisation which would transform the 80 or so parameter list into six independent groupings. Letting

$$\theta = \frac{T - TLo}{Thi - TLo} = \frac{T - TLo}{\Delta T} \quad \ldots . (2)$$

$$\tau = \frac{t}{t*} \quad \text{with } t* = 24 * 3600 \text{ sec}$$

$$G' = \frac{G}{G*} \;, \quad \text{both G and G* contain incident angle modifiers Ib, Id}$$

then inserting these definitions into eqn(1) and dividing by the total load, L, one obtains

$$\frac{MCp\Delta T}{L}\;\frac{d\theta}{d\tau} = \frac{F'' \, Ac \, \eta_o \, G^* \, t^*}{L}\left\{G' + \frac{U\Delta T}{\eta_o G^*}\,(\theta a - \theta)\right\}^+$$

$$+ \; p(t)\,\frac{\dot{m}Cp^*\;\Delta T\, t^*}{L}\,\{\theta c - \theta\} + q(t)\frac{\overline{\dot{m}Cp}\;\xi\;\Delta T\, t^*}{L}\,\{\theta rm - \theta\}$$

$$+ \left(\frac{UAs}{MCp}\right)\frac{MCp\,\Delta T\, t^*}{L}\,(\theta as - \theta) \qquad \ldots . \; (3)$$

Defining,

$$Ma = \frac{F''\,Ac\,\eta_o\,G^*\,t^*}{L}$$

$$Ke = \frac{U\Delta T}{\eta_o G^*}$$

$$Rs = \frac{MCp\Delta T}{L} = \frac{M\int_{TLo}^{Thi} Cp(T)\,dT}{L}$$

$$Ls = \frac{UsAs}{MCp}$$

$$p'(\tau) = \frac{p(t)\,\dot{m}Cp^{*}\ell\,\Delta T\,t^*}{L}$$

$$q'(\tau) = q(t)\,\overline{\dot{m}Cp}\;\xi\;\Delta T\,t^* \qquad \ldots . \; (4)$$

the non-dimensional differential equation to be solved becomes

$$Rs\dot{\theta} = Ma\,\{G' + Ke\,(\theta a - \theta)\}^+ + p'(\tau)\;(\theta c - \theta) + q'(\tau)\;(\theta rm - \theta)$$

$$+ \; RsLs\;(\theta as - \theta). \qquad \ldots . \; (5)$$

The daily solar fraction, Ed, then is determined from the integration of eqn(5),

$$Ed = \int_0^1 (p'(\tau)(\theta c - \theta) + q'(\tau)(\theta rm - \theta))\,d\tau \qquad \ldots . \; (6)$$

and from the daily solar fraction the monthly solar fraction, E.

In principle, the solution for E from eqns(2-6) would appear straightforward, provided that the reference parameters, Thi, TLo, G* and G*t* have been well chosen, and therein lies the major difficulty. The full explanation of the various choices can be found in Kenna (3,4,5). What complicates the choice is the fundamental difference between an *open loop* system, e.g. dhw where water is heated, then dumped to waste, and a *closed loop* system where the storage fluid is recirculated. This difference is accentuated because in the dhw system all the energy above the mains cold temperature, Tc, is useful while for the sh system the energy must be above Tminu before it is useful. So, while there are clearly different but intuitively obvious choices for Thi and TLo for either the dhw system alone or the sh system alone, the proper choice for a *combined* system is less straightforward.

A further difficulty remains for the best choice of (G*t*) in the Ma number and G* for the Ke number. Space does not permit us here to elaborate on the subtle choices available except to say that the first choice (energy in Joules) can be best related to

$$G^*t^* \stackrel{\Delta}{=} \Phi\,(TLo)\,Htilt \qquad \ldots . \; (7)$$

where $\overline{\phi}(TLo)$ is the utilisable energy above the threshold radiation based on TLo and Htilt is the total energy available on the tilted surface. Recall that $\overline{\phi}(TLo)$ is defined by

$$\overline{\phi}(TLo) = \frac{\frac{1}{N\ Days} \sum_{1}^{N\ Days} \sum_{tsr}^{tss} (Gtilt - U/\eta_o\ (TLo - Ta))^{+}}{\sum_{1}^{N\ Days} \sum_{tsr}^{tss} Gtilt} \quad \ldots . (8)$$

In a similar fashion G* for use in the Ke definition is best related to

$$G^* \stackrel{\Delta}{\equiv} \phi(TLo)\Big|_p \overline{Gp} \quad \ldots . (9)$$

where $\overline{Gp}$ is the peak average irradiation level plus ambient energy gain multiplied by the average $\phi(TLo)$ evaluated from the hour of maximum (i.e. peak) irradiance during the day.

In Table 1 are listed the "best" choices of the reference parameters examined to date. For either the dhw alone or the sh system alone Kenna has used data for a total of 6 European stations leading to a correlation of the form

$$E = \frac{a(Rs)\ Ma}{b(Rs) + MaK^*} \quad \ldots . (10)$$

$$\text{with } K^* = Ke + c(Rs)\ Ma \quad \ldots . (11)$$

where "(Rs)" denotes an Rs parameter dependence. Here 12 average monthly values of Rs, Ma and Ke give rise to 12 monthly solar fractions, Ei, i = 1-12. The correlations for each of the three system types are seen in Table 2. Using the data gathered from simulations under the previous contract for combined dhw and sh systems a similar correlation was devised. The results in the form of E/Ma versus MaK* for all three types of systems (dhw, sh, and combined) for several values of the Rs parameter appear as Fig.1. The yearly solar fraction, E, determined from eqns.10 and 11 agree to within ±5% of the results from hour by hour simulations for either the dhw alone or sh alone but only to within ±9% for the combined system. Thus, some work is still needed to improve the fit in the last case.

2.2 Implications for Alternative Storage Devices

Let us now briefly re-examine the results described in the previous meetings in terms of the non-dimensional parameters Ma, Ke and Rs with one eye on the correlation appearing in Fig.1.

The Ma parameter is clearly the ratio of utilisable energy based on TLo on the tilted surface to total load. The dominant influences here are the optical efficiency times area, η_oAc, and the utilisable energy on the tilt, $\overline{\phi}(TLo)$ Htilt, together with the load. From the correlation, eqn(10), it is desirable to size the system at an Ma value exceeding unity but it is wiser and probably less expensive to use improved collectors leading to a lower Ke because the gain in performance by increasing Ma beyond unity is relatively small. Further a large Ma is best achieved by choosing a parallel auxiliary system thereby reducing Tminu and thus increasing $\overline{\phi}(TLo)$.

The single most influential parameter on system performance is Ke and hence the quality of collector, U/η_o. Contained in this Ke parameter, too, is the influence of the heat distribution sizing parameter by way of the reference $\Delta T = Thi - TLo$. For sh systems alone Thi, see Table 1, depends on the heat emitter effective capacity rate, $\dot{m}Cp\xi$. The low temperature, TLo, on which the utilisable energy on the tilted plane is based, $\overline{\phi}$(TLo) Htilt, is a function of the method of auxiliary injection. For a parallel auxiliary system TLo is the building set point temperature, Trm, as all energy above this level is useful, hence TLo = Tminu = Trm + ΔTon with perhaps a $5^{o}C$ switch-on differential for practical implementation. For a series auxiliary, the store fluid temperature is boosted in series by a boiler to some level T* in order to power the heat emitter. The return temperature, Tr, is therefore considerably above room temperature, leading to a minimum useful temperature, Tminu = Tr on which the utilisable energy is based, i.e. TLo = Tminu. Thus, the series auxiliary is characterised by both less utilisable energy on the tilt, ϕ(TLo) Htilt, and a lower average peak rate of energy capture, ϕ(TLo) Gp. Therefore Ma is lower and Ke is larger leading to a poorer system performance compared with the parallel auxiliary system. The same effect occurs in the combined system.

For dhw alone the influence of a lower demand temperature, Thi = Td, is the crucial temperature level. Taken together Ke represents the reference collector loss rate, $(U/\eta_o)\Delta T$, to incident irradiation, G*. Solar fractions above 50% simply cannot be achieved unless Ke < 1.0, a result which applies equally well to all three systems.

Examination of Table 2 indicates that there is found to be much weaker dependence of system performance on the storage sizing parameter, Rs, evident from the expressions for a(Rs), b(Rs), c(Rs). Thus the storage parameter, Rs, is seen to play only a secondary role, compared with the influence of Ma and Ke on the performance.

The one factor which may alter this observation is the influence of the demand pattern. For dhw alone, an evening peak demand pattern (as used in this study) will increase the sensitivity to the Rs parameter. Similarly a continuous but not constant sh load pattern as used in this study will decrease the sensitivity to Rs. It should be obvious that the best system performance is obtained if the energy collected is put to immediate use.

Two further aspects of the Rs parameter are worth mentioning. Firstly, the Rs parameter is seen to be independent of the collector area and, instead, related to the reference ΔT and load, L. Therefore, the relative storage mass per collector area, i.e. $kg\,m^{-2}$, is really *unsuitable* as a reference parameter. Secondly, it should be clear that for pcm storage the product $MCp\,\Delta T$ has to be replaced by the energy storage density between Thi and TLo, and not an arbitrary temperature trajectory. A quick calculation on the ΔT valid for the "typical" average house conditions reveals that the range TLo $\sim 20^{o}C$ to Thi $\sim 35^{o}C$ is the more appropriate range so that the low melting temperature pcm materials such as Glauber's salt and Calcium Chloride hexahydrate possess energy storage densities equal to and at best 3 times higher than water. This helps explain why the paraffin wax pcm with a low melting temperature performs as well or better than water and not much worse than either $CaCl_2$ or Glauber's salt.

3. Future Work on a Design Method

In summary, the non-dimensional form of the fully mixed store equation, eqn(5), together with a proper choice of reference parameters (Thi, TLo, G*t*, G*) is an extremely powerful tool for assessing the sensitivity of system performance to fundamental parameter variations. The correlations derived to date for dhw alone and sh alone and the combined system demonstrate the universality of this method. The degree of fit is very encouraging. A list for future work includes a deeper examination of monthly irradiation correlations on the tilt, the demand patterns and the experimental evidence that the Rs parameter, i.e. ΔT, has been well chosen. Lastly, work is required to establish the correlation validity to climates outside Northern Europe, e.g. the U.S.A.

References

1. Marshall R.H. — Modelling of Thermal Storage for Solar Heating Systems. Proceedings of the EC Contractors Meeting, Athens, Greece, 11-13 Nov. 1981, D. Reidel Pub. Company for the CEC, p.p.118-123, 1981. (Also to appear as a Final Report to the CEC in expanded form, 1982).

2. Duffie J.A., Beckman W.A. — Solar Engineering of Thermal Processes, John A. Wiley and Sons, 1980.

3. Kenna J. — A Parametric Study of Open Loop Solar Heating Systems, Proc. ISES Solar World Forum, Brighton, England, August 1981, Pergamon Press 1982.

4. Kenna J. — A Parametric Study of Closed Loop Solar Heating Systems, submitted for publication in Solar Energy, April 1982.

5. Kenna J. — The Cardiff SEU Design Methods, HELIOS, Number 15, editor C. M. Johansson, Cardiff Solar Energy Unit, July, 1982.

Nomenclature

Ac	collector area, m^2
a,b,c	Design Method correlation constants
Cp	specific heat capacity, $J\,kg^{-1}\,K^{-1}$
dhw	domestic hot water
E	solar fraction
F"	collector flow factor
G	irradiance incident on tilt, $W\,m^{-2}$
G*	the reference irradiance level, $W\,m^{-2}$
$\overline{Gp}$	monthly daily average peak irradiance, $W\,m^{-2}$
Htilt	energy on the tilt, $J\,m^{-2}$
Ib,Id	beam and diffuse incident angle modifiers
Ke	ratio of collector loss rate to average peak rate of energy available
L	monthly average daily load, J
Ls	non-dimensional loss parameter ratio of loss energy to stored energy
$\dot{m}$	flow rate, $kg\,s^{-1}$

M	storage mass, kg
Ma	ratio of energy available on the tilt to load
NTU	number of transfer units ($UA/\dot{m}Cp$)
$p'(\tau)$	the non-dimensional dhw load pattern
$q'(\tau)$	the non-dimensional sh load pattern
Rs	the ratio of energy stored to load over the range TLo-Thi
t,t*	time and reference time, s
tsr,tss	the sunrise and sunset times, s
T	store temperature, K
Ta	ambient temperature, K
Tc	mains water temperature, K
T*	emitter required inlet temperature to meet the average load rate, K
Tminu	minimum useful temperature, K
Tr	emitter return temperature, K
Trm	room temperature, K
Thi	reference high temperature, K
TLo	reference low temperature, K
U	collector loss coefficient (FrU_L), $W\,m^{-2}\,K^{-1}$

Greek

ξ	load heat exchanger effectiveness
η_o	collector efficiency at ambient temperature ($Fr\ \tau\alpha$)
ϕ	ratio of utilisable energy (or rate) to total energy or rate
θ	non-dimensional temperature
ΔT=Thi-TLo	reference temperature interval
τ	non-dimensional time

Subscripts

a	ambient
ℓ	dhw loop
s	store

Superscript and other

*	denotes a reference value
$(\ )^+$	positive values only

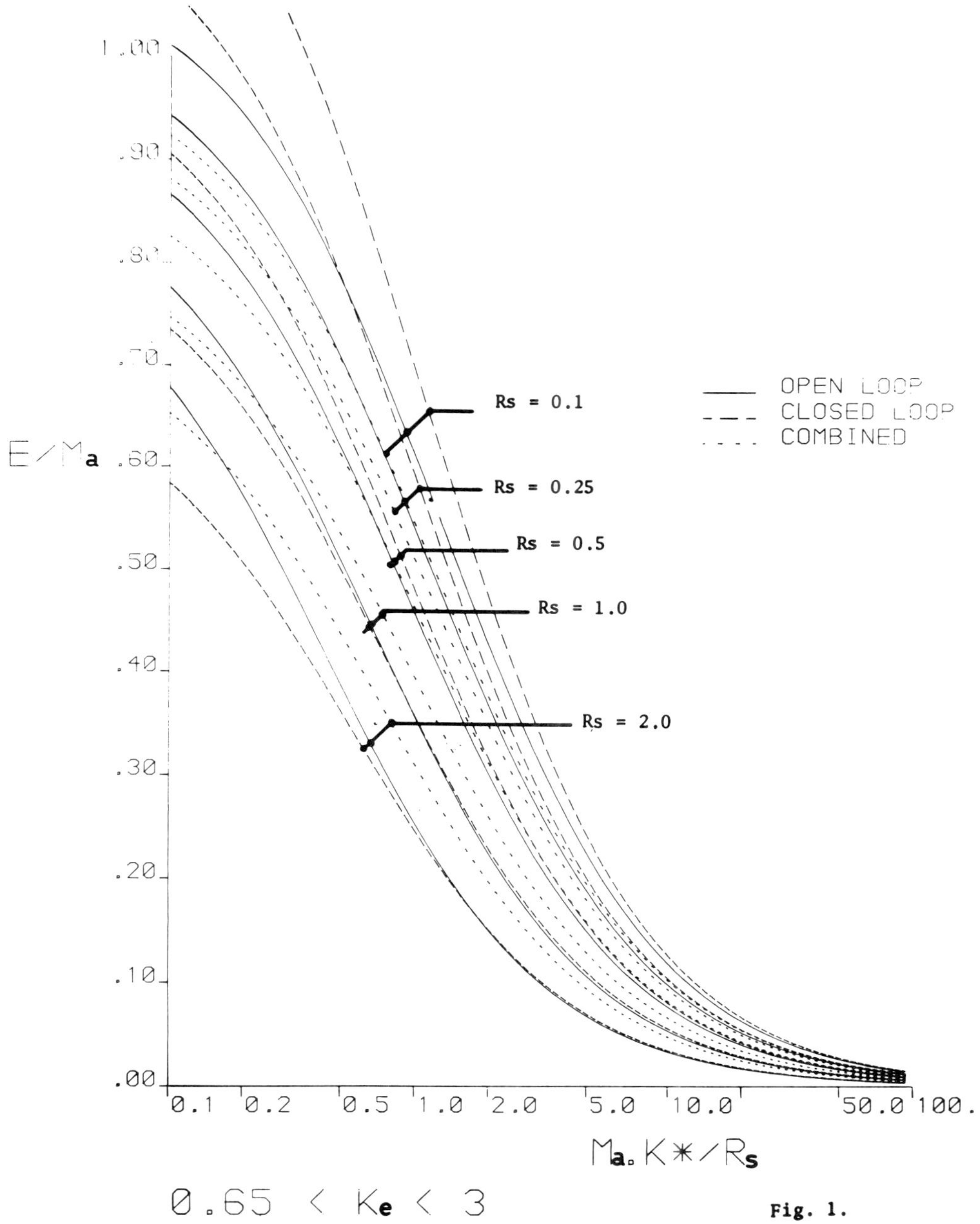
1.00
.90
.80
.70
.60
.50
.40
.30
.20
.10
.00
E/Ma
OPEN LOOP
CLOSED LOOP
COMBINED
Rs = 0.1
Rs = 0.25
Rs = 0.5
Rs = 1.0
Rs = 2.0
0.1
0.2
0.5
1.0
2.0
5.0
10.0
50.0
100.
Ma.K*/Rs
0.65 < Ke < 3

Fig. 1.

	DHW ONLY	SH ONLY	COMBINED DHW & SH
t^*	1 day (in seconds)	same	same
G^*	$\{I_b G_b + I_d G_d + \frac{U}{\eta_o}(T_a - T_{LO})\}^{+}_{max}$ sum daily maximums take monthly average	same	same
ΔT	$T_{hi} - T_{LO}$	$(T_{hi} - T_{LO})$	$L/(\dot{m}C_{p_{eff}}{}^* t^*)$
T_{hi}	T_d	$T_{LO} + \frac{UA_B}{\dot{m}C_p \xi}(T_{RM} - \overline{T}_a)$	$T_{LO} + \Delta T$
T_{LO}	T_c	$(T_{minu}, T_{RM})_{max}$	$f\ \overline{T}_c + (1-f)\ T_{minu}$ $f = DHWL/L$ $\dot{m}C_{p_{eff}}^* = f\ \dot{m}C_{p_\ell} + (1-f)\ \overline{\dot{m}C_p \xi}$ $\dot{m}C_{p_\ell}^* = \frac{DHWL}{t^*(T_d - T_c)}$
M_a	$F''\ A_c\ \eta_o\ H/L$ $H = \int_{t_{sr}}^{t_{ss}} (I_b G_b + I_d G_d)\, d\tau + \frac{U}{\eta_o}(\overline{T}_a - T_c)(t_{ss} - t_{sr})$	$F'' A_c \eta_o \overline{\phi}(T_{LO}) H_{tilt}$ $H_{tilt} = \int_{t_{sr}}^{t_{ss}} (I_b G_b + I_d G_d)\, d\tau$	same same
K_e	$\frac{U}{\eta_o}(T_{hi} - T_{LO})/G^*$	same	same
R_s	$\frac{MC_p \Delta T}{L}$	same	$\frac{M \int_{T_{LO}}^{T_{hi}} C_p(T)\, dT}{L}$ (for pcm's)
L_s	$\frac{U_s A_s}{MC_p}$	same	similar

TABLE I. - THE REFERENCE PARAMETERS

CORRELATION (Used on a monthly basis)

1. DHW ONLY $0.1 \leq R_s \leq 1.0$

$$E = a\,M_a/(b + M_a K^*)$$

$$K^* = K_e/(1.0 + 0.11\,K_e) \qquad K_e > 0.65$$

$$K^* = (K_e + C\,M_a)/(0.69 + 0.87\,K_e) \qquad K_e \leq 0.65$$

$$a(R_s) = 1.6\,R_s/(1.0 + 1.74\,R_s)$$

$$b(R_s) = 1.39\,R_s/(1.0 + 1.31\,R_s)$$

$$c(R_s) = 0.063\,R_s/(1.0 + 0.24\,R_s)$$

2. SH ONLY $0.1 \leq R_s \leq 2.0$

$$E = A\,M_a/(b + \frac{M_a K^*}{R_s})$$

$$K^* = K_e/(1.0 + 0.11\,K_e) \qquad K_e > 0.65$$

$$K^* = (K_e + c\,M_a)/(0.69 + 0.87\,K_e) \qquad K_e \leq 0.65$$

$$a = 1.67/(1.0 + 1.74\,R_s)$$

$$b = 1.13/(1.0 + 0.55\,R_s)$$

$$c = 0.136\,R_s/(1.0 + 1.23\,R_s)$$

3. Combined DHW and SH

$$E = a\,M_a/(b + \frac{M_a K^*}{R_s})$$

$$K^* = K_e/(1.0 + 0.11\,K_e) \qquad K_e > 0.65$$

$$K^* = (K_e + c\,M_a)/(0.69 + 0.87\,K_e) \qquad K_e \leq 0.65$$

$$a = 1.64\,R_s/(1.0 + 1.74\,R_s)$$

$$b = 1.26\,R_s/(1.0 + 1.0\,R_s)$$

$$c = 1.0\,R_s/(1.0 + 0.75\,R_s)$$

TABLE II. - THE DESIGN METHOD CORRELATION

PHYSICAL MODEL FOR THE STUDY OF MASS AND ENERGY TRANSFERS IN THE NON-SATURATED LAYER OF SOIL LOCATED ABOVE A SOLAR ENERGY STORAGE

Authors :J.C.BENET, G.DELLA-VALLE, P.JOUANNA, D.KOUFOGIANNIS, C.SAIX

Contract number : E.S.A-S-051-F(S)

Duration : 30 months 1 July 1981 - 1 December 1982

Total budget : 656 810 FF CEC contribution 328 000 FF

Head of project : Pr.P.JOUANNA

Contractor : Laboratoire de Génie Civil-Université Montpellier II

Address : Place Eugène Bataillon 34060 - MONTPELLIER - France.

Summary

The aim of the project is to study the mass and energy transfers in a non-saturated layer of soil placed above a saturated layer.

The execution of the test embankment is described: saturation zone simulation, stockpiling the earthfill, preliminary compaction tests, placing and compacting a layer, monitoring compaction.

Then the placing of the different probes is given: thermal conductivity and temperature, capillary succion, gas pressure, settlement measurements, water content.

Initial values of the state variables have been measured. After a one-month period of rest for homogeneisation of the water content, the test embankment will be ready to begin the one-year test programme of experimentation.

1. Introduction

The aim of the project is to study the mass and energy transfers in a non-saturated column of soil placed above a saturated layer of soil used as a storage zone for solar energy. The fields of the different state variables such as the specific weight of solid, liquid water, water vapour and air, succion and temperature obtained on this physical model will be compared with the values obtained by a theoretical model, previously developed in the laboratory, thus obtaining a test of reliability of this theoretical model. In the case of good agreement of both results, this theoretical model could be used for predicting the behaviour of solar heat storage under or in non-saturated soil masses, for different types of soils and various boundary conditions.

The final choice of the physical model, which is a large column of soil more than 6m in diameter placed above a concrete tank filled with saturated sand, has been discussed previously (1).

The detailed description of the project of the test zone and the different state variables to be measured, was given at the last contractors'meeting (2).

The present paper describes the execution of the test zone and gives the initial state of the different variable fields as measured in situ.

2. Execution of the test zone - Fig.1 and Fig.2

2.1 Saturation zone simulation

The saturation zone is simulated by a reinforced concrete tank filled with saturated sand; water is distributed throughout the sand by concrete pipes placed at the top of the concrete slab, with fine holes protected by a glass fibre membrane acting as a filter to prevent penetration of sand in the water circuit. The sand placed above has been slightly compacted. No transition zone between the sand and the earthfill material has been necessary, transition requirements between these two materials being satisfied.

Heating resistances have been placed in the concrete slab, in the same manner as in heated slabs. Great care has been attached to regulating the heating device by a thermocouple placed in the sand layer and to the safety devices as well. Electrical tests have shown that the circuit functions satisfactorily.

The water is not allowed to saturate the sand during the first stage of the experiment consisting of a one-month rest period for the earth embankment.

2.2 Stockpiling the earthfill

Due to poor weather conditions throughout the fall and winter period preventing extraction of earth in the borrow pit, supplying and stockpiling the silt were performed during the first days of spring. Stockpiling has been essential to ensure an homogeneous water content for all the material at least in the central portion of the trial zone, this stockpile being protected from atmospheric influences by a plastics sheet. Stockpiling has also been useful to secure a sufficient rate of supply of earth during the compaction process.

2.3 Preliminary compaction tests

The first two layers were performed independently of the rest of the earth embankment in order to study carefully the choice of the different compaction parameters: exact water content, type of roller and thickness of layer before and after compaction, number of passes,...

The technique of placing the different measurement probes was also studied at the same time. Moreover control tests were also perfected.

2.4 Placing and compacting a layer

The earth was supplied from the stockpile to the earthfill by a mechanical shovel, avoiding passing over the central portion of the earthfill, thus preventing overcompaction of the earth material. In this central region the earth was chosen at the right water content, close to the optimum Proctor water content. The thickness of the layer was fixed at 21 cm before compacting to obtain a thickness of 17 cm after compaction. After 6 passes of a 700 kg vibrating roller, the dry density of the material ceased to increase. Before placing the next layer, the surface of the fill was scarified with a handscarifier in the central portion and a mechanical sacrifier around it. This is important to prevent the formation of a crust between each layer, which could cause heterogeneous profiles of permeability.

2.5 Control tests during compaction

The most important parameters during the compaction process were checked as follows:

i- The water content was controlled essentially by two procedures. Immediate values of the water content after compacting a layer were obtained using a neutron detector He3-3411/B. The scattering of the results is well known and a comparison with conventional water content measurements on the site or in the laboratory was necessary.

ii- The dry density of the soil in a compacted layer was also checked by two procedures. The same apparatus 3411/B using a backward diffusion process of gamma rays was used; the values obtained for wet density were found to be very close to the values obtained using a classic densitometer. The dry density was obtained using the water content measured in the laboratory.

iii- The thickness of the layer was measured after each two roller passes using a surveying technique.

3. Placing the measuring probes

The state of the soil in the central zone of the test fill will be monitored using the following measurement probes:

3.1 The thermal conductivity and the temperature are measured by a probe described in reference(2), fig.2. Such a probe was placed every two layers. The first was placed at the top of the second layer, but the following ones were placed inside the earth before compaction to ensure a better bond between the probe and the surrounding earth material.

3.2 The capillary succion is measured by classic cells using a porous ceramic disc, as described in reference(2), fig.3. They were placed in small grooves at the top of two layers, the face of the porous disc being in contact with the compacted soil. The increasing slope of the connecting copper tubes ensures satisfactory bleeding of air bubbles from the circuit.

3.3 The gas pressure is measured by probes described in reference(2), As noted hereabove they were placed in small grooves but with tubes having a decreasing slope to ensure that condensed water flows out.

3.4 Special probes to measure the settlement or heaving of the soil have been also placed in the fill, at the top of each two layers.

3.5 To monitor the water content variations, aluminium tubes were finally placed horizontally to receive a neutron Solo 20 probe, as developed by the CEA at Cadarache, the method using the results of the thermal conductivity probe to measure the evolution of the water content not being accurate enough. Placing these tubes vertically would have been difficult during compaction with a great risk of obtaining very scattered values of the dry density around tubes.

The probes to measure the partial water vapour pressure will be placed later in the two top layers of the soil.

4. Initial conditions

The earthfill was entirely completed on the 22nd of May 82 and is resting at the moment waiting for homogeneisation of water content. The initial state of the variables is given in figure 3 for the dry density and the water content, in figure 4 for the capillary succion and in figure 5 for the temperature. The entire earthfill is protected against climatic variation by a plastics sheet during this homogeneisation period. After this period will begin the one-year programme of experimentation.

References

(1) E.C Solar Energy R&D Program - Project A
Coordination meeting of CEC contractors - Contract ESA-S-051-F(S)
21-23 January 1981- Brussels

(2) E.C Solar Energy R&D Program - Project A
Coordination meeting of CEC contractors - Contract ESA-S-051-F(S)
11-12-13 Nov.1981 - Kavouri.

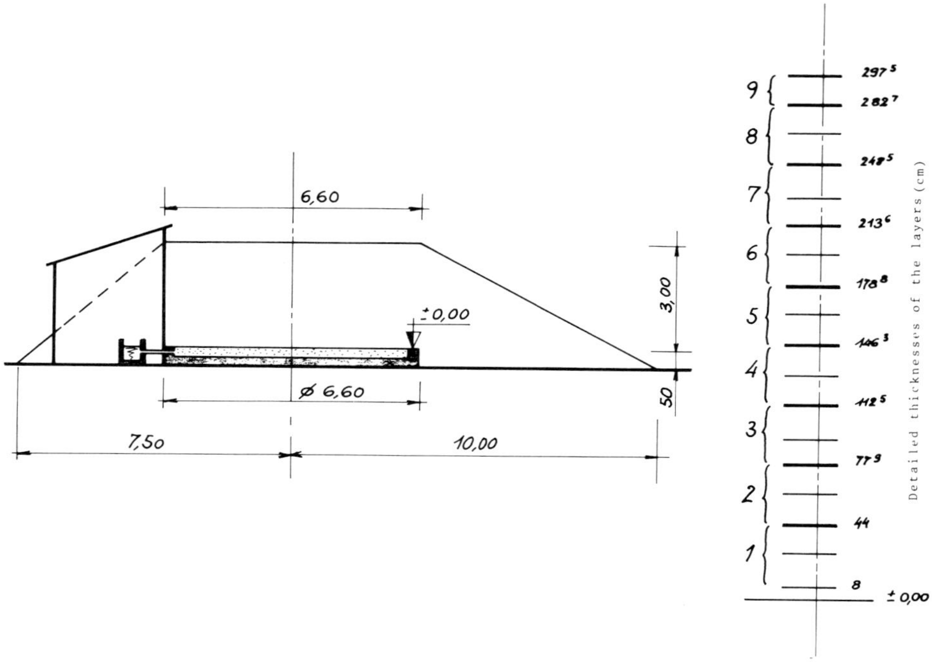

Fig. 1 - Test embankment - cross section

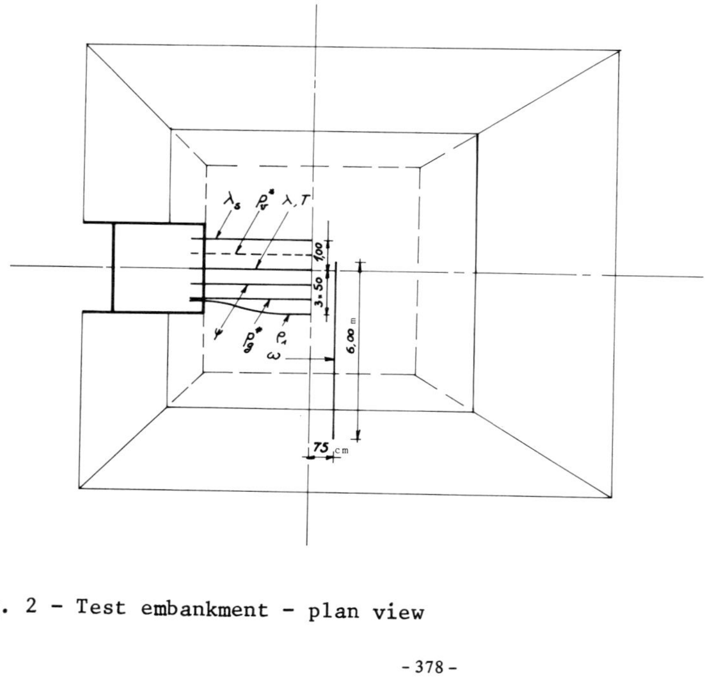

Fig. 2 - Test embankment - plan view

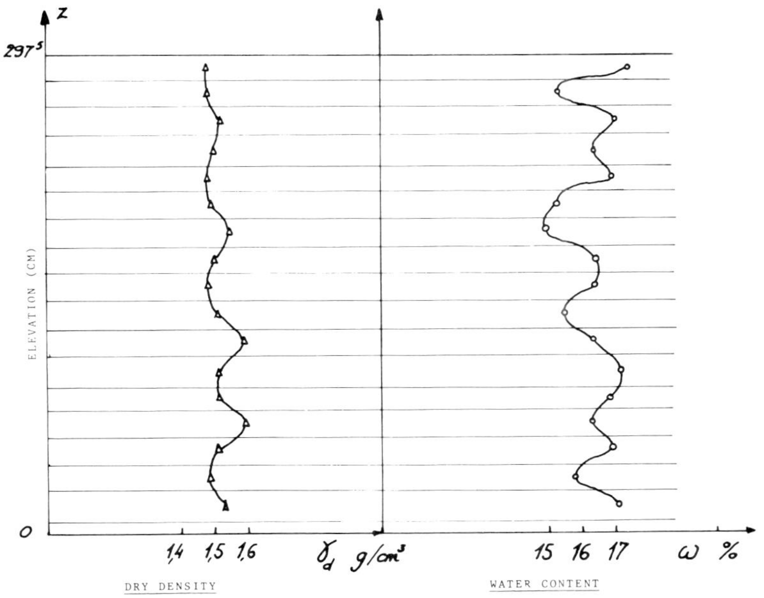

Figure 3

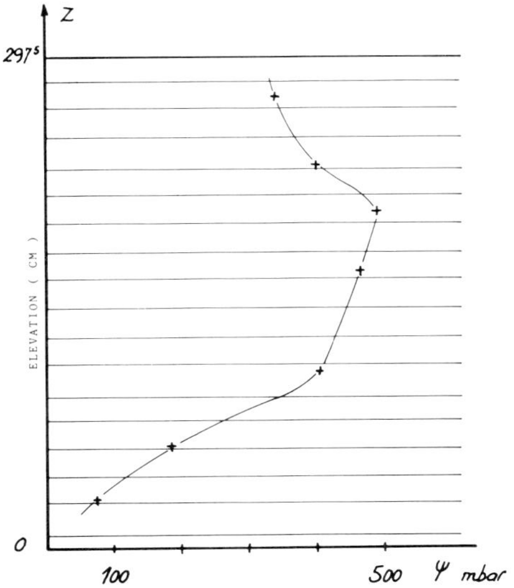

Figure 4 - Capillary succion

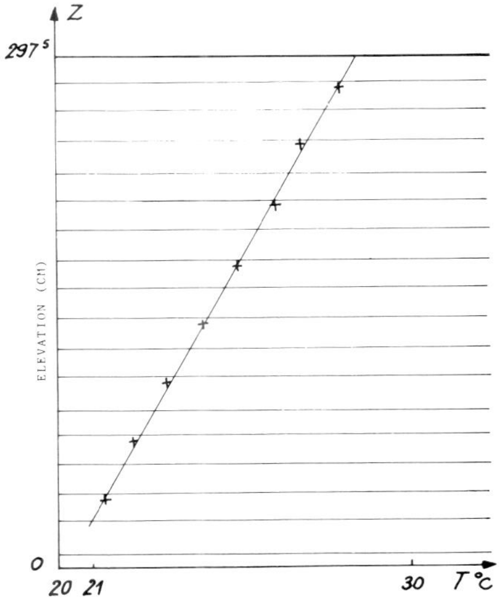

Figure 5 - Temperature

LONG TERM SOLAR HEAT STORAGE THROUGH UNDERGROUND WATER TANKS FOR THE HEATING OF HOUSING

Authors : M. Cucumo, V. Marinelli, G. Oliveti, A. Sabato

Contract n° : ESA/S/049/I

Duration : 36 months from 1/7/1980 to 1/7/1983

Total budget : 400,000,000 LIT CEC contribution 75,000,000 LIT

Head of Project: Prof. eng. Valerio Marinelli

Prof. eng. Giuseppe Oliveti

Organization : Dipartimento di Meccanica, Facoltà di Ingegneria, Università della Calabria

Address : 87030 - Arcavacata di Rende (Cosenza) - Italia

Summary

This project consists in the development of design methods of solar plants for heating of housing by means of the interseasonal storage of solar energy through water tanks located under or above ground. In this report the two-dimensional transient conduction PACI computer code for cylindrical tanks is described and some results are commented on.

1. Introduction

this project has three phases: the first consists in the development of calculation methods and codes necessary to correctly design solar plants which provide all heat needs of residential housing, by means of the interseasonal storage of solar energy through underground or above ground water tanks;the second phase consists in the construction of a prototype plant, and the third and last phase in the analysis of operation data from this plant and the comparison with design predictions (1),(2).

We at present are in an advanced stage of first phase of the project: five computer codes on a total of seven have been up to now completely developped, ad the last two are in a good stage of progress. SM, PROTESE, PAPI and PASFE codes have been described previously (3),(4); in this report PACI code is described and obtained calculation results are shown and commented on.

2. PACI computer code

PACI is a two-dimensional numerical computer code able to evaluate the thermal behaviour of cylindrical tanks located underground or above the ground. The water in the tank is represented by a single node, while the shell of the tank and the surrounding is subdivided in radial and axial nodes, up to a maximum of 800 nodes. The code evaluates, at each time step, the temperature distributions of all nodes. in Fig. 1 a typical nodalization is shown.

In the case of a tank exposed to the external air, the air is considered as a node with infinite thermal capacity, and its time temperature profile is specified as an input datum in the code, according to methereological conditions.

The nodal equation system has been solved by means of a Cholesky technique, which transforms the coefficient matrix of unknowns in a vector having a dimension lower than the half of the matrix dimension.

3. Comparisons between spherical and cylindrical tanks

It has been considered interesting to compare results obtained with PACI code, for cylindrical tanks, to results of spherical tanks having the same volume of cylinders, and evaluated with a previousley developped code for spherical geometry, PASFE code.

In fig. 2 the temperature profiles of underground tanks of 800 m^3, having a spherical and a cylindrical geometry, are compared, for two different values of the insulator thickness: 0, and 20 cm. The figure shows that the cylinders have a water temperature lower than that of spheres, with a maximum difference of about 5 °C. This is primarily due to the different S/V ratio for the two geometries equal, in this case, to 0.52 for the sphere and to 0.596 for the cylinder.

Per cent heat losses versus time are compared in Fig. 3 which shows, in the case of insulated tanks with 20 cm of polyurethane, larger losses for the cylinder by 5-6% after first year of transient.

According to the S/V ratios, such differences should by equal to 14.5%: this means that a favourable geometry effect exists for the cylinder, due to the larger values of radial internodes thermal resistances.

This last point is confirmed from the graph of Fig. 4, where the heat losses for unit area versus time are plotted for the two geometries.

In Fig. 2, 3, and 4, results obtained with PAPI code, valid for plane walle, are also plotted as reference data.

4. Comparison between underground and above ground cylindrical tanks

In Fig. 5 two temperature profiles are shown, for a cylindrical tank located underground and one located above ground in the open air. From this figure it is evident that an underground tank with t=10 cm and S=196 m^2 matchs the design requirements of a maximum temperature of about 90° C and a minimum temperature of about 50° C.

Also an above ground tank with t=15 cm and S=186 m^2 matchs these requirements.

This means that the tanks of solar plants proposed by us may placed also in the open air, and have not necesserely to be buried.

The selected underground tank presents a total cost of Lit 194,500,000 (Lit 96,500,000 tank + 98,000,000 collectors), while the above ground tank has a total cost of Lit 165,000,000 (Lit 72,000,000 tank + Lit 93,000,000 collectors).

These sums can be recovered in 11 and 10 years of plant operation, assuming an increasing of oil price of 20% for year.

5. Calculation models in progress and future actions.

A threedimensional (x,y,z,) transient conduction computer code for the design of parallelepiped tanks is under development (PARE code).

A simulation code for the entire solar plant, including the components and regulation systems modelling, is under development (COSIMI code). This code will furnish, at each hour, the temperature distributions in all components and the actual hourly thermal loads.

The construction of a prototype plant will be initiated, starting the second phase of the project. This has been until now delayed for several reasons.

6. Conclusions.

The design phase of this project has reached a good stage, with the development of PACI transient conduction code which permits a realistic modelling of cylindrical tanks and their correct design.

Design studies performed up to now have indicated that three different solutions are possible and valid, from a thecnical and also economic point of view: tanks may be located underground, or placed above ground, in the open air, or may be half-buried, according to different conditions of space availability. Above ground tanks cost less then buried tanks for the cost of excavation, but both solutions appear actractive economically. In a future generation plant the tank may be embodied within the structure of the building and an underground floor may be utilized as storage water tank.

7. References.

1) Long term solar heat storage by inground water tanks for the heating of civil buildings.
M. Cucumo, V. Marinelli, G. Oliveti, S. Orlando, A. Sabato.
Bruxelles Contractors Meeting; January 1981, Report EUR 7343 EN

2) Long term solar heat storage through water tanks for the heating of housing.
M. Cucumo, V. Marinelli, G. Oliveti, A. sabato.
Dipartimento di Meccanica, Università della Calabria, Progress Report n. 1 (1981).

3) Long term solar heat storage through underground water tanks for the heating of housing.
M. Cucumo, V. Marinelli, G. Oliveti, A. Sabato.
Athens Contractor Meeting, November 1981, Solar Energy Applications to Dwellings, Series A, vol. 1, Reidel Publishing Co.

4) Long term solar heat storage through water tanks for the heating of housing.
M. Cucumo, V. Marinelli, G. Oliveti, A. Sabato.
Dipartimento di Meccanica, Università della Calabria, Progress Report n. 2 (1981).

5) Sulle perdite di calore di recipienti interrati contenenti liquidi caldi.
M. Cucumo, V. Marinelli, G. Oliveti, A. Sabato.
XXXVI Congresso Nazionale ATI, Viareggio, ottobre 1981.

6) Sulle perdite di calore di recipienti sferici interrati ed esposti all'aria ambiente, contenenti liquidi caldi.
M. Cucumo, V. Marinelli, G. Oliveti, A. Sabato
XXXVII Congresso Nazionale ATI, Padova, ottobre 1982.

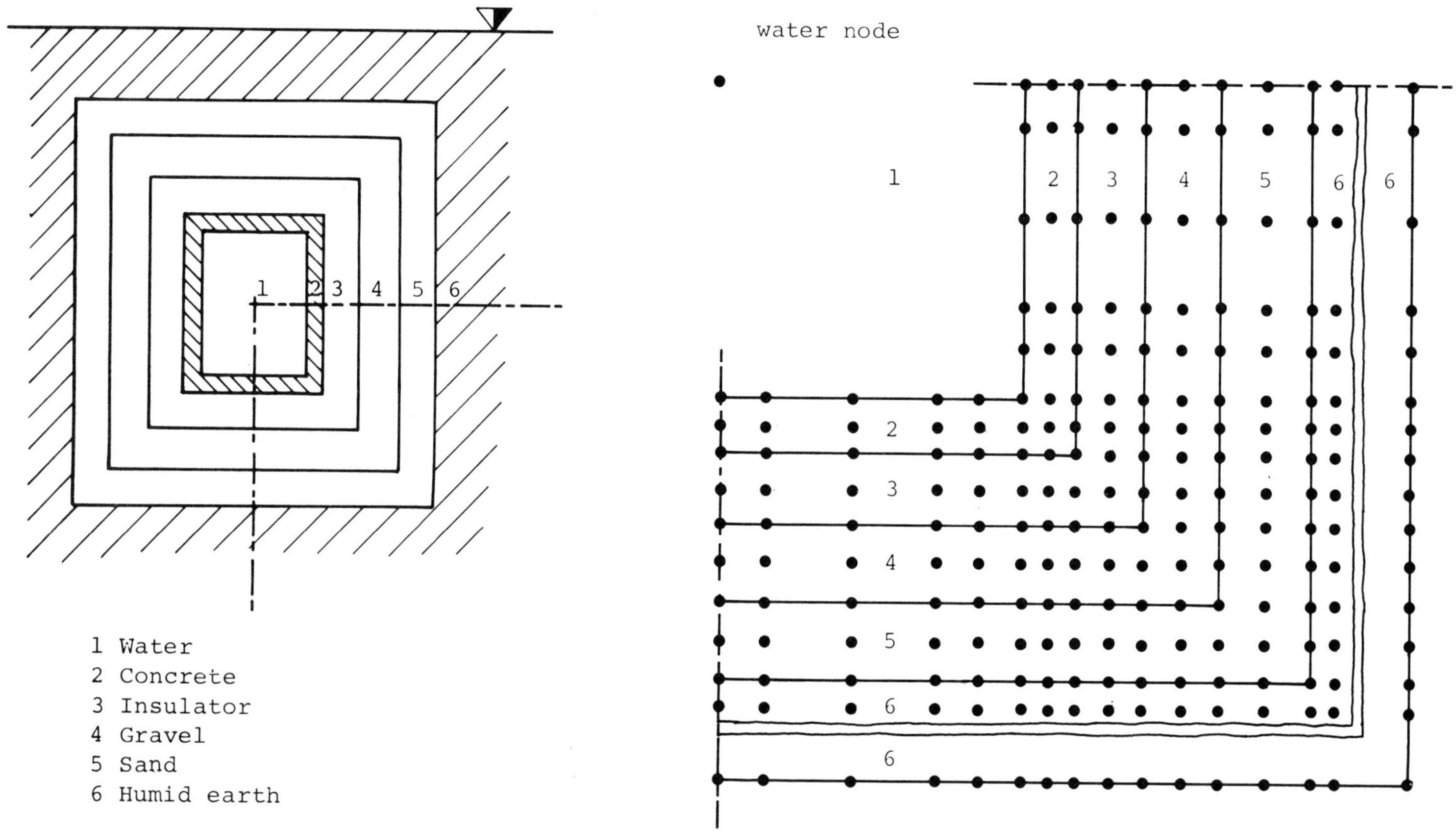

Fig 1- Typical nodalization of cylindrical underground tanks in PACI code.

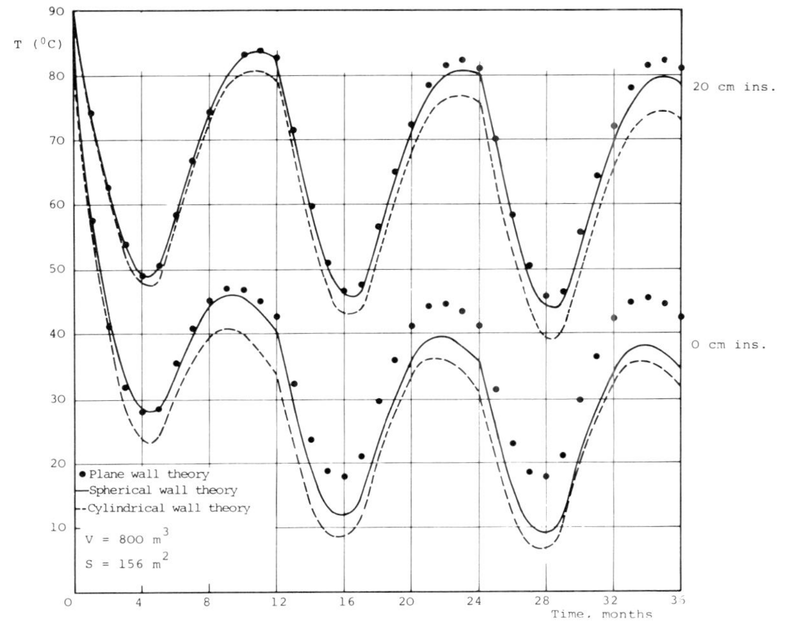

Fig. 2 - Temperature profiles of underground water tanks calculated by plane, spherical and cylindrical wall theory

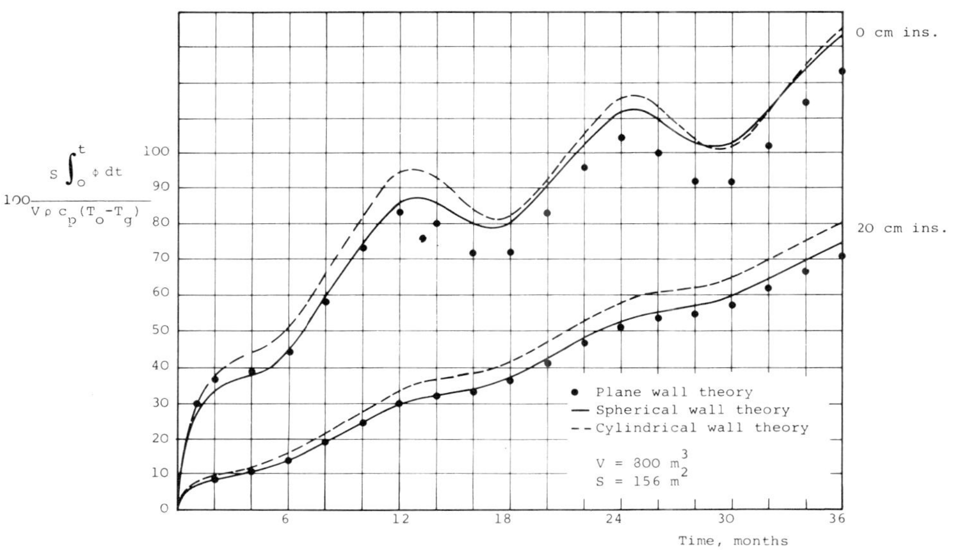

Fig. 3 - Comparison between plane wall theory, spherical wall theory and cylindrical wall theory : % heat losses to the earth versus time

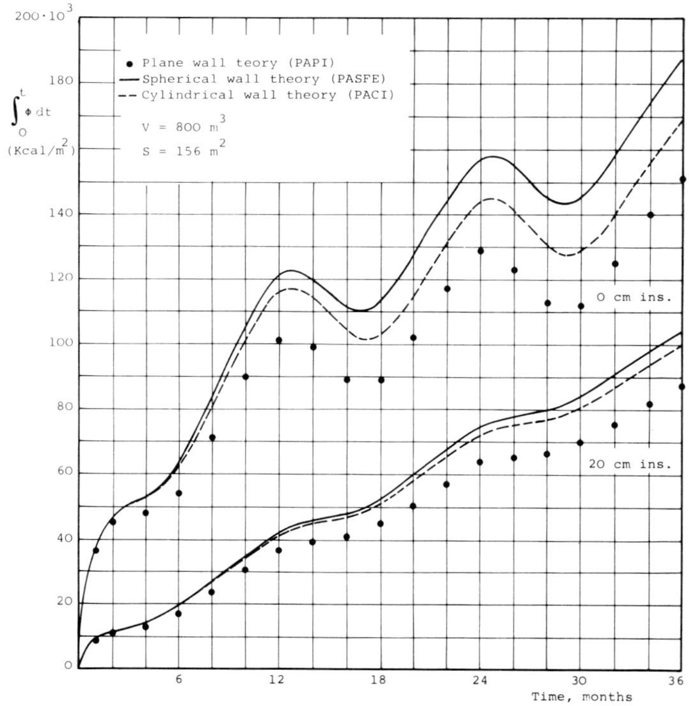

Fig. 4 - Comparison between plane wall theory, spherical wall theory and cylindrical theory : heat loss per unit of area versus time

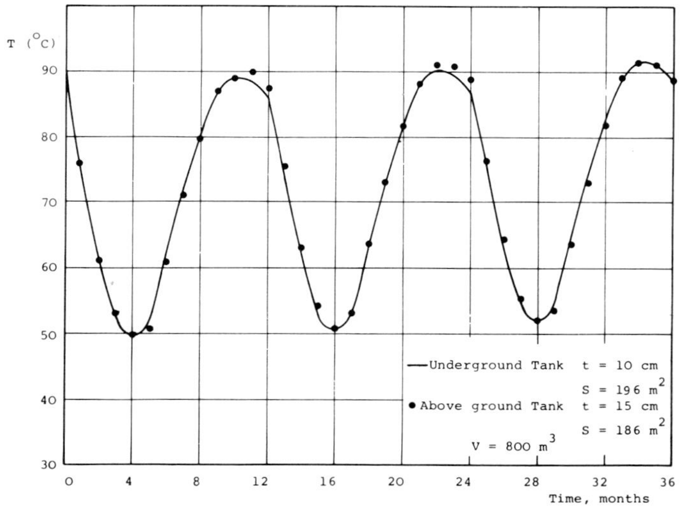

Fig. 5 - Temperature profiles for underground and above ground cylindrical tanks

THE HELIOGEOTHERMAL DOUBLET : A REAL SCALE TEST FACILITY

Author : P. IRIS

Contract number : ESA/S/119/F

Duration : Dec. 81 to Dec. 83

Total budget : 1 879 028 FF   CEC Contribution : 620 000 FF

Head of project : P. Iris

Organization : A.R.M.I.N.E.S.

Adress : A.R.M.I.N.E.S.
60 Boulevard Saint-Michel
75272 PARIS CEDEX 06
France

Summary

Aquifers generally remain at a constant temperature and represent a very good cold source for heat pumps. A system is proposed in order to make possible a large thermal use of aquifers, specially in urban areas : the "heliogeothermal doublet" based on the seasonal thermal recharge of the energy extracted during winter, by solar energy collected during summer. A feasability study of this system has been done in the case of 224 collective appartments which are now under construction near Paris, over a regional aquifer. The design of the system is described in the Proceedings of the EC Contractor's Meeting held in Athens (Greece), 11-13 Nov. 81. The present work concerns the scientific control of the realisation, the design of an acquisition data system and the interpretation of the global system with numerical models for the first year of heating which will begin in October 1982.

1. Introduction

The project deals with the thermal utilisation of "cold" aquifers at low depth (from about 10 to 200 m) by heat pumps. The system of the "heliogeothermal doublet" is proposed to make possible a large development of the thermal use of aquifers even in urban areas. During winter, water is pumped from a well at 13°C and injected back in a second well after being cooled at 4°C by the extraction of heat with the heat pump. During summer, the system is reversed, the cold water at 4°C is pumped back from the second well, heated with climatic energy collected at low temperature (solar radiation and heat exchange by natural convection) and injected at the first well at its initial temperature (13°C). It permits to maintain a constant temperature at the production well each winter and to avoid any thermal disturbance in the aquifer with time. This system is now under construction in the case of 224 collective appartments at Aulnay-sous-Bois near Paris. The aquifer is a sandy aquifer which lies under the northern part of the Parisian Basin at 80 m depth and represents a very attractive target for collective buildings in the region.

2. Description of the system (for memory)

The design of the complete system is described in the proceedings of the Athens EC Contractor's Meeting (11-13 Nov. 1981).

The major components are the following :

2 heat pumps for space heating (thermal power 500 kW for a nominal peak demand of the building of 891 kW).
1 heat pump for domestic hot water (thermal power 160 kW).
2 wells at 85 m depth screened between 65 and 85 m (diameter 220 mm).
1275 m^2 of aerosolar collectors in batteries, characterised by a high coefficient of heat exchange (K = 45 W/m^3.°C).
1 heat distribution system consisting in a single network of heating floors and radiators (maximum temperature 45°C).
5 accumulation tanks (5 m^3 each) for domestic hot water mainly heated during the night.

The regulation is the following :

during the heating period the heat exchanger on the aquifer is in serie with the aerosolar collectors to give the heat at the cold source of the heat pumps. When the exterior air temperature is to low the heat exchanger works alone (tex $<$ 3°C) when it is high enough (tex $>$ 10°C) the collectors work alone, the rest of the time they work together. It permits to limit the extraction of heat from the aquifer and to maintain the heating capacity and the COP of the heat pumps at a high level during the whole heating period.

During summer, the collectors work during the night to give energy to the cold source of the heat pump for domestic hot water, during the day they are used to recharge the thermal capacity of the aquifer.

The global thermal balance of the system is the following (calculated by models with simplified methods) :

Space heating needs : 1384 MWh.
Hot water needs : 457 MWh (winter) + 328 MWh (summer) = 785 MWh.
Heat extracted from the aquifer : 705 MWh.
Heat extracted form the collectors : winter : 660 MWh.
summer : 931 MWh.

Total needs : 2169 MWh.
Heat pump consumption : 548 MWh.
Auxiliary consumption (circulation pumps, geothermal pumps, etc...) : 72 MWh.
Oil consumption : 25 MWh.
COP of the heat pumps during the whole season : 3.9.
COP of the heat pumps with the auxiliaries : 3.4.
Oil saving : 315 tons of oil.
Payback of the investment compared to classical oil heating system : 10 years.
Overcost of the system for 224 apartments (compared to oil heating system) : 3 616 000 FF.

3. General aspect of the work under contract

The objective of the work is to interpret the behaviour of the global system, working at a real scale. There are different parts in the work.

1) The first part consists in the scientific monitoring of the system during its construction (Dec. 81-Nov. 82) with different tests on it :

- pumping and injection test on the production wells,
- chemical tests on the water of the aquifer,
- heat pumps test (control of the performances, test of the regulations),
- Aerosolar collectors test (control of the parameters of the collectors),
- design and installation of the data acquisition and data treatment system.

2) Design of a numerical model of simulation of the heating system taking into account at a one hour time step, the underground storage, the collectors, the heat pumps and the heat exchangers. This model which will be able to simulate different systems using heat pumps and storages, will be calibrated on the experimental measurements collected during the heating and recharging periods. It will permit to interpret the bahaviour of the system and to simulate it under other strategies (temperature level, regulation, etc...) in order to optimize it for the future. By comparison, it will also permit to test the validity of simplified methods of calculation of performances of heat pumps systems. A 3 dimensional finite difference numerical model of simulation of the thermal behaviour of the underground medium will be adapted to the particular case of the heliogeothermal doublet.

3) During the first year of heating (1982-1983) data will be collected and interpreted as follows :

- continuous data acquisition and treatment
- control of the behaviour of the heating system during the heating period (winter), and the recharge period (summer),
- calibration of the models of simulation,
- energetical, technical and economical balances of the system,
- simulation under optimized configuration - conclusions on the possibility to developpe the system at an industrial scale.

4. Work already done

4.1 Control of the realisation

At the present time the construction of the buildings is terminated, excepted the indoors installations. The heat pumps and the aerosolar collectors are to be connected.

The two wells of production are drilled and different tests have been performed :

- chemical tests have confirmed that no major problems should occur if the water is not oxygenized, which makes necessary to chose the geothermal network.
- Pumping tests have been performed on each well.
- The specific flowrate is 14 $m^3/h/m$ for the production well and 10 $m^3/h/m$ for the injection well of winter (the specific flowrate is the flowrate for which the level in the well drops from 1 m). The static level of the water in the wells is about 15 m depth, the screens are located between 68 and 85 m depth.
- Injection tests have also been performed on each well during four hours : the stabilisation of the dynamic level has been obtained no clogging appeared, the performances of the wells being exactly symetrical in injection and pumpage on the hydraulical point of view.

Those different tests have been performed at flow rates of 80 m^3/h and 100 m^3/h. The maximum flow rate will be about 50 m^3/h according to the thermal capacity of the heat pumps. Those first results are very good, the wells have the expected characteristics. Now we are defining the way to maintain their capacity during the exploitation of the system.

- The heat pumps and the aerosolar collectors will be tested as soon as they will be connected to the network.

4.2 Design of the data acquisition system

The data acquisition system will permit to know during the whole period at a one hour time step the energetical balance of every component of the system and the temperature at each node. The acquisition system is also adapted to make "zooms" at smaller time steps on particular parts of the system.

All the data will be collected at a time step of one minute (or less if necessary) and averaged or integrated on one hour before being registered on a tape recorder. The autonomy of the system is of one week, which permits to follow precisely the installation without treating too many data. About 80 parameters will be registered in such a way :

- climatic data (wind velocity, solar insulation, outside temperature, wet and dry, precipitations),
- temperatures (inside temperature, temperature in the aquifer, heat pumps, heat exchangers, network),
- flow rates (hot lock, cold lock, geothermal flow rate),
- energy balances (by instantaneous products of $\Delta\theta$ and flow rates, and integration),
- auxiliary energy, electricity consumption (with transductors of power and integration),
- The data will be treated each week on micro-computer and stored for mathematical treatments (calibration of numerical models).

The system will be ready in October 1982.

4.3 Mathematical model of simulation of the heating system

The numerical model "systherm" will permit to simulate precisely complex systems of heating, with solar collectors heat pumps, and heat storages. It has a modular structure in order to be able to simulate various configurations : first the user describs the system of heating, and according to this description the program couples the equations representing the different elements of the system. A particular modul which is interactive with the others simulates the regulation of the system. The time step can be chosen by the user.

Two prototyps of the model have already been performed in Basic and in Fortran with the following elements :

- solar collectors,
- heat exchanger (water-water),
- heat pump (water-water),
- distribution network,
- envelopp,
- storage.

The model will be calibrated on the measurements.

4.4 Work to be done

The work to be done is explained in 3. The next months will permit to install and to test the data acquisition and treatment system.

The collectors and the heat pumps will also tested (in October) before the beginning of the heating period which will start at the end of October.

First results will be communicate in December 1982.

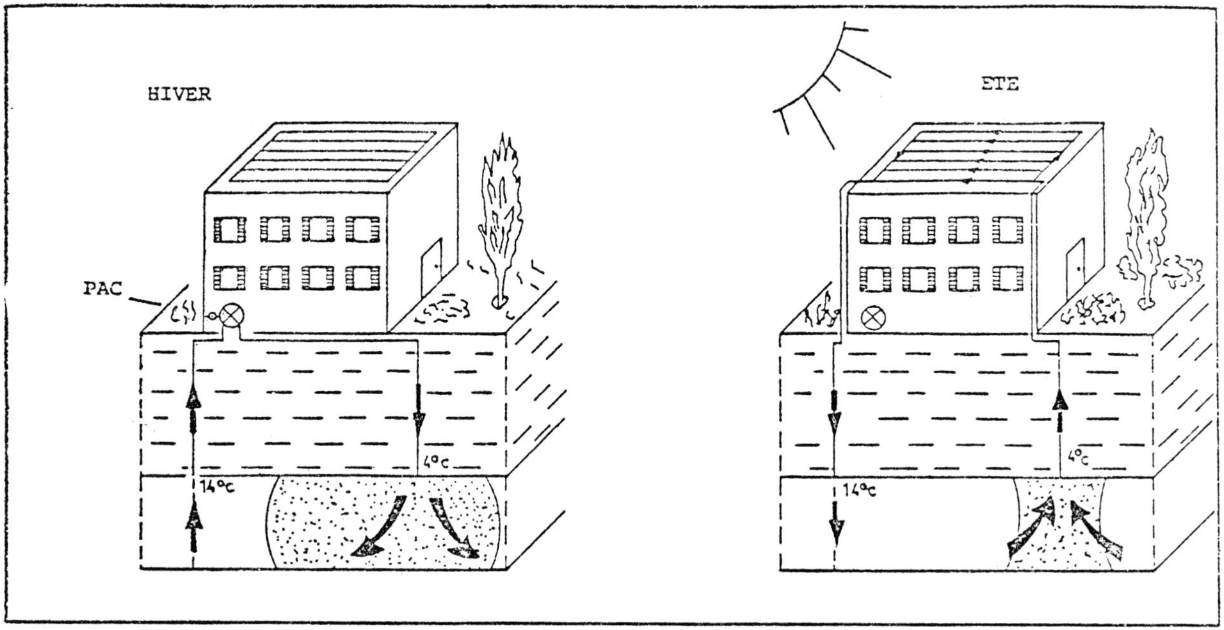

Fig. 1 : Principle of the heliogeothermal doublet

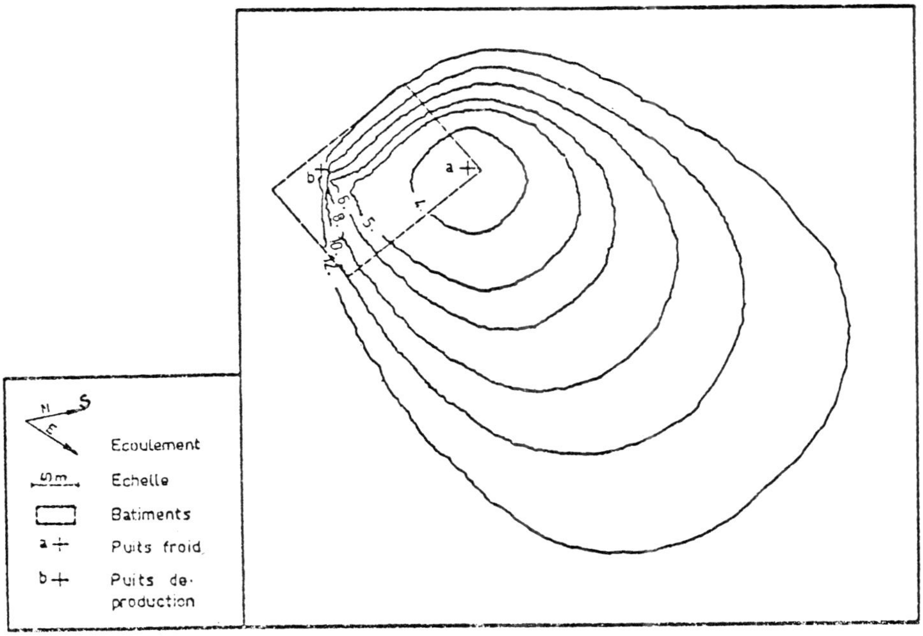

Fig. 2 : Isotherms (°C) in the aquifer after 20 years in the case of a doublet without recharge (initial temp. 13.5°C, result of a numerical simulation)

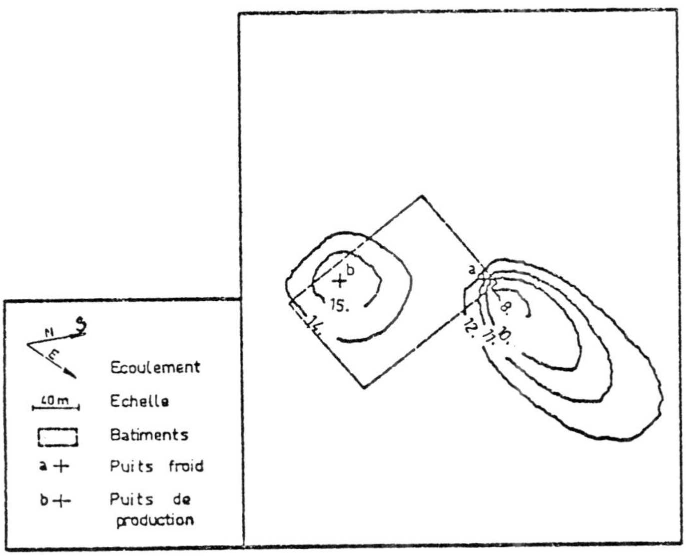

Fig. 3 : Isotherms (°C) in the aquifer after 20 years in the case of a doublet with recharge (heliogeothermal doublet)

Fig. 4 : View of a well during drilling

FIELD TEST TO INVESTIGATE THE PERFORMANCE OF AN UNDEEP PROTOTYPE SEASONAL HEAT STORAGE SYSTEM WITH A HEAT CAPACITY FOR 100 SOLAR HOUSES USING THE SOIL AS THE STORAGE MEDIUM

Authors : A.J.Th.M. Wijsman, J.W. de Feijter (DSML)

Contract number : ESA/089/NL

Duration : 36 months, 1 July 1980 - 30 June 1983

Total budget : Dfl. 2.665.270,-- CEC contribution: Dfl. 550.000,--

Head of project : Ir. C. den Ouden

Contractor : Technisch Physische Dienst TNO-TH (Institute of Applied Physics)

Address : P.O. Box 155, 2600 AD DELFT, The Netherlands

Subcontractor : Delft Soil Mechanics Laboratory

Address : P.O. Box 69, 2600 AB DELFT, The Netherlands

Summary

The project contains the detailed engineering design, the execution and the investigation of the performance of a system for the seasonal storage of heat in a water saturated sandy soil with a heat capacity for approximately 100 houses.
The aims of the project are:
- gathering practical experience with the actual realization of a large scale seasonal heat storage system;
- monitoring the overall behaviour of the soil storage system;
- verifying the applicability of small tests and computation methods.

Recently developed techniques in geotechnical engineering will be used to install the soil storage heat exchanger and the circumferential impermeable barrier. The heat storage will be charged and discharged according to the calculated thermal behaviour of a group of 100 solar houses with seasonal heat storage in the soil.

This project is a continuation of the research project 'the use of soil as a storage medium for seasonal storage of solar energy' under contract number 516-78-1 ESN.

The result of the first design work was a too expensive heat storage reservoir design. This caused a delay in the start of the construction of the reservoir. We reported about that during the Athens meeting. In the meantime extra studies have been carried out to reduce the cost of the reservoir.

In this paper the results of this studies and the new design of the reservoir are given.

1.1 Introduction

Within the first 4-years programme on solar energy a study was carried out on 'the use of soil as a storage medium for seasonal storage of solar energy' by the Delft Soil Mechanic Laboratory (DSML) as maincontractor and as subcontractors by the Institute of Applied Physics (IAP) and by Philips. This study delivered the theoretical ins and outs of a total solar heating system with long term storage in soil for a group of solar houses both for the technical and economical behaviour. To complete the picture of such a system with actual data on heat balances, costs and geotechnical realization it is necessary to construct and investigate a heat storage in soil on real scale.

In the test set-up (see fig. 1) the heat storage reservoir in the soil is on real scale and has a heat storage capacity for 100 solar houses. The charging and discharging of the heat storage will be according to the thermal behaviour of a group of 100 solar houses with seasonal heat storage. This thermal behaviour is calculated by a computer simulation model. The heat input/output of the seasonal heat storage reservoir is realized by a controlled boiler/cooler.

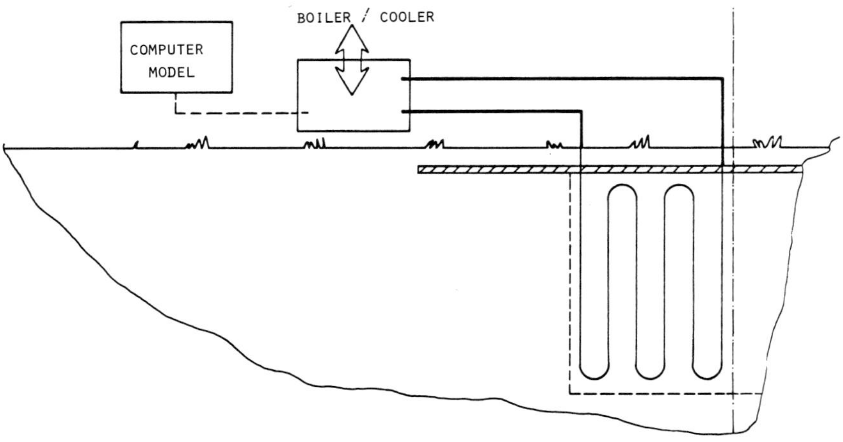

FIG. 1: TEST SET-UP OF THE PROJECT

The project contains the following phases:

a. Detailed engineering design of resp. the heat storage system and the simulation of the heat input/heat output.
b. Preparation of the execution of the construction.
c. Supervision of the soil storage installation.
d. The testing (check) of the system.
e. Execution of the fieldtest and the evaluation of the testresults.
f. Reporting and transfer of know-how.

The result of the first design work was a too expensive heat storage reservoir design. To bring these costs down extra studies have been done. The new storage reservoir design will be discussed.

1.2 Description of the test set-up

A short description will be given of the principle of the heat storage reservoir resp. the principle of the heat input/output simulation.

a. The heat storage reservoir in the soil (figure 2)

In principle the heat storage reservoir in the soil consists of a layer of soil (up to a depth of 20-30 m) with a heat exchanger. The reservoir is not bounded by walls. Only at the top the reservoir is furnished by an insulation foam layer, at the other sides the soil itself acts as insulation. In water saturated porous soil a vertical screen at the edge of the reservoir can be necessary to reduce the heat losses by convection of the groundwater and by groundwater movement.

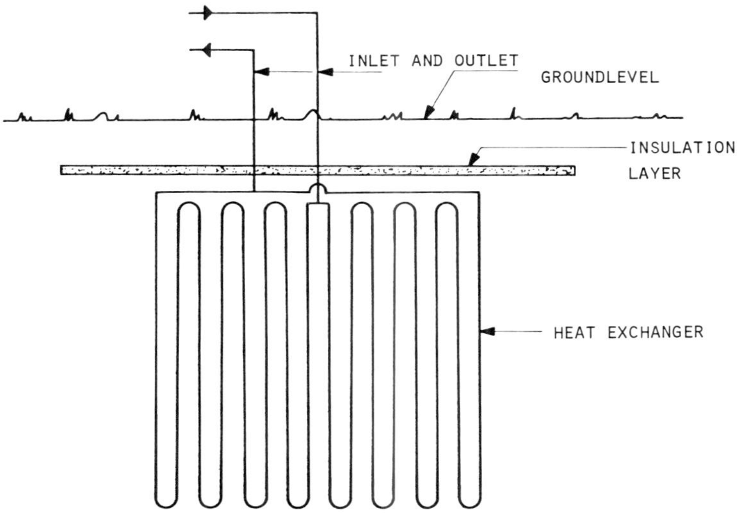

FIG. 2: PRINCIPLE OF THE HEAT STORAGE RESERVOIR IN THE SOIL

The heat exchanger will consist of vertical (flexible) tubes which are, as it were, stitched into the soil from the surface by a specially shaped lance. Because of temperature stratification in the seasonal heat storage reservoir in the charging mode the hot transfer fluid enters the heat exchanger in the centre of the reservoir and leaves the heat exchanger at the edge. In the discharging mode the fluid direction is opposite.

b. The simulation of the heat input/output of the reservoir

For a group of solar houses with seasonal heat storage in the soil a computersimulation model is available to predict under given weather conditions the several heat flows in the total system. The so calculated heat flows to and from the seasonal heat storage reservoir are realized by a controlled boiler respectively a cooler.

The above mentioned simulation differs from our first idea: amplification of the heat input/output of a pilot plant with a rate of 100.

1.3 Design results

a. The heat storage reservoir

Extra studies have been done to bring the cost of the reservoir down:
- the heat exchanger design has been changed
- the vertical screen can be omitted
- the thickness of the top insulation layer has been decreased.

The new reservoir design will be discussed now.

The project will be executed in a new district of the town Groningen in the northern part of The Netherlands. An extensive investigation of this site has been made to determine the soil properties. The results are given in figure 3.

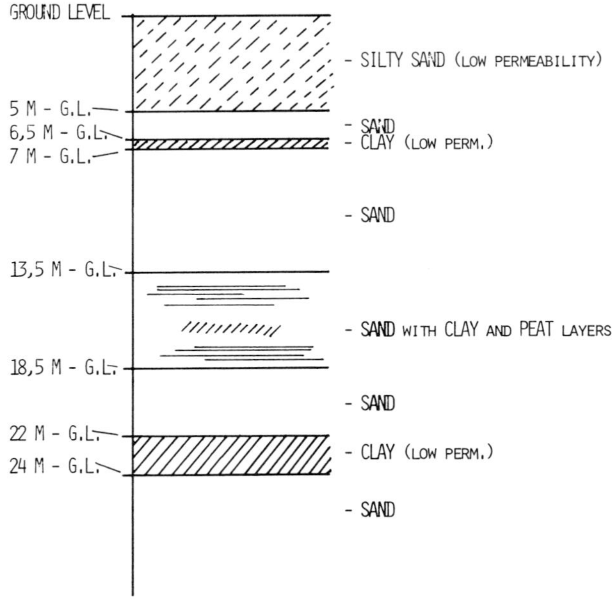

FIG. 3: MAIN SOIL COMPOSITION IN GRONINGEN

The soil can roughly be described as:

Water saturated sand with thick layers of clay and some thin layers of peat. The mean heat capacity is 2.7 MJ/m^3K, the thermal conductivity lies between 1.5 and 2.2 W/mK and the permeability between $5 * 10^{-12}$ $2 * 10^{-11} m^2$ (with impermeable layers in between).

For a total heat capacity of 60.000 MJ/K (for 100 houses) the reservoir should have a volume of about 23.000 m^3 (diameter 38 m, depth 20 m).

The vertical screen is omitted. The clay layers (low permeability) in the subsoil of Groningen (see fig. 3) obstruct the vertical flow of groundwater caused by free convection. So these heat losses will be restricted. A study of the local natural groundwater flow (horizontal) showed a maximum head of 1 m/km. The influence of a clay screen with low permeability around the storage has been determined with the 3-D computer simulation programme of the DSML for heattransport in saturated soils. The calculations have shown that the efficiency of the heat storage reservoir improves with 6% by using the formerly designed screen. For the unscreened storage the heat losses caused by free convection and natural groundwater flow are of the same order of magnitude.

The decision has been made to omit the screen in the final design because the investment for the screen is to high in comparison with the expected extra heat output.

The heat exchanger consists of vertically inserted strings (to a depth of 20 m); in one string the fluid goes downwards, in the other leg upwards. The strings are of flexible polybutene tube (20 x 16 mm). The density of the strings: 1 per 3.2 m^2; the string width can go down to 0.50 m without a significant decrease in heat exchange rate (see fig. 4). This string width requires a less heavy vibrating apparatus for the insertion of the strings.

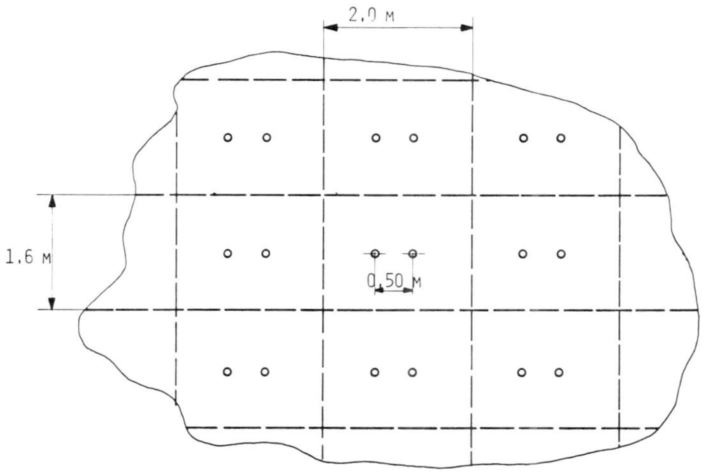

FIGURE 4. TOP VIEW ON THE STRINGS

The number of strings is 360, which is half of the number in the first design. This size of heat exchanger became possible by the use of a buffertank of 100 m^3 between the district network and the heat exchanger in the soil (see fig. 5 and 6). The fluid flow in the soil heat exchanger is half of the flow in the district network; the buffertank acts as a short term heat storage. The buffertank is burried in the soil in the middle of

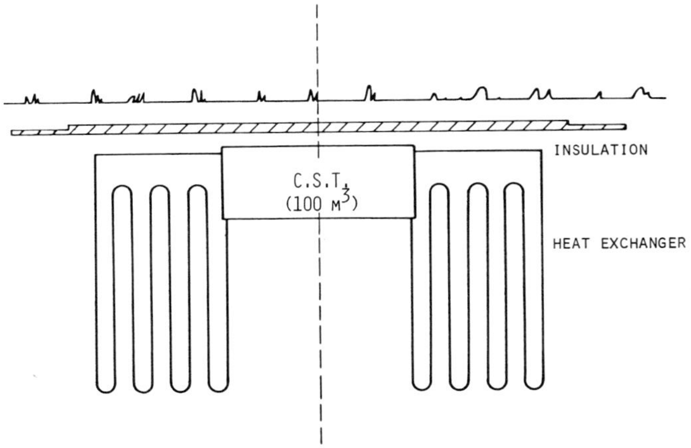

FIG. 5: SCHEME OF THE SEASONAL HEAT STORAGE RESERVOIR WITH A CENTRAL SHORT TERM STORAGE RESERVOIR (C.S.T.)

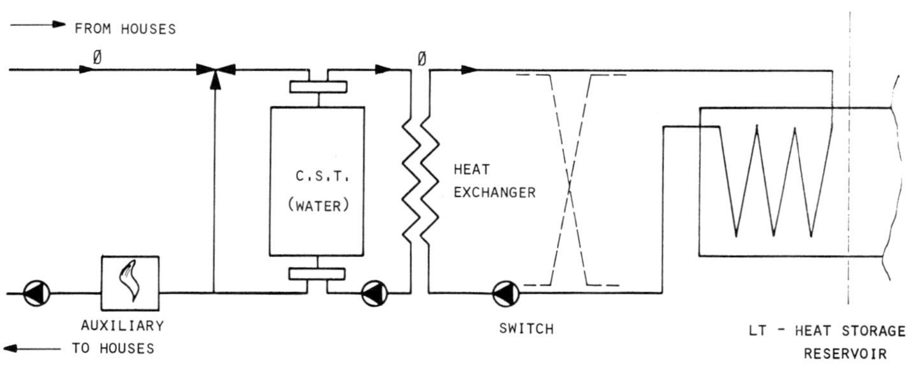

FIG. 6: SCHEME OF SEASONAL HEAT STORAGE SYSTEM

the reservoir. The volume of the buffertank followed from the maximum temperature allowable in the system at extreme conditions.

Because of temperature stratification in the seasonal heat storage reservoir in the charging mode the hot transfer fluid enters the heat exchanger in the centre of the reservoir and leaves the heat exchanger at the edge. In the discharging mode the fluid direction is opposite. From a study at the Delft University it appeared, that this improves the performance of the seasonal heat storage with more than 10%. Further improvement is expected from a special strategy, whereby the heat exchanger is divided into three concentric rings.

The interconnection network at the top has been changed; the interconnection of the strings is more concentrated at the centre of the reservoir near to the buffertank. De-airing of the heat exchanger strings can happen now separately.

The top insulation layer has been changed. In the first design 0.30 m foam glass was planned. Now in the top insulation layer two different insulation materials are used (see fig. 7). The lower layer (0.40 m) of expanded clay grains contains the interconnection network between the strings. On top of this layer there is a plastic film and a sand layer. The upper layer is formed by 0.10 m foam glass. An 1 m thick layer of fertile soil covers the top insulation layer. The effective insulation value of the top layer is lower than in the first design.

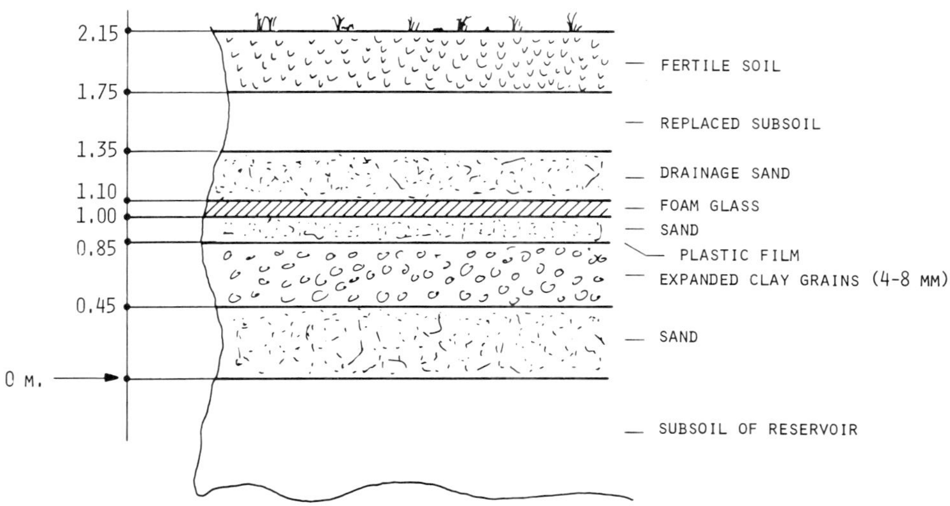

FIG. 7: THE TOP INSULATION LAYER

b. The simulation of the heat input/output

For a group of 100 solar houses, each with a collector area of 25 m^2 and a heat demand at design conditions of 6 kW, the heat input to and the heat output from the storage reservoir is simulated. For this simulation a computer simulation model for a group of solar houses with seasonal heat storage in the soil is used. The with this model calculated heat input/output of the seasonal heat storage reservoir is realized by a controlled boiler/cooler. The maximum boiler capacity will be 850 kW, the maximum cooler capacity 600 kW.

The above mentioned simulation procedure will be proceded by a four months period consisting of 10 weeks constant heat input, 2 weeks without heat input/output and 4 weeks of heat output. By these step responses several characteristics of the seasonal heat storage system can be derived.

1.4 Instrumentation of the heat storage reservoir

In the soil temperatures are measured in three directions from the centre of the reservoir. One direction is full instrumentated. The low groundwater flow rate permits a less dense instrumentation in the other two directions. In figure 8 and 9 the total instrumentation is shown. The instrumentation in the soil:

18 vertical and 2 horizontal temperature lances
4 vertical and 4 horizontal heat flux meters
12 water pressure gauges
3 tubes for density measurements
6 observation wells
9 beacons for vertical deformations
4 vertical beacons for horizontal deformations

The output signals of the temperature, the heat flux and the pressure sensors are registered by the central data acquisition system in the boiler house.

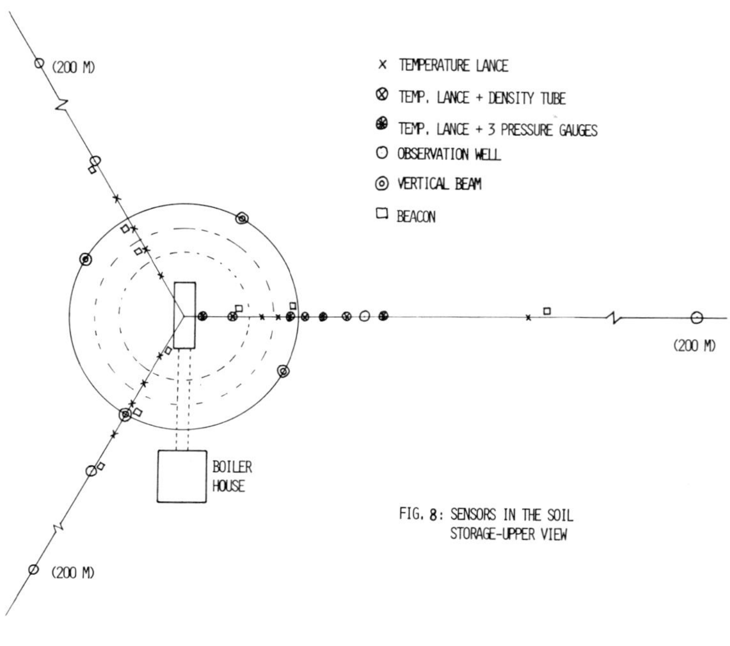

FIG. 8: SENSORS IN THE SOIL STORAGE-UPPER VIEW

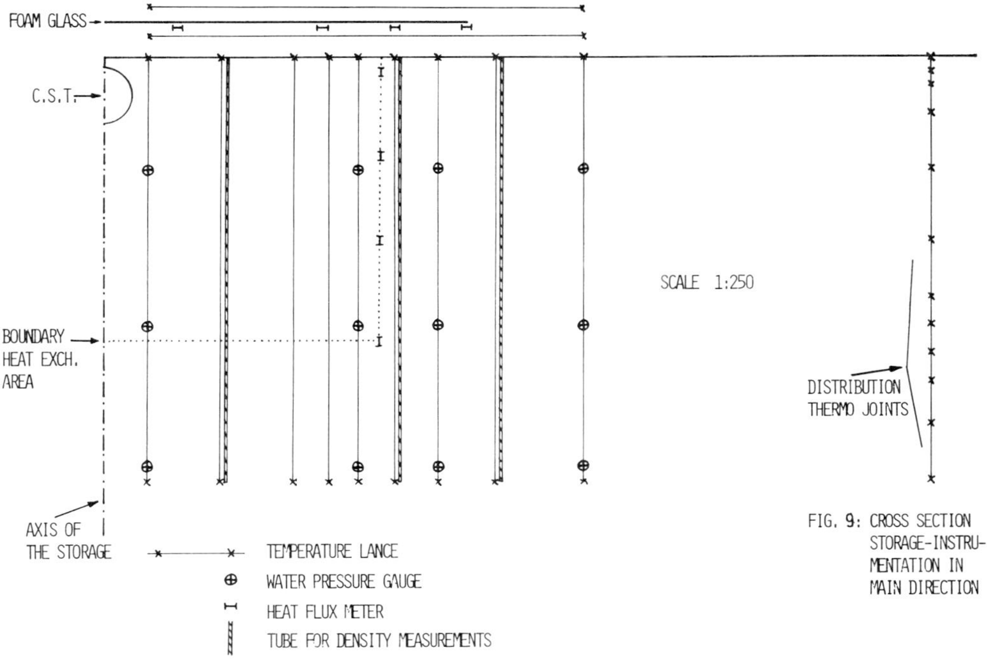

FIG. 9: CROSS SECTION STORAGE-INSTRUMENTATION IN MAIN DIRECTION

1.5 State of the art and future work

The extra studies resulted in a new design of the seasonal heat storage reservoir. The costs are estimated at 1.2 million Dutch guilders (instrumentation is not included); the first design was 2.3 million. The cost includes material and cost for Labour. The cost of this new design implies a cost of about Dfl. 50 per m^3 of soil (~ 20 ECU/m^3 soil).

In last April the conformation for the start of the construction of the seasonal heat storage reservoir was given. According to the planning the construction of the reservoir and the whole test-set-up will be finished at December 1982. In January 1983 the first experiment of four months will be started. At the end of 1983 a decision will be made about the connection of the group of solar houses to the reservoir.

2. Conclusions

A second, less expensive detailed design of the heat storage reservoir and the simulation of the heat input/output has been made. The preparation of the execution of the construction is almost finished. It is expected, that in January 1983 the experiments can start.

SEASONAL HEAT STORAGE IN UNDERGROUND WARM WATER STORES (CONSTRUCTION AND TESTING OF A 500 M^3 STORE)

Authors : Kurt K. Hansen, Preben N. Hansen

Contract number : ESA-S-162-DK (G)

Duration : 20 months 1 November 1981-30 June 1983

Total budget : Dkr. 1.000.000 CEC contribution : 50%

Head of project : Preben N. Hansen

Contractor : Thermal Insulation Laboratory

Address : Technical University of Denmark
Building 118
DK - 2800 Lyngby
Denmark

Summary

The project contains the projection, the construction and the testing of the efficiency of a 500 m^3 warm water store. The aims of the project are:

- gathering practical experience with the actual realization of the pilot plant;
- monitoring the overall behaviour of the storage system;
- verification and modification of the digital computer program.

In this paper the results of the first phase, the examination of the soil conditions on the site of the store and the projection of the storage system, are given.

1. Introduction

On the basis of the preliminary studies done by Preben N. Hansen et al. [1], [2], [3], [4] concerning the heat losses from the store and on the basis of a work [5] done by the Danish firm Dipco Engineering, Virum, concerning possible principles of construction, projection of a 500 m^3 store is carried out. The site of the store is on our University campus, see figure 1.

The project contains the following phases:

a. Detailed examination of the soil conditions on the site of the store.
b. Using the results from a) detailed projection is performed.
c. Execution of the construction. The construction is financed by the Danish Ministry of Energy.
d. Experimentation. Tests are made to determine the storage efficiency.
e. Verification and modification of the digital computer program.
f. Reporting.

The results of phase a. and b. will be given in this paper.

2. Examination of the soil conditions

The soil conditions are determined by the Danish Geotechnical Institute. Two continuous borings are made to collect samples for identification of the soil types in order to determine the porosity of the soil in the field. In the laboratory the water content, the dry density, the shear strength, the grainsize distribution, the degree of saturation and the thermal properties are determined.

In one of the boreholes a standpipe is installed to check the case of a secondary groundwater pressure (the primary groundwater is situated more than 40 m below the surface of the ground). In the other borehole a tube with five temperature sensors is placed. A photograph of the boring is shown in figure 2.

2.1 Determination of the thermal properties

To determine the thermal conductivity of moist soil a diagram is given in the literature [6]. The diagram is shown in figure 3 and the entry values are the quartz-content, the dry density and the degree of saturation. Depending on "course-grained" or "fine-grained" soil in the example in the diagram the conductivity is found to be 1.15 W/m^{o}C or 0.90 W/m^{o}C.

It should be mentioned that some of the conductivities found by the diagram are checked in the laboratory using the instrument shown in figure 4.

To determine the heat capacity of moist soil a diagram is given in the literature [7]. The diagram is shown in figure 5 and the entry values are the porosity and the degree of saturation. From the knowledge of the thermal conductivity and the heat capacity, the thermal diffusivity α is found by the relation $\alpha = \lambda/C$.

The results are given in figure 6. The soil can be described roughly as a 4-5 m thick layer of clay. Below this layer of clay is water saturated sand (different degrees of saturation). The thermal conductivity lies between 0.8 W/m^{o}C and 2.2 W/m^{o}C, and the thermal diffusivity lies between 17.0 m^{2}/year and 31.4 m^{2}/year.

3. Projection of the storage

The store is digged into the ground ending up with a pyramidal geometry (see figure 7). The ground surface is made waterproof by use of a plastic liner. (No insulation materials are used on this surface). The top is heat insulated and protected against evaporation and against the weather by use of a plastic liner.

The heat input is generated by an oil burner simulating one or more different types of solar collectors.

4. Conclusion

A detailed examination of the soil conditions and a detailed projection of the storage have been made. The construction of the storage has started this month.

5. Literature references

[1] Hansen, P.N., Analytical description of the heat losses from underground thermal seasonal heat stores. Paper presented at the U.S. Department of Energy Conference "Seasonal Thermal Energy Storage", October 19-21, 1981, Seattle, Washington, U.S.A.

[2] Hansen, P.N., Seasonal Heat Storage in Hot Water Stores. Paper presented at the International Conference on Numerical Methods in Thermal Problems, July 2-6 1979, Swansea.

[3] Hansen, P.N., Varmetab fra store varmelagre, (Heat losses in big heat stores). 1979. Thermal Insulation Laboratory, Technical University of Denmark.

[4] Lawaetz, H. and Hansen, P.N., Solvarmesystem med sæsonlagring - et projektforslag, (A Solar Energy System using Seasonal Heat Storage - a Proposal). Report No. 78-39, 1978. Thermal Insulation Laboratory, Technical University of Denmark.

[5] Dipco Engineering, Preliminary study of the construction principles to be used building seasonal hot water stores in the ground (in Danish). December 1979. Thermal Insulation Laboratory, Technical University of Denmark.

[6] The Royal Norwegian Council for Scientific and Industrial Research and the Public Roads Administration's Committee on FROST ACTION IN SOILS, No. 1-17. Oslo 1970-1976.

[7] Balstrup, T., Varmepumpeanlæg - Varmeovergangsforhold i jord, (Heat pumps - heat flows in soil). Technological Institute, Copenhagen, 1977.

TECHNICAL UNIVERSITY OF DENMARK

Figure 1. The site of the seasonal heat storage.

Figure 2. A photograph of the boring.

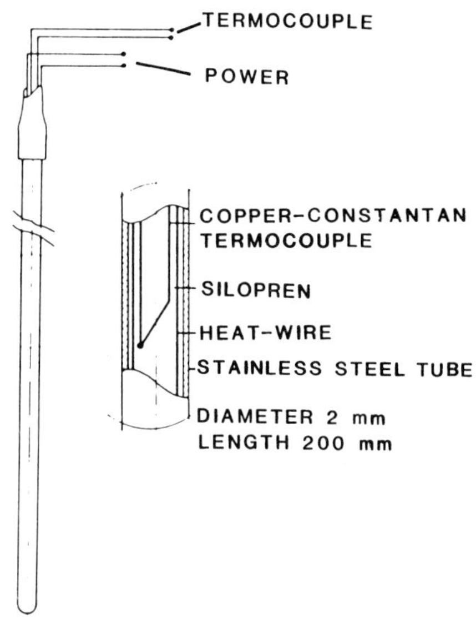

Figure 4. Sketch of a probe for measuring the coefficient of thermal conductivity.

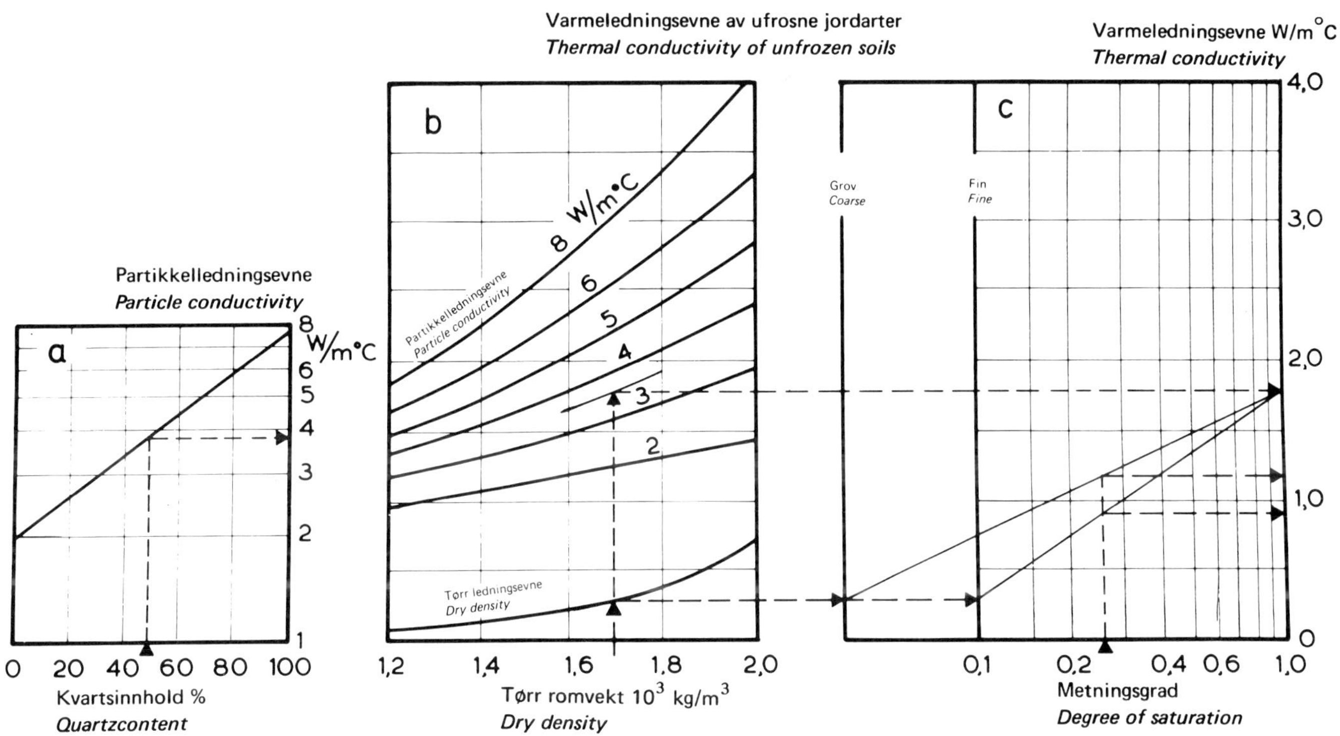

Figure 3. Thermal conductivity of unfrozen mineral soils. Reference [6].

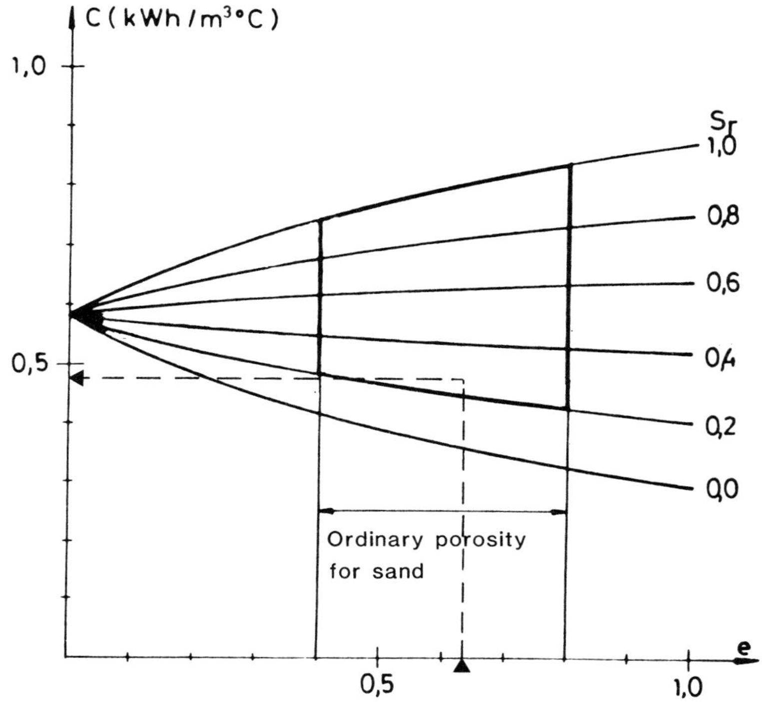

NOTE :

C is heat capacity

e porosity

S_r degree of saturation

Figure 5. Determination of the heat capacity. Reference [7].

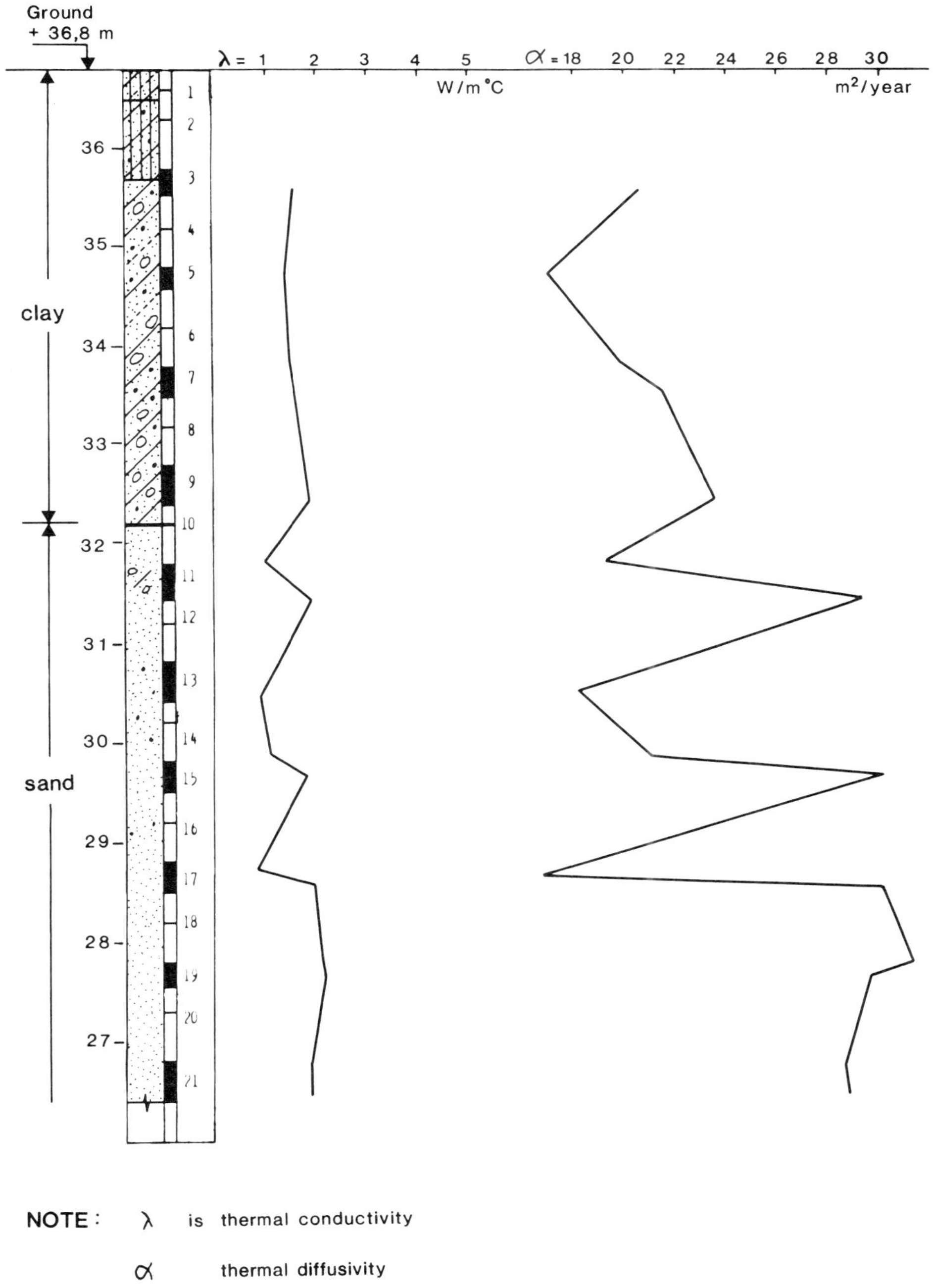

Figure 6. Schematic soil strata and the thermal properties.

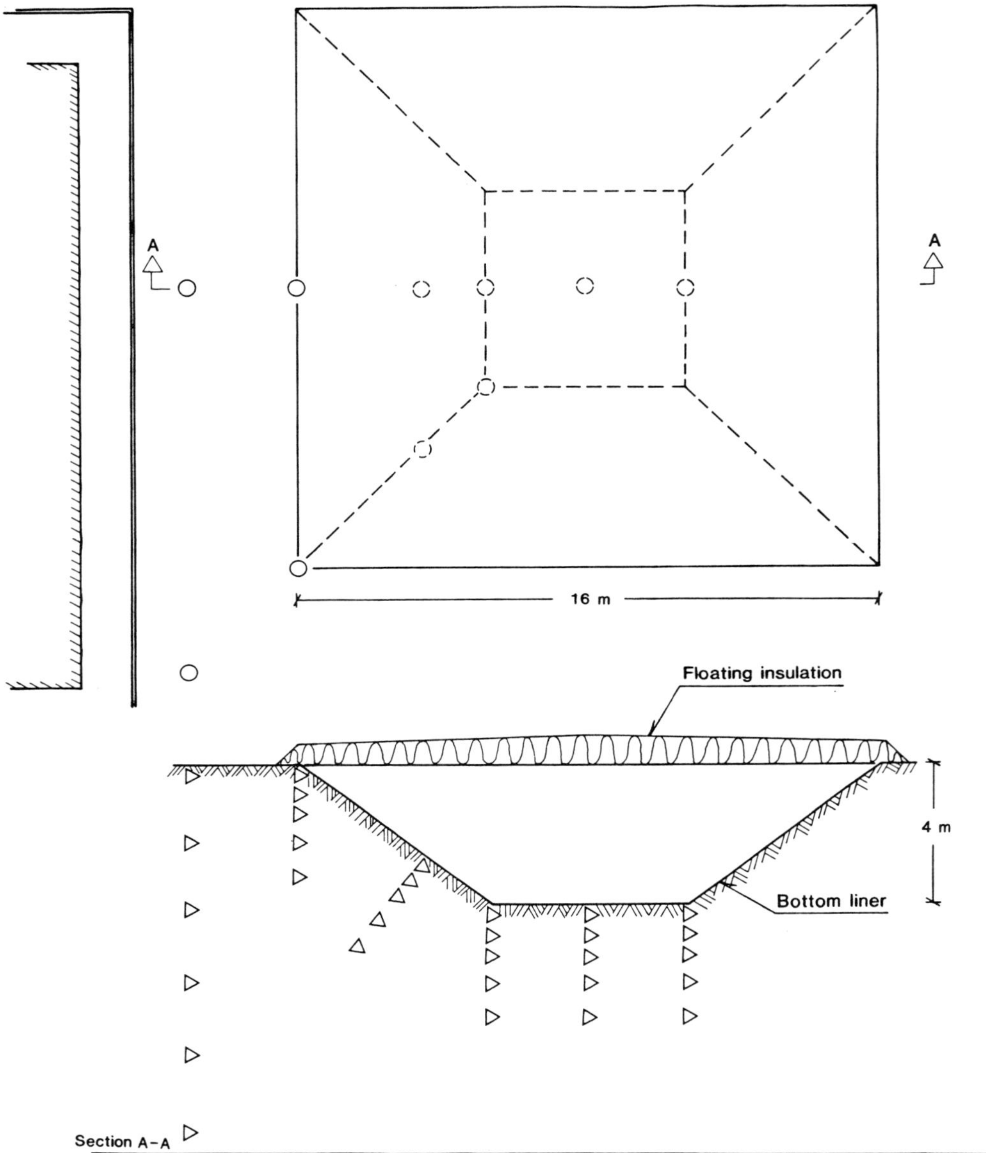

NOTE:

○ Boring with temperature sensors

▷ Temperature sensor

Figure 7. The 500 m³ heat storage

LIST OF PARTICIPANTS

ADNOT, J.	Centre d'Energétique Ecole des Mines de Paris 60, Bd St. Michel F - 75272 PARIS CEDEX 06 Tel. 1/329 21 05 Telex
ALPER, B.	ETSU AERE Harwell Building 156 UK - DIDCOT, Oxon OX11 ORA Tel. 0235/83 46 21 Telex 214250 atomha
ARANOVITCH, E.	Joint Research Centre Commission of the European Communities Ispra Establishment I - ISPRA (Varese) Tel. 322/78 97 68 Telex 38042/38058 euratom
BAKER, N.	Energy Conscious Design 44 Earlham St. UK - LONDON WC 2 Tel. 240 05 66 Telex 8952165
BANNARD, J.	NIHE National Institute for Higher Education EIR - LIMERICK Tel. 061/43644 Telex 6959 nihe ei
BAUMANN, R.	Gesamthochschule Kassel FB 12 Postfach 101380 D - 3500 KASSEL Tel. 0561/804 5310/12/13 Telex
BOOSE, C.A.	Metal Institute TNO P.O. Box 541 NL - 7300 AM APELDOORN Tel. 055/77 33 44 Telex 36395 tno ap

BOUGARD, J.	Faculté Polytechnique de Mons Service de Thermodynamique 31, Boulevard Dolez B - 7000 MONS Tel. 065/33 81 91 Telex
BOURDIN, F.	Saint Gobin Vitrage Centre de Développement Industriel 15, rue Rouget de Lisle F - 92400 COURBEVOIE Tel. 1/334 30 31 Telex
CECCHI-PAONE, F.	FIDIMI Cons. 68 V. Arte I - 00144 ROME Tel. 592 57 41 Telex
COHEN, R.	Cranfield Institute of Technology UK - BEDFORD Tel. 0234/750 111 Telex
COURTIER, A.	Corning-France 44, avenue de Valvins F - 77210 FONTAINEBLEAU-AVON Tel. 6/422 49 15 Telex 690 432
COWAN, I.J.	Institute for Industrial Research and Standards Ballymun Road EIR - DUBLIN 9 Tel. 01/37 01 01 Telex 25449 iirs ei
DEBLESSER, Y.	Faculté Polytechnique de Mons Service de Thermodynamique 31, Bd. Dolez B - 7000 MONS Tel. 065/33 81 91 Telex
DEN OUDEN, C.	Project Leader - Project A Institute of Applied Physics TNO P.O.Box 155 NL - 2600 AD DELFT Tel. 015/56 93 00 Telex 38091 tpd dt nl

DURAND, A. | Les Maisons Bruno Petit
21, rue des Capucins
F - 92190 MEUDON
Tel. 1/534 75 05
Telex

DUTRE', W.L.E. | Katholieke Universiteit Leuven
Lab. Warmteoverdracht en Reaktorkunde
300 a, Celestijnenlaan
B - 3030 HEVERLEE (Leuven)
Tel. 016/23 49 31
Telex

FITZGERALD, D. | Civil Engineering Dept.
The University
UK - LEEDS LS2 9JT
Tel. 0532/43 17 51 x 260
Telex

ERBAGGI, C. | CTIP Solar SpA
Ple Giulio Douhet 31
I - 00144 ROMA
Tel. 06/59 021
Telex 610078 ctip rm

FERRARO, R. | Energy Conscious Design
44 Earlham Street
UK - LONDON WC2H 9LA
Tel. 01/240 05 66
Telex 8952165 create g

FLISI, U. | Montepolimeri CSI
viale Lombardia 20
I - 20021 BOLLATE (MI)
Tel. 02/350 12 01
Telex 310679

FUHSE, W. | FES
Fuhse Energie System GmbH
Schulstrasse 37
D - 7997 IMMENSTAAD
Tel. 07545/6984
Telex

HANSEN, K. | Thermal Insulation Laboratory
Technical University of Denmark
Building 118 - Lundtoftevej 100
DK - 2800 LYNGBY
Tel. 02/88 35 11
Telex 37529 dthdia dk

HAUGLUSTAINE, J.-M.	Laboratoire de Physique du Bâtiment Univ. de Liège, Fac. Sciences Appl. 15, avenue des Tilleuls, Bât. D1 B - 4000 LIEGE Tel. 041/52 01 80 x 416 Telex 41746 enviro b
HAUSE, R.	Dornier-System GmbH An der Bundesstrasse 31 D - 7990 FRIEDRICHSHAFEN Tel. 7545/83703 Telex 07-34359
HAYDEN, J.	NIHE National Institute for Higher Education EIR - LIMERICK Tel. 061/43 644 Telex 6959 nihe ei
HOUGHTON-EVANS, W.	Stephen George and Partners 170 London Road UK - LEICESTER LE2 IND Tel. 0533/55 18 07 Telex
JOUANNA, P.	Laboratoire Génie Civil Université Montpellier 2 Place E. Bataillon F - 34060 MONTPELLIER Tel. 67/63 13 62 Telex 490 944 ustmont f
LEBENS, R.	Ralph Lebens Associates 4 Tottenham Mews UK - LONDON W1 Tel. 01/636 71 72 Telex 24170 maytel g ref. RL1
LEPOIVRE, J.-P.	AFME - Agence Française pour la Maîtrise de l'Energie Route des Lucioles - Sophia Antipolis F - 06560 VALBONNE Tel. 93 74 79 79 Telex comessa 461 357 f
LEWIS, J.O.	Energy Research Group - School of Architecture - University College Dublin Richview EIR - DUBLIN 14 Tel. 01/69 64 16 Telex 25278 ucd ei

LYALL, A.
Waterloo Grille Co Ltd
14, Parsons Road
UK - BENFLEET, Essex SS7 4PT
Tel. 03745/4121
Telex 99324 watloo g

MARIE, J.-P.
Plan Construction
Ministère Urbanisme et Logement
1, rue François 1er
F - 75008 PARIS
Tel. 225 99 19
Telex passy b 610 835 f

MARINELLI, V.
Dipartimento di Meccanica
Università della Calabria
Facoltà di Ingegneria
I - 87030 ARCAVACATA DI RENDE (CS)
Tel. 0984/83 80 73
Telex

MARSHALL, R.
Solar Energy Unit
University College Cardiff
Newport Road
UK - CARDIFF CF2 1TA
Tel. 0222/44 211 x 7058
Telex 498635

MECHEL, F.P.
Fraunhofer Institut für
Bauphysik, Stuttgart
Koenigstraessle 70
D - 7000 STUTTGART
Tel. 0711/76 50 08
Telex

MERGES, V.
M.B.B.
Dep. RT 346
P.O.Box 80 11 69
D - 8000 MUENCHEN 80
Tel. 089/6000 3366
Telex 5287-0 mbb d

MEUNIER, F.
CNRS, Laboratoire de Thermodynamique des
Fluides - Bât. 502
Campus d'Orsay
F - 91405 ORSAY
Tel. 6/941 82 50
Telex facors 692 166 f

MOON, J.E.
Solar Energy Unit
University College Cardiff
P.O. Box 78
UK - CARDIFF, CF1 1XL Wales
Tel. 0222/44 211
Telex 498635

NAJAT AL-SALEH,	Sussex University Maths and Physics Building Falmer UK - BRIGHTON Tel. 606 755 x 611 Telex
NICOLAY, D.	Commission of the European Communities Directorate General Information Market and Innovation P.O. Box 1907 - JMO B4/072 L - 2920 LUXEMBOURG Tel. 43011 x 2946 Telex 3423/3446 comeur lu
OLIVE, G.	Consultant Ingénieurs Conseil Gilles Olive 16, rue Nansouty F - 75014 PARIS Tel. 580 99 52 Telex
OLSEN, L.	Thermal Insulation Laboratory Technical University of Denmark Building 118 - Lundtoftevej 100 DK - 2800 LYNGBY Tel. 02/ 88 35 11 Telex 37529 dth dia dk
O'MALLEY, P.	NIHE (MAC Dept.) National Institute for Higher Education EIR - LIMERICK Tel. 061/43 644 Telex 6959 nihe ei
PALZ, W.	Commission of the European Communities Directorate General Science, Research and Development 200, rue de la Loi B - 1049 BRUSSELS Tel. 735 11 11 Telex 21877 comeu b
PHILLIPS, V..	Maths and Physics Sussex University Falmer Building UK - BRIGHTON Tel. 606 755 x 611 Telex
PREUSS, H.-P.	Fraunhofer Institut für Solare Energiesysteme Oltmannsstrasse 22 D - 7800 FREIBURG Tel. 0761/40 50 46 Telex

ROSSI, M.
SpA Termomeccanica Italiana
V. del Molo 1
I - 19100 LA SPEZIA
Tel. 0187/50 31 51 - 50 32 41
Telex 270 171 tmi sp i

SABATO, A.
Università della Calabria
Dipartimento di Meccanica
Facoltà di Ingegneria
I - 87036 ARCAVACATA DI RENDE (CS)
Tel. 0984/83 80 73
Telex

SCHMID, J.
Fraunhofer Institute for
Solar Energy Systems
Oltmannsstrasse 22
D - 7800 FREIBURG
Tel. 0761/40 50 46
Telex

SIDOROFF, S.
Société SOREIB
29, rue de Longchamp
F - 92200 NEUILLY-SUR-SEINE
Tel. 1/722 83 49
Telex

STEWART, D.
Fulmer Research Institute Ltd
Hollybush Hill, Stoke Poges
UK - SLOUGH SL2 4QD
Tel. 02816/2181
Telex 849374

SVENDSEN, S.
Thermal Insulation Laboratory
Technical University of Denmark
Bldg 118, Lundtoftevej 100
DK - 2800 LYNGBY
Tel. 02/88 35 11
Telex 37529 dthdia dk

TURRENT, D.
Energy Conscious Design
44, Earlham Street
UK - LONDON WC2 9LA
Tel. 01/240 05 66
Telex 8952165 create g

VANDENDAEL, Y.
Institut de Mécanique Appliquée
Université Libre de Bruxelles
50, Av. F.D. Roosevelt
B - 1050 BRUSSELS
Tel. 2/649 00 30
Telex 23069 unilib b

VAN DIJK, H.A.L.	TPD TNO/TH Institute of Applied Physics P.O. Box 155 NL - 2600 AD DELFT Tel. 015/56 93 00 Telex 38091 tpd dt nl
VAN GALEN, E.	TNO Institute of Applied Physics Stieltjesweg 1 NL - 2628 CK DELFT Tel. 015/56 93 00 Telex 38091 tpd dt nl
VAN LIT, J.A.	ERA Bouw b.v. Postbus 62 NL - 2700 AB ZOETERMEER Tel. 079/16 67 40 Telex 31133 era nl
VAN LOEIJ, J.	Fakt. Toegep. Wet. Afd. Bouwkunde Vrije Universiteit Brussel Pleinlaan 2 B - 1050 BRUSSEL Tel. 02/641 29 22 - 641 29 21 Telex
VELLUET, P.	ARMINES - Centre d'Energétique Ecole des Mines de Paris 60, Boulevard Saint-Michel F - 75006 PARIS Tel. 329 21 05 x 480 Telex
VERDONSCHOT, J.	TNO/TH Institute of Applied Physics P.O. Box 155 NL - 2600 AD DELFT Tel. 015/56 93 00 Telex 38091 tpd dt nl
WIJSMAN, A.	TNO/TH Institute of Applied Physics P.O. Box 155 NL - 2600 AD DELFT Tel. 015/56 93 00 Telex 38091 tpd dt nl
WOOD, R.J.	Cranfield Institute of Technology UK - CRANFIELD, Bedfordshire MK43 OAL Tel. 0234/75 01 11 Telex